Food Stabilisers, Thickeners and Gelling Agents

To Katie

Food Stabilisers, Thickeners and Gelling Agents

Edited by

Alan Imeson

FMC BioPolymer, UK

WILEY-BLACKWELL

A John Wiley & Sons, Ltd., Publication

This edition first published 2010

Blackwell Publishing was acquired by John Wiley & Sons in February 2007. Blackwell's publishing programme has been merged with Wiley's global Scientific, Technical, and Medical business to form Wiley-Blackwell.

Registered office
John Wiley & Sons Ltd, The Atrium, Southern Gate, Chichester, West Sussex, PO19 8SQ, United Kingdom

Editorial offices
9600 Garsington Road, Oxford, OX4 2DQ, United Kingdom
2121 State Avenue, Ames, Iowa 50014-8300, USA

For details of our global editorial offices, for customer services and for information about how to apply for permission to reuse the copyright material in this book, please see our website at www.wiley.com/wiley-blackwell.

Library of Congress Cataloging-in-Publication Data

Food stabilisers, thickeners, and gelling agents / edited by Alan Imeson.
p. cm.
Includes bibliographical references and index.
ISBN 978-1-4051-3267-1 (hardback : alk. paper)
1. Hydrocolloids. 2. Food additives. 3. Food–Analysis. 4. Food–Composition. 5. Food industry and trade. I. Imeson, A. (Alan)
TP456.H93F66 2010
664′.06–dc22 2009016433

A catalogue record for this book is available from the British Library.

Set in 10/12 pt Times by Aptara® Inc., New Delhi, India
Printed and bound by CPI Group (UK) Ltd, Croydon, CR0 4YY

C9781405132671_231025

The manufacturer's authorized representative according to the EU General Product Safety Regulation is Wiley-VCH GmbH, Boschstr. 12, 69469 Weinheim, Germany, e-mail: Product_Safety@wiley.com.

Contents

A colour plate section falls between pages 50 and 51

Preface

Stabilisers, thickeners and gelling agents are inherent in almost all living organisms. They determine a number of critical functions including moisture binding and control, structure and flow behaviour that enable organisms to thrive in a natural environment. For use in food products, the functional materials are carefully extracted from various natural raw materials and incorporated into foods to give the structure, flow, stability and eating qualities desired by consumers.

These additives include traditional materials such as starch, a thickener obtained from many land plants, and gelatine, an animal by-product giving characteristic melt-in-the-mouth gels. Cellulose, the most abundant structuring polymer in land plants, seed gums and other materials derived from sea plants extend the range of polymers. Recently approved additives include the microbial polysaccharides of xanthan, gellan and pullulan. With stringent regulations in place governing the use of additives, it is unlikely that many new polymers will be approved and researchers must employ the current range of products to deliver the range of attributes needed for their particular food products.

Hydrocolloids have a profound impact on food properties when used at levels ranging from a few parts per million for carrageenan in heat-treated dairy products to high levels of Acacia gum, starch or gelatine in jelly confectionery. The correct application of these materials is a fascinating topic that continues to engage the attentions of many expert researchers. Over recent years, investigative techniques have shed more light on the fine structure of the polymers to enhance the understanding of network formation and how they combine with other polymers. These structures determine a number of properties in finished foods, such as emulsion stability, the long-term suspension of fortified beverages using 'fluid gels' and for giving rich, creamy eating qualities.

Calorie-dense materials such as fats and oils may be replaced with 'structured water' to give healthy, reduced-calorie foods with excellent eating quality. Some fibres are currently being studied for their effects on satiety and the reduction of daily energy intake. In addition to the functional attributes, future acceptance and, possibly, positive endorsement may derive from the recognition that soluble and insoluble fibres contribute many physiological benefits to the natural function and well-being of the body.

This book is highly practical and directed to all those involved in various sectors of the food industry. Although it is particularly valuable for product and process developers, marketing personnel will appreciate the value of these highly functional materials and it will help people involved in ingredient procurement appreciate that these materials are often complex functional additives. The information is easy to read and assimilate. New students will find chapters presented in a standard format, enabling key points to be located quickly.

Those with more experience will be able to compare and contrast different materials and gain a greater understanding of the interactions that take place during food production. This concise, modern review of hydrocolloid developments will be an invaluable teaching resource and reference text for all academic and practical workers involved in hydrocolloids in particular and food development and production in general.

Acknowledgements

In commending this book to readers, I must pay tribute and thank all the authors and contributors to the chapters in this book. Their great enthusiasm and commitment over the long period needed to complete this project has produced an excellent series of chapters linking the structure and function of the polymer in nature to a wide range of properties needed for high-quality foods.

These excellent contributions summarise the current state of knowledge on the use of these materials in food. The authors have used data, diagrams and figures made available by many suppliers to the hydrocolloid industry. The cooperation and support from major manufacturers and suppliers have been essential in producing this book. These companies continue to enhance and extend their product ranges, to actively investigate new applications for their products, to provide detailed support and direction to new customers and to contribute to new publications.

I would also like to acknowledge the support of the publishers, Wiley-Blackwell, who have continued to encourage and support this project, despite several setbacks, before this successful conclusion.

Finally, I must recognise the great inspiration, encouragement, support and tolerance from my family, particularly my wife, Hazel, whilst I have nurtured this project to a successful outcome.

Alan Imeson

Contributors

Peder Andersen
Technologist, FMC BioPolymer, Sandvika, Norway

William R. Blakemore
Principal, Celtic Colloids Inc., Topsham, ME, USA

Sarah M. Brejnholt
Research Scientist, CP Kelco, Lille Skensved, Denmark

Greg Buliga
Technology Growth Leader, FMC BioPolymer, FMC Corp., Princeton, NJ, USA

Sandra J. Caputo
Technical Service, Regulated Industries, Ashland Aqualon Functional Ingredients, Wilmington, DE, USA

Mary Jean Cash
Technical Service, Regulated Industries, Ashland Aqualon Functional Ingredients, Wilmington, DE, USA

Hiroto Chaen
Plant Manager, Okayama Plant II, Hayashibara Co., Ltd., Okayama-shi, Japan

Ross Clark
Distinguished Research Fellow, CP Kelco, San Diego, CA, USA

Therese Fjæreide
Project Leader, FMC BioPolymer, Sandvika, Norway

Olav Gåserød
Team Leader Alginate Commercial Technology, FMC BioPolymer, Sandvika, Norway

Alan R. Harpell
Senior Scientist, FMC BioPolymer, FMC Corp., Princeton, NJ, USA

Trond Helgerud
Alginate Commercial Technology Manager, FMC BioPolymer, Sandvika, Norway

Alan Imeson
Account Manager, FMC BioPolymer, Epsom, Surrey, UK

Gregory R. Krawczyk
Senior Research Scientist, FMC BioPolymer, FMC Corp., Princeton, NJ, USA

Christian Klein Larsen
Senior Development Chemist, FMC BioPolymer, Sandvika, Norway

Jenny M. Mayes
Technical Manager, Arthur Branwell & Co. Ltd., Epping, Essex, UK

Jean-Marc Parry
Reasearch & Development Manager, Kalys SA., Bernin, France

Dennis Seisun
President, IMR International, San Diego, CA, USA

Paul Sheldrake
Market Manager, Avebe Food, Sheffield, UK

Paul Stevens
Technical Service and Application Development Manager, Rousselot NV, Gent, Belgium

Graham Sworn
Group Manager Gums and Systems Development, Danisco, Paris, France

Francis Thevenet
Technical Director, Colloides Naturel International, Rouen, France

Domingo C. Tuason
Senior Engineering Associate, FMC BioPolymer, FMC Corp., Princeton, NJ, USA

Raymond Valli
Senior Research Scientist, CP Kelco, San Diego, CA, USA

Willem Wielinga
Consultant, Taegerwilen, Switzerland

Rudy Wouters
Head of Food Application Technology, BENEO Group, Tienen, Belgium

1 Introduction

Dennis Seisun

ABSTRACT

Food stabilisers, thickeners and gelling agents are obtained from a wide range of natural raw materials including microorganisms, land and sea plants and animal connective tissues. They control moisture and provide structure, flow, stability and eating qualities to food products. Approvals for food use and purity criteria are closely controlled by regulation. Commercial applications are determined by the combination of properties provided by these materials including the current significant market drivers of price and availability coupled with consumer and retailer preferences. Future developments with hydrocolloids will recognise the value of nutritional and therapeutic benefits in addition to the functional attributes.

1.1 INTRODUCTION

'Little known and yet ubiquitous in virtually all processed foods'. This statement summarises the role of stabilisers, thickeners and gelling agents in today's food industry. It is virtually impossible to list the multitude of functions that these additives have in making foods look, feel and taste like they do. Such a list is virtually impossible, not only because it is so long and all encompassing, but also because it is changing and growing all the time. New uses and functions for these unique ingredients are constantly being found.

1.1.1 Scope

Within the food industry, stabilisers, thickeners and gelling agents are often more simply referred to as food hydrocolloids. The hydrocolloids traditionally used in food thickening and gelling include, but are not limited to, the following: agar, alginates, arabic, carrageenan, cassia tora, carboxymethyl cellulose, gelatin, gellan, guar, karaya, konjac, locust bean gum, methyl cellulose, hydroxypropylmethyl cellulose, microcrystalline cellulose, pectin, starches, tara, tragacanth and xanthan.

This book deals with all of these materials in a structured fashion starting with raw materials, followed by the production process and ending with application-related information. Chemical structure and conformation, viscosity and gelation charts and many food formulations are included for each ingredient in separate chapters. Readers will obtain a good overview of scientific, technical and commercial aspects for each material.

1.1.2 Definition

All the stabilisers, thickeners and gelling agents covered in this book are also known and described as 'food hydrocolloids' implying that functional properties are obtained by mixing them with water. A strict definition of a hydrocolloid is, however, difficult. Ask ten scientists what is a hydrocolloid and it is likely that ten different answers will be obtained. These could include statements as follows:

- A colloidal substance obtained from . . .
- A material that hydrates in water . . .
- A colloid forming a suspension and not a true solution in water . . .
- A synonym for gum (e.g. guar gum, locust bean gum, gum arabic . . .)
- A macromolecule, such as a carbohydrate polymer or a protein, that is water soluble . . . .

Many of these ingredients are carbohydrates but at least one important hydrocolloid, gelatin, is a protein. Most are agricultural derivatives but some are biotechnology derived, and gelatin, of course, is an animal product. This volume presents some scientific information but focuses more on the 'real world' of application-related data that will be of most benefit to food technologists and food formulators. Issues of functional properties, synergy, production and raw materials are most relevant to readers, but molecular structure and chemical definition have also been covered.

1.1.3 Classification

The hydrocolloids are treated in alphabetical order in this book, but readers should bear in mind that there are several methods of grouping them. Raw material origin has been used to classify them, for example as seaweed extracts, seed gums, fermentation products or plant exudates. Their general functional properties may also be used to classify them as thickeners, stabilisers or gelling agents. More recently, their commercial availability and price stability have been used as a differentiating factor. Those that are commercially steady in availability and price include the cellulose derivatives and fermentation products, such as xanthan and gellan gum. On the other hand, notorious for their instability in terms of price and availability are guar gum, locust bean gum and gum arabic. More recent developments may allow a further aspect of classification and differentiation, namely nutritional and therapeutic function. Future research may uncover yet more functions for these versatile food wonders.

1.1.4 Differentiated grades

Many, if not all, of the thickening and gelling agents for food are available in a wide range of differentiated grades. Starch is a good example. There are literally hundreds of different food starches based on different raw materials and production process conditions. A double-derivatised, waxy maize starch is totally different from, for example, a pregelatinised potato starch. Cellulose derivatives, such as carboxymethyl cellulose, microcrystalline cellulose, methyl cellulose and hydroxypropylmethyl cellulose, come in a virtually limitless range of differentiated grades depending on the degree of substitution and other processing factors. A large number of 'new' and 'differentiated' properties have been and continue to be developed for hydrocolloids that fall under an 'umbrella category'; for example methyl cellulose, hydroxypropylmethyl cellulose and microcrystalline cellulose can be produced in a multitude of grades to suit a wide range of specific functional needs. Xanthan is offered in different mesh sizes, rapidly hydrating, brine tolerant and/or as a clarified grade. New versions are constantly being developed and assure the specialty future of at least part of this market.

1.2 FUNCTIONAL PROPERTIES

The following is a brief overview of the key functional properties for which these ingredients are used. Nutritional properties are relatively new and nutraceutical or health-enhancing properties are even more recent. Further work is sure to advance the use of hydrocolloids beyond modification of the rheology of foods.

1.2.1 Viscosity

Viscosity is probably one of the most widely used properties. In this respect, hydrocolloids are often used in systems where the oil or fat content has been reduced or eliminated through substitution with water. The hydrocolloid thickens water, which, in turn, replaces the fat or oil to give a product with similar properties to the full-fat food. A typical application for this function is reduced-fat salad dressings. In other cases, the thickened water simply adds body, texture and mouthfeel to a food such as table syrups, particularly low-calorie syrups.

1.2.2 Stability

If oil or fat is partially removed from a formulation and is replaced with thickened water, an emulsion is usually formed. Often the function of the hydrocolloid is to stabilise the emulsion, to prevent separation and, in the case of frozen foods, to control ice crystal formation. New technology and new ingredients have been developed specifically to address the problem of ice crystals in frozen foods, but hydrocolloids will continue to play a role. Virtually every ice cream product sold in retail outlets is stabilised with carrageenan, locust bean gum and/or guar gum. Low-fat salad dressings, discussed above, also benefit from emulsion-stabilising properties.

1.2.3 Suspension

If insoluble particles are included in the thickened product then separation and settling should be eliminated or at least minimised. Some hydrocolloids create solutions with a 'yield point' that will keep particles immobilised in suspension. Salad dressing is a good example of this and xanthan gum is the typical hydrocolloid to supply this functionality.

1.2.4 Gelation

One of the key texturising aspects of hydrocolloids is the ability to gel and solidify fluid products. For example, in gelled milk desserts, even low levels of carrageenan will form a solid milk gel. Other classic gelling agents are pectin, gelatin and agar. Many others, however, will form a gel under specific conditions. Certain grades of alginates form gels with calcium ions. Xanthan and locust bean gum do not gel individually but together they display synergy and form a strong cohesive gel. Methyl cellulose and hydroxypropylmethyl cellulose are unusual in forming solutions that reversibly thicken or gel when heated. The food industry has a myriad of gelling applications ranging from soft, elastic gels to hard and brittle gels.

1.2.5 Nutritional and nutraceutical

There is already a wide use of some hydrocolloids, arabic and guar gum, for example, as sources of *soluble* dietary fibre. Much research has been conducted in the nutraceutical benefits of hydrocolloids. Potential benefits range from cholesterol reduction to cancer risk prevention. Their use in weight loss programmes is already widespread and likely to expand further.

1.3 REGULATORY ENVIRONMENT

1.3.1 Background

The use of hydrocolloids in food has been steadily evolving. Pectin, agar, starches and gelatin have been used for centuries. They are amongst the few hydrocolloids that are sold directly to consumers at the retail level. These are the hydrocolloids with which the consumer is most familiar and comfortable. Until the 1980s and 1990s, gelatin was amongst the preferred and most label-friendly hydrocolloids, but this changed rapidly and dramatically with the advent of bovine spongiform encephalopathy (BSE), otherwise known as mad cow disease. Gelatin provides an example of the changing fortunes for individual hydrocolloids in the marketplace.

Many of the hydrocolloids in use today were developed long before regulatory approvals and constraints were imposed on use levels or in specific applications. Alginates, agar and carrageenan, for example, were extracted from seaweeds some of which were eaten as a basic food. Red seaweeds have long been used as a food. In Ireland, 'carrageen' was used to gel dairy products centuries ago. Extracts from such seaweeds were therefore deemed safe for use in food. This principle of ingredients and extracts thereof that are 'generally recognised as safe' (GRAS) is still in use today albeit under more rigorous review. Many of the more recent texturising options offered to food technologists derive from differentiated hydrocolloids based on materials that are currently approved. Carrageenan, for example, can be modified in many ways. There is lambda, kappa and iota carrageenan. There is a refined version produced through alcohol precipitation or specific precipitation with potassium chloride (the 'gel press' process) and there is a semi-refined carrageenan produced through a much simpler process. Semi-refined carrageenan made its way into the texturising world only in the 1980s and 1990s, whereas refined carrageenan has been offered for 60–70 years or more.

1.3.2 Legislation

Regulatory approval of a food ingredient is critical. Without approval of the appropriate government bodies, the additive has no market or function in food. Nowadays, obtaining regulatory approval for a new ingredient is a very costly and time-consuming process. The approval process itself is evolving and has opened up opportunities for differentiated versions of existing products. This has allowed for continued improvement and innovation to be offered to food formulators.

Labelling is a key factor in marketing any food ingredient. The rules governing food label nutritional information have changed significantly. There has been little impact on hydrocolloids other than fibre claims. There are literally hundreds of differentiated food starches. On the food label, however, starches are simply segmented into modified or native starches. Dozens of grades of carboxymethyl cellulose are produced, but they are all simply called carboxymethyl cellulose or cellulose gum. In Europe, E-numbers have been established for all hydrocolloids. Often, however, for marketing reasons, food processors elect to replace the E-number with the accepted name of the additive. For example, locust bean gum, guar gum and carrageenan are often declared instead of E410, E412 and E407, respectively.

Regulatory authorities strictly control the approval of food additives. Chemical modifications are generally not allowed with the exception of starches, cellulose derivatives and propylene glycol alginate. Physical and enzymatic modification, however, is allowed. Physically modified pectin, sold under the name Slendid™ by CP Kelco Division of JM Huber, is an example. Some new hydrocolloids are brought to market under the GRAS designation, such as tara gum, konjac and pullulan. Gellan gum is one of the last hydrocolloids to go through a full food additive petition on a global scale. Its approval took many years and tens of millions of dollars of research and lobbying effort. Nearly 20 years after its approval, gellan gum has not yet reached commercial volumes that justify the cost of bringing it to market. These high stakes and high risks probably mean that no new hydrocolloid will be taken through a full approval process in the foreseeable future. Cassia gum has recently been approved in France (August 2008) and more widespread approval in the EU is expected soon. Approval for human food in the USA is still pending.

Genetic engineering could offer a tremendous opportunity for new functional developments in hydrocolloids. Bio-fermentation products such as xanthan and gellan gum could be manipulated to provide specific functional properties not now available. Seaweed, seed or other agricultural raw materials could be genetically enhanced. Giant, rapidly growing kelp could be programmed to produce more than alginates. The time for a carob tree to reach maturity and give a commercially viable yield could be reduced from the required 12–15 years. All these scenarios for improved production currently have a lid tightly sealed over them by consumer concerns. This is not to say that future generations and future nutritional conditions will not radically change.

1.3.3 Consumer concerns

Readers of this book must bear in mind one fundamental concept in terms of markets for food hydrocolloids: 'The perception of consumers is the reality of hydrocolloid producers.' In other words, whatever the scientific facts, it is the consumers' perceptions and resulting action or inaction that dictates the commercial future of any food ingredient, including hydrocolloids. In terms of perception and label image, pectin probably has the best and most friendly image. Seaweed extracts such as agar, carrageenan and alginates are also 'label friendly'. Cellulose derivatives have a 'variable' image in the mind of consumers. Terms such as carbohydrate gum and vegetable gum, used in the USA to describe methyl cellulose and hydroxypropylmethyl cellulose, are very label friendly. On the other hand, a name such as carboxymethyl cellulose has a very chemical connotation, but its synonym, cellulose gum, is much more label friendly. This is the main reason why EU authorities were lobbied to allow cellulose gum on the label.

Until the late 1980s and early 1990s, gelatin had an excellent label image, but that changed dramatically with the advent of BSE or mad cow disease as it is more commonly known.

Consumers made a perceptual link between gelatin and BSE. Gelatin consumption suffered dramatically and most large food formulators made strong efforts to replace or eliminate the use of gelatin. All the scientific evidence indicates that there is no risk of the BSE prion being found in gelatin, but consumer concerns and their effect on buying behaviour are often more emotional than rational and scientific.

1.4 COMMERCIAL ENVIRONMENT

1.4.1 Global market

The global market value for all food hydrocolloids during the period April 2007 to March 2008 is estimated at US$4.2 billion. This market value is calculated at the basic producer level. Once markups by service companies and distribution channels have been factored into the equation, the food hydrocolloid market is worth around US$5.0–5.5 billion per year. Gelatin and starches are by far the largest in terms of value, as indicated in Table 1.1.

The price of gelatin and starch is lower than the average of many other hydrocolloids. This means that, in terms of volume, gelatin and starches account for an even higher proportion of the total than indicated in Table 1.1. Pure gellan gum is one of the most expensive hydrocolloids at US$14–15/lb (US$31–33/kg). Native starches are at the other end of the scale and can be purchased for US$0.15–0.20/lb or US$0.33–0.44/kg. Based on price and value calculations, the total volume of hydrocolloids in 2007–2008 was around 1.7–1.8 million metric tonnes. Growth rates vary depending on the hydrocolloid. Not surprisingly, large volume items such as starches are growing at a low rate of 1.0–1.5%. Modified starches in the USA, however, are growing much faster at 4–5%. Overall, the average growth for hydrocolloids is estimated at 2.5–3.0%.

Table 1.1 Global hydrocolloid estimates (April 2007 to March 2008, estimated annual market).

Hydrocolloid	Price (million US$)	Value (%)
Starches	1307	27
Gelatin	823	24
Pectin	443	10
Carrageenan	416	9
Xanthan	225	6
Agar	209	5
Arabic	153	5
Locust bean gum	134	4
Carboxymethyl cellulose	146	3
Alginates	125	2
Guar	76	2
Microcrystalline cellulose	60	1
Methyl cellulose/hydroxypropylmethyl cellulose	57	1
Other	42	1
Total	4216	100
Excl. starch	2909	73

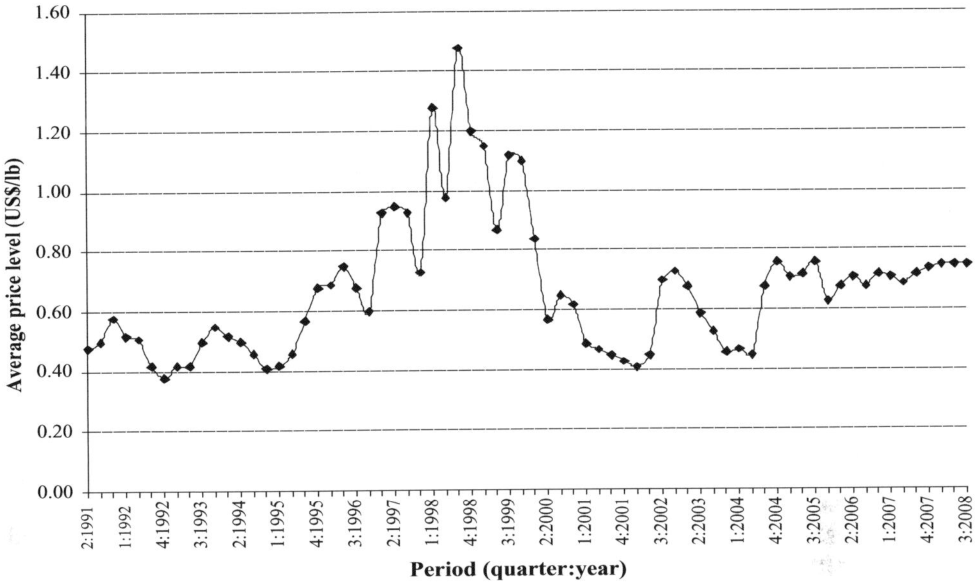

Fig. 1.1 Guar price history.

1.4.2 Cost

These food ingredients are cost-effective and generally stable in price. Overall, the cost of texturising, stabilising or gelling a food formulation has been steady or even declined over the last 30–40 years. Hydrocolloid producers have improved the efficiency of production, reduced costs and even improved performance such that lower use levels are needed. The suppliers of raw materials have also tried to keep costs down. There are, however, severe fluctuations in price and availability in some of these products. Prices in 2008 have shattered the image of stability which most of these ingredients had achieved. Dramatic price increases and even shortages were experienced for several hydrocolloids in 2008 and expected to continue well into 2009.

1.4.2.1 *Guar gum*

Guar gum, for example, is notoriously cyclical in price as indicated in Fig. 1.1.

The main cause for the cyclical nature of guar availability and price is its concentrated geographic source. Virtually all the guar gum in the world originates in the Indian subcontinent. If the monsoon rains in that region are poor, there is a poor guar crop and tightness in world supply.

1.4.2.2 *Locust bean gum*

Locust bean gum is another hydrocolloid notorious for periodic tightness in supply and correspondingly high prices. The cycle (8–10 years) of locust bean gum price and availability is longer than that of guar gum (3–4 years), but more severe when it does impact the market

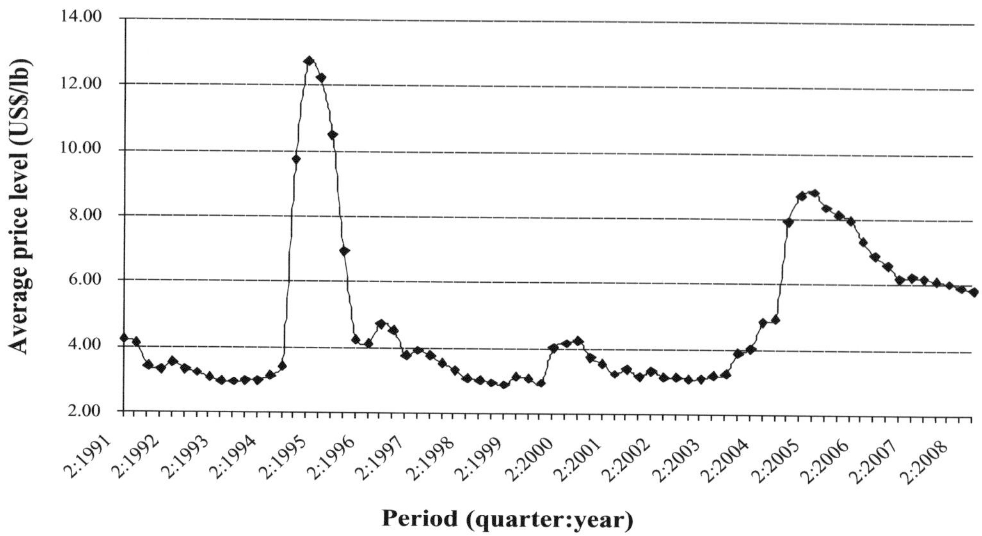

Fig. 1.2 Locust bean gum price history.

(Fig. 1.2). It is impossible to predict when another shortage will occur, but Mother Nature will ensure that periodically there will be a poor crop. In 1994, locust bean gum prices went up to US$18–20/lb (US$40–44/kg). In 2004, prices again rose more than 100%. They have since been steadily declining, until the next locust bean gum crisis!

1.4.2.3 *Acacia gum or gum Arabic*

Acacia gum or gum arabic serves as a final example of volatile price and availability. Much of the world's gum arabic originates in one country of the world, namely Sudan. A bad season, combined with political unrest, resulted in dramatic price increases in 2004. It must be noted that these increases were preceded by a steady period of decline for the previous 6–8 years as indicated in Fig. 1.3.

Gum arabic availability and price will remain dependent on conditions in the Sudan despite efforts in other countries, such as Nigeria and Chad, to diversify the availability of raw material.

1.4.2.4 *Xanthan gum*

Xanthan gum is one of the more versatile of the hydrocolloids and demand for it has grown rapidly. In contrast to gum arabic, xanthan gum has been a paragon of price predictability with a small exception in the late 1990s when demand outstripped supply for a short period (Fig. 1.4). The number of new producers of xanthan gum entering the market makes it unlikely than another tight supply situation will occur in the foreseeable future. In the last 16 years that IMR has been tracking quarterly hydrocolloid prices, the cost of xanthan gum has *declined* from $5.65/lb in 1991 to $2.10 in 2007. This is an average price drop of 5.7% per year. Some recovery in xanthan and several other hydrocolloid prices started at the end of 2007 and accelerated as 2008 events drove up the price of most hydrocolloids.

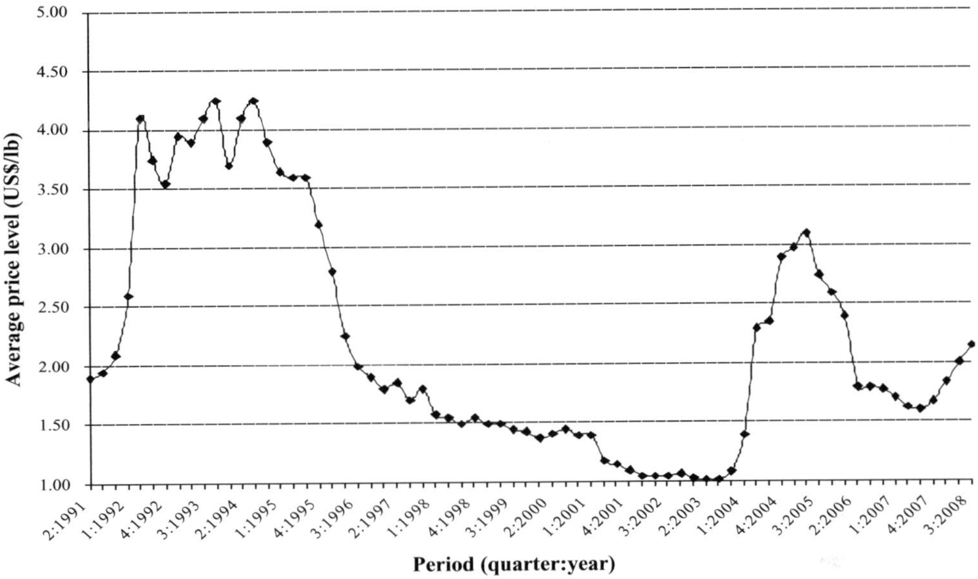

Fig. 1.3 *Acacia* gum price history or Gum arabic price history.

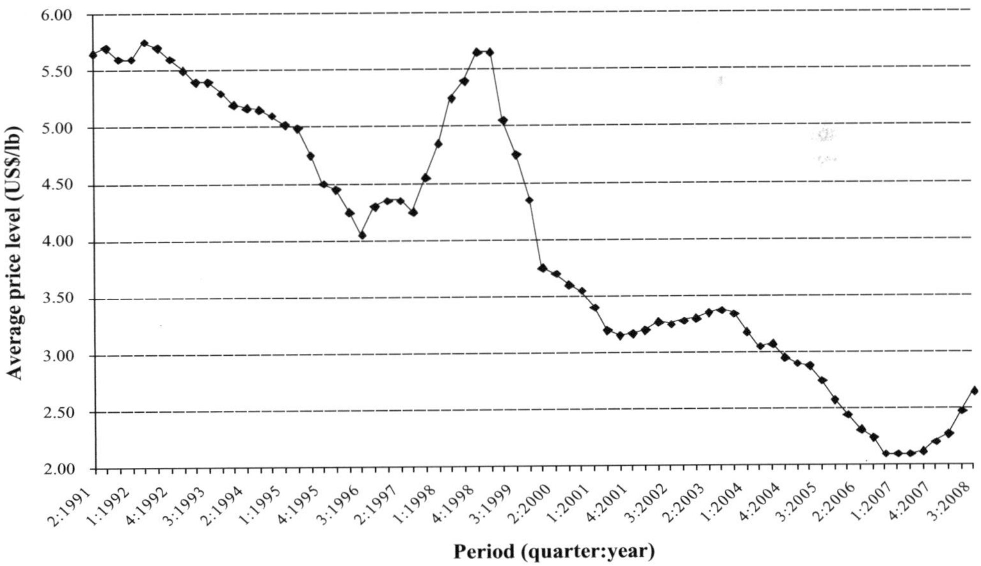

Fig. 1.4 Xanthan gum price history.

1.5 FUTURE DEVELOPMENTS

Processed foods are here to stay. As the world continues down the path of modernisation, consumption of processed foods will increase and the corresponding demand for texturising agents will also increase. Four key factors will drive growth in processed foods:

- Convenience
- Quality
- Nutrition
- Cost

In developed and developing countries alike, leisure time is becoming more limited and more precious, particularly in families with two working partners. Neither person is willing or able to spend the time to cook meals 'from scratch'. Even countries with more traditional cooking practices, such as France and Italy, are inexorably going towards higher consumption of processed foods.

Quality has always been an issue but even more so in a world where consumers approach zero tolerance for risk and/or variation in quality. Specifications for purity and performance are getting tighter. The performance of ingredients, such as hydrocolloids, becomes all the more critical when their use level is very low.

The aspect of nutrition has become paramount in the mind of consumers. This concept has been taken one level higher with the development of 'functional foods' which fulfil some therapeutic role at the same time as being a food. Added fibre, calcium, vitamins and various other elements have become a standard in many foods. Some hydrocolloids provide benefits of soluble or insoluble fibre. Gum arabic has achieved the status of being a food ingredient rather than a food additive in the EU. Others are being evaluated with a promising outlook for providing other therapeutic benefits.

2 Acacia Gum (Gum Arabic)

Francis Thevenet

ABSTRACT

Acacia gum, also known as gum arabic, is a natural gum exudate obtained from acacia trees in the 'African sub-Sahelian zone'. The gum has a highly branched compact arabinogalactan structure which gives a low-viscosity solution together with a central protein fraction that provides good emulsification properties. The powder hydrates readily in water and concentrations up to 40–50% can be handled easily. Key food applications include a range of confectionery products, flavoured oil emulsions and capsules and health foods as a source of soluble fibre with prebiotic properties.

2.1 INTRODUCTION

Acacia gum, also known as gum arabic, is a natural, vegetable exudate from acacia trees known since antiquity and used for thousands of years in foods as an additive and ingredient, in the pharmaceutical industry and for technical purposes.

There are various species of acacia trees with more than 700 spread across the world in Africa, Australia, India and South America.

The botanical definition of acacia gum was specified at the 53rd Joint FAO/WHO Expert Committee session in 1999 as being 'a dried exudate obtained from the stems and branches of *Acacia senegal* (L.) Willdenow or *Acacia seyal* (fam. *Leguminosae*): synonyms: Gum arabic, Acacia gum, arabic gum, INS No. 414'. This new definition has been approved by the Codex Alimentarius Commission (ALINORM 99/12.A). In Europe, acacia gum is listed as a food additive by the European Directive 98/86/CE and 2008/84/EC with the number E414. Acacia gum (gum arabic) is also listed by the US Code of Federal Regulations number 21CFR184.1330 and accorded Generally Recognised as Safe (GRAS) status. Worldwide, these natural gums from *A. senegal* and *A. seyal* species are approved for pharmaceutical applications and listed in the US National Formulary and European pharmacopeia (01/2009: 307).

To secure the supply of gum, international associations have made many technical and economic efforts to extend plantations and promote the collection from acacia trees in the sub-Saharan zone of Africa (Dondain and Phillips, 1999). In place of the original three main gum-producing countries, namely Sudan, Nigeria and Chad, gum is now collected from 12 different countries from Senegal to Ethiopia, forming the so-called 'African Gum Belt'.

Thirty years ago, 90% of the gum arabic used in food was from *A. senegal* species. *A. seyal* grades were kept only for technical purposes, such as glue, textiles, inks and printing

(Meer, 1980). As a result of the development of new purification processes, research on specific properties and the promotion of new applications, the market for gum from *A. seyal* species is increasing every year. With new amendments of the food and pharmaceutical regulations in favour of *A. seyal* gum, the worldwide consumption of gum from *A. seyal* species is now about the same as gum from *A. senegal*. This is a very important factor for the stabilisation of the gum market.

Acacia gum is the third largest hydrocolloid additive and ingredient used by industry in terms of volume: worldwide consumption was around 57 000 tons in 2008. The market for acacia gum is growing every year because it is a multifunctional food additive with properties as a texturising agent, film former, emulsifier and stabiliser, with numerous health benefits as a soluble fibre with prebiotic effects, and reinforced by the natural, non-modified vegetable origin.

2.2 ORIGIN AND PURIFICATION PROCESS

All *A. senegal* and *A. seyal* gums used in the food industry as additives or ingredients are collected from native sources or plantations in Africa. Acacia trees grow in a geographical belt (gum belt) located at the border of the desert from the western part of Africa in Senegal to the eastern part in Ethiopia (Imeson, 1992).

The gum exudes naturally after tapping the tree during the dry season from December to May. Nodules are hand-collected by farmers. One tree produces only about 400 g of gum per year (Vassal, 1985; Coppen, 1995).

A. senegal and *A. seyal* gums are normally harvested from different areas in the same country. *A. senegal* and *A. seyal* trees are very different in terms of shape, size, colour and thorns and are not easily confused. In gum-producing countries, the main gum manufacturers have subsidiaries equipped with laboratories where the quality of the gum is checked before shipment to purification plants.

Raw gum from the same botanical origin is a blend of gum nodules with different mesh sizes, containing vegetable and mineral impurities and fluctuating bacteriological contamination. Using dry purification steps, such as kibbling, sieving and pulverisation, the level of impurities can be slightly reduced but bacteriological contamination cannot be improved. Most of the time, raw gum does not meet international food or pharmaceutical specifications. Consequently, the dry methods of purification have been substituted by purification in aqueous solution, which is much more efficient (see Fig. 2.1). The gum is fully dissolved in water and all the impurities removed by a cascade of filtration steps giving levels of insoluble matter in the finished product as low as 0.02%. Bacteriological contamination is reduced by a plate heat exchanger and the gum syrup is concentrated to between 25% and 35% and dried, giving a level of contamination in the powder as low as 500 total germs per gram or below.

During solubilisation and purification, the thermal conditions are critical. Acacia gum contains proteins which are very important for the emulsifying properties but sensitive to heat denaturation. The selection of temperature and selection of time are key parameters for determining the quality of the gum.

Different processes are used for recovering purified, powdered acacia gum from the syrup. Roller drying is used to produce a gum in powder form with good hydration properties, but it has reduced emulsifying properties due to the drastic thermal treatment during the drying step. Regular spray drying is also used, which gives the gum good physical qualities and functional properties. Recently, spray drying has been improved by using a multi-stage

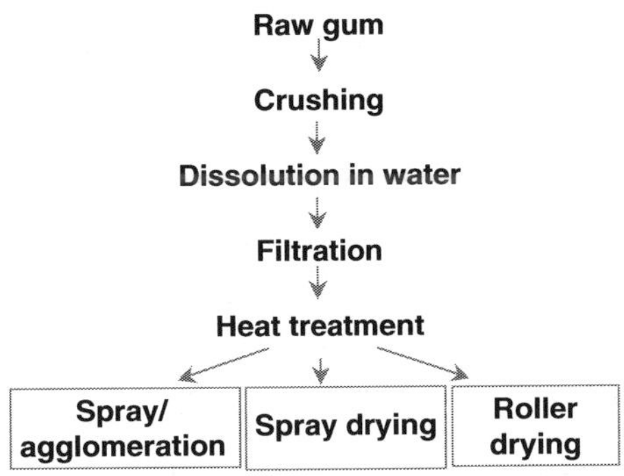

Fig. 2.1 Purification process of acacia gum.

drying process where fine particles of gum produced during drying are recycled at the top of the dryer. Agglomerated gum particles are obtained, keeping the entire properties of the raw gum, but containing no dust or particles below 75 μm and giving unique hydration and dissolution properties in water, without any lump formation, up to the maximum level of solubility of 45–50%.

Such a purified gum is sold with general specifications according to the local food or pharmaceutical regulations together with specific criteria related to the proposed application of the gum, such as viscosity, colour, microbiology and functionality tests. As a result of the integration between the producing countries and processing plants, full traceability 'from the tree to the finished product' is guaranteed.

2.3 CHEMICAL STRUCTURE

Acacia gum is a highly branched arabinogalactan polysaccharide with a high molecular weight developing a low viscosity in water.

All molecules of acacia gum contain the same sugars: galactose, arabinose, rhamnose and glucuronic acids (Jurasek *et al.*, 1993) partially neutralised with calcium, potassium, sodium and magnesium salts. In the 1960s and 1970s, Anderson and Farquhar (1974) conducted full taxonomy studies on many exudates from various acacia species. With new techniques of gel permeation, flow field-flow fractionation and multi-angle laser light scattering, Phillips, Fenyo and Muller have been able to clarify the complex structure of acacia gum (Connolly *et al.*, 1987; Picton *et al.*, 2000; Al-Assaf *et al.*, 2005b).

An acacia gum molecule is not a unique structure. Gel permeation and size-exclusion chromatography have shown that both *A. senegal* (*Vulgares* species) and *A. seyal* (*Gummiferae* species) have at least two fractions with different molecular weights. For both exudates, the highest molecular weight fraction contains the majority of the proteins but represents a minority percentage of gum.

The 'wattle blossom' structure (Fincher *et al.*, 1983; Connolly *et al.*, 1988) represents the highly branched compact structure of acacia gum from *A. senegal*. Arabinogalactans are attached to a protein skeleton forming the arabinogalactoprotein (AGP) fraction. The polysaccharide fraction is composed of a linear chain of β[1,3]-linked galactose. In position [1,6], this chain is branched with side chains of galactose and arabinose. Rhamnose,

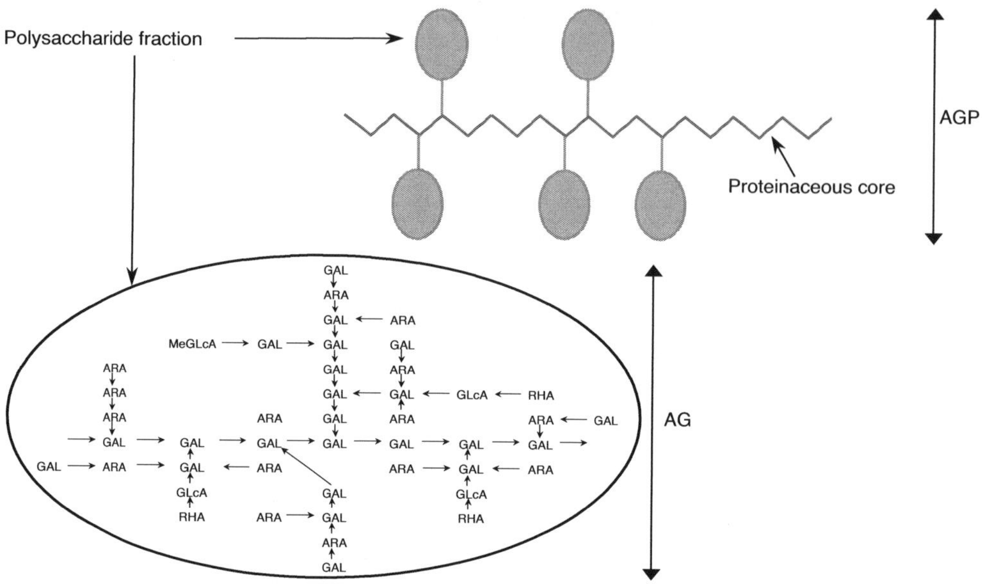

Fig. 2.2 Schematic representation of *A. senegal* gum molecule: the 'wattle blossom' model (Fincher *et al.*, 1983). (Reprinted, with permission, from the *Annual Review of Plant Physiology*, Volume 34, © 1983 by Annual Reviews, www.annualreviews.org.)

glucuronic acid or methyl glucuronic acid units are found as chain terminations in the Arabinogalactan (AG) fraction (Fig. 2.2) (Street and Anderson, 1983).

From Table 2.1, we can compare the main physical parameters of various samples of the two species approved for food and pharmaceutical applications (Islam *et al.*, 1997; Al-Assaf *et al.*, 2005a; Hassan *et al.*, 2005). A simple fingerprint of the molecule is the value of the optical rotation (Biswas and Phillips, 2003) which easily demonstrates from which acacia species the gum has been collected. *A. senegal* gives laevorotatory rotation of about $-30°$ and *A. seyal* shows dextrorotatory rotation of about $+50°$. For further structural information, sugar composition is determined by HPLC after acidic gum hydrolysis. This qualitative sugar composition is now part of the Food Monograph and Pharmacopeia. The ash content results from the cations associated with the uronic acids. The presence of uronic acids in acid and salt forms gives *A. senegal* and *A. seyal* gums buffer properties which make the gum solutions pH stable after the addition of acids or bases.

The nitrogen content of *A. senegal* gum is about double that of *A. seyal* gum. It has been found that the protein fraction, which contains the same amino acids in *A. senegal* and *A. seyal* gums, is usually linked to the high-molecular-weight fraction. It has been shown that the protein in *A. seyal* is also much less available than in *A. senegal* (Anderson and McDougall, 1987). The lower content and lower availability of the protein fraction explain why *A. seyal* gum is less efficient for emulsion stabilisation and, consequently, it is not used for emulsions where long-term stability is required.

A comparison of molecular weight clearly shows a higher molecular weight for the *A. seyal* gum and, in general, for gums from the *Gummiferae* species. Despite this higher molecular weight, the intrinsic viscosity of *A. seyal* gum is lower than that of *A. senegal* gum at 15 mL/g compared to 20 mL/g. This indicates a more compact structure for *A. seyal* (Flindt *et al.*, 2005).

Table 2.1 Comparison between the compositions of *A. senegal* and *A. seyal* gum.

	***Acacia senegal* (*Vulgares* species)**	***Acacia seyal* (*Gummiferae* species)**
Sugar composition after hydrolysis		
Galactose (%)	44	38
Arabinose (%)	27	46
Rhamnose (%)	13	4
Glucuronic acids (%)	14.5	6.5
4-0-Methylglucuronic acid (%)	1.5	5.5
Specific rotation (degrees)	−30	+51
Intrinsic viscosity (mL/g)	16–24	13–17
Nitrogen (%)	0.29	0.14
Ash (%)	3.93	2.87
	***Acacia senegal* (Var. *senegal*)**	***Acacia seyal* (Var. *seyal*)**
Average molecular weight ($\times 10^5$)	3.3–9.4	8.4–35.6
Molecular weight peak 1 ($\times 10^6$)	1.6–3.0	2.06–9.28
Mass peak 1 (%)	9–17	6–29
Molecular weight peak 2 ($\times 10^5$)	2.6–4.0	7.2–12.8
Radius of gyration (nm)	17–30	22–36

Source: Al-Assaf *et al.* (2005b), Hassan *et al.* (2005), Islam *et al.* (1997).

Compared to other water-soluble polysaccharides with a similar molecular weight, acacia gum exhibits very low viscosity in water. At 1% concentration, guar, xanthan and lambda carrageenan develop viscosity around 3000–5000 mPa s (Brookfield viscometer). To reach such viscosities, acacia gum has to be dissolved at concentrations of 40–45%. The low viscosity in solution is due to the globular, highly branched structure of acacia gum which hinders the formation of cross-links or hydrogen bonding with water. Rheological behaviour of acacia gum solutions is Newtonian up to 25% concentration and then becomes pseudo-plastic (Sanchez *et al.*, 2002). This highly branched structure for both acacia gums makes the product highly resistant to hydrolysis in acidic media and to degradation in extreme thermal conditions and by enzymes.

2.4 APPLICATIONS

2.4.1 Confectionery

Acacia gum is not considered a thickening hydrocolloid when dissolved in water at low concentrations up to 30%. However, when used in sucrose or sugar-free systems at high levels of dry solids, acacia gum provides a unique texture to confectionery products (Edwards, 1995). Acacia gum is used in a wide range of finished products including moulded candies, jujubes, pastilles with sucrose or polyols, for coated and non-coated chewy products, for sugar-free hard candies and in different tableting processes where binding properties are needed.

2.4.1.1 *Moulded candies*

For moulded candies made with sucrose, acacia gum is used at different levels depending upon the texture required.

For a hard texture, acacia gum from *A. senegal* is used alone at high concentrations in the finished confectionery. A typical formula contains 35% acacia gum, 30% sucrose, 25% glucose and about 10% water plus flavouring and colouring. Because of the occasional shortage of acacia gum, modified starch has partially substituted the natural gum. And yet, compared to modified starch, it is recognised that hard candy made with acacia gum lasts longer in the mouth, does not stick to the teeth and provides a unique flavour release.

To produce moulded candies with a softer texture, acacia gum is used with other gelling agents, such as gelatine. The combination of the hard texture from acacia gum with the flexible, soft gel from gelatine gives a wine gum-type texture. A typical formula is based on 15% acacia gum from *A. seyal* or *A. senegal* with 5% gelatine, 24% sucrose, 44% glucose, colour and flavour, and the remaining part being 12% moisture.

It is important to note that since acacia gum is all natural, vegetable in origin and a source of fibre, it benefits from a unique marketing image which is used in claims on packaging. In Scandinavian countries, sugar-free gumdrops have always been popular by claiming reduced calorific values and anti-cariogenic properties. Acacia gum is the only ingredient used to guarantee these claims. A formula contains 45–50% of acacia gum (*A. senegal*), with the remaining components being polyols, sorbitol, mannitol or maltitol, sweeteners, liquorice or flavours and, sometimes, ammonium chloride.

The production of moulded candies needs specific equipment which involves the following:

- Efficient stirring/dissolving systems to prepare a 70–75% dry solids syrup of acacia gum and sugars.
- A thermal exchanger followed by a vacuum chamber to fully deaerate the thick syrup.
- An automatic depositing system (Mogul) to mould the syrup at 72–75% dry solids into printed starch trays.
- Ovens for drying the trays at 35–45°C for 2–3 days.
- Machines for removing pastilles from the moulds and for cleaning and glazing the pastilles with sugar or oil.

2.4.1.2 *Chewy confectionery*

A chewy product is a slightly whipped, soft confection containing 74% sucrose and glucose, 5% hydrogenated vegetable fat, flavour and acid. Specific textures are obtained by including 1% gelatine for aeration and 1% gum from *A. seyal* for a long-lasting, cohesive chew. These chewy products can also be used as centres and then sugar coated.

2.4.1.3 *Sugar-free hard candies*

Because of the water-binding properties of acacia gum, the addition of a low level of 2–5% of gum in the formulation of a sugar-free hard candy, based on sorbitol, maltitol or mannitol, slightly increases the amount of residual water by 1–3% after cooking and, therefore, decreases the cooking temperature between 5 and 15°C. Hygroscopicity of the candy is reduced, recrystallisation of polyols is avoided and wrapped sweets are not sticky.

2.4.1.4 *Tableting*

Tableting includes different techniques: direct compression, wet granulation and making lozenges which are a type of tablet. Agglomerated acacia gum from *A. seyal* and

A. senegal is used as a binder in these different processes to make food and pharmaceutical products.

For direct compression, purified and agglomerated acacia gum is mixed with the other powders having the same mesh size before filling the die.

In wet granulation, a solution of acacia gum is added to the powders to make a slurry which is dried and sieved to produce a free-flowing material which is then compressed.

For lozenges, two binders are combined: acacia gum and gelatine are dissolved in water and used to bind the flavoured icing sugar. The paste is then sheeted, cut to shape and dried in an oven. The lozenges are usually flavoured with mint and have an old-fashioned traditional appearance with a rough surface.

2.4.2 Coating and panning

Coated confections, also called dragées, are one of the oldest forms of confectionery. Different types of centres, such as chocolate, lentils, almonds, nuts, jellies, liquor centres and chewing gum, are coated with sugar, polyols or chocolate (Lynch, 1992). Depending upon the centre, the nature of the coating and the required texture, three different processes are used: hard coating, soft coating and chocolate coating (see Fig. 2.3). These processes involve many different steps. Numerous layers of syrup are applied and dried with more than 100 for a hard sugar coating. The complete coating process may last more than 1 day.

Acacia gum is used for its film-forming properties to improve the physical and mechanical parameters of the centres and make the hard- and soft-coating layers more effective. Gum from *A. seyal* is used for this application.

In sugar-coated products, acacia gum is mainly used during the gumming step. An almond or a chocolate lentil contains between 30% and 55% fat. If the centre is not sealed with a flexible film, fat migration or blooming will occur, with cracking and an oxidised taste on the surface of the sugar-coated dragées. In a rotating-pan coating system, by applying a syrup of acacia gum, sucrose and glucose to the centre, fat migration and shell cracking is avoided. In addition, the mechanical resistance of the centre is improved, especially at the corners which tend to break during the rotation of the pan. The surface of the centres will be more even, allowing further sugar layers to adhere more easily. An example gumming syrup formula contains 60% water, 20% acacia gum, 15% sugar and 5% glucose. This syrup is sprayed over the centre and dried by dusting with granulated sugar and then a 50:50 blend of icing sugar and fine-mesh acacia gum. After complete crystallisation overnight, the centre is perfectly sealed and ready to be hard or soft coated.

For multiple-layered sugar coating, the hard-coating layer obtained after forced-air drying will be softer and less brittle if acacia gum is added at low levels of between 3% and 5% to the 80% dry solids content sucrose syrup. Acacia gum is known to act as a plasticiser for sucrose by reducing the formation of large sugar crystals.

In soft coating, a 70% dry solids syrup of a sugar–glucose mix is sprayed to wet the centres and then dried by dusting with icing sugar. Acacia gum can also be added in powder form to reinforce the coating layers.

For chocolate coating, the addition of 30–35% acacia gum solution is used to polish and shine the final layers of chocolate. This shiny effect is sensitive to air humidity and is protected by further layers of shellac.

Recently, sugar-free coated products appeared on the market and they continue to grow rapidly. Consumers are looking for sugar-free dragées because of their low calorific value, non-cariogenic effect and unique release of flavour due to the cooling effect of the polyols.

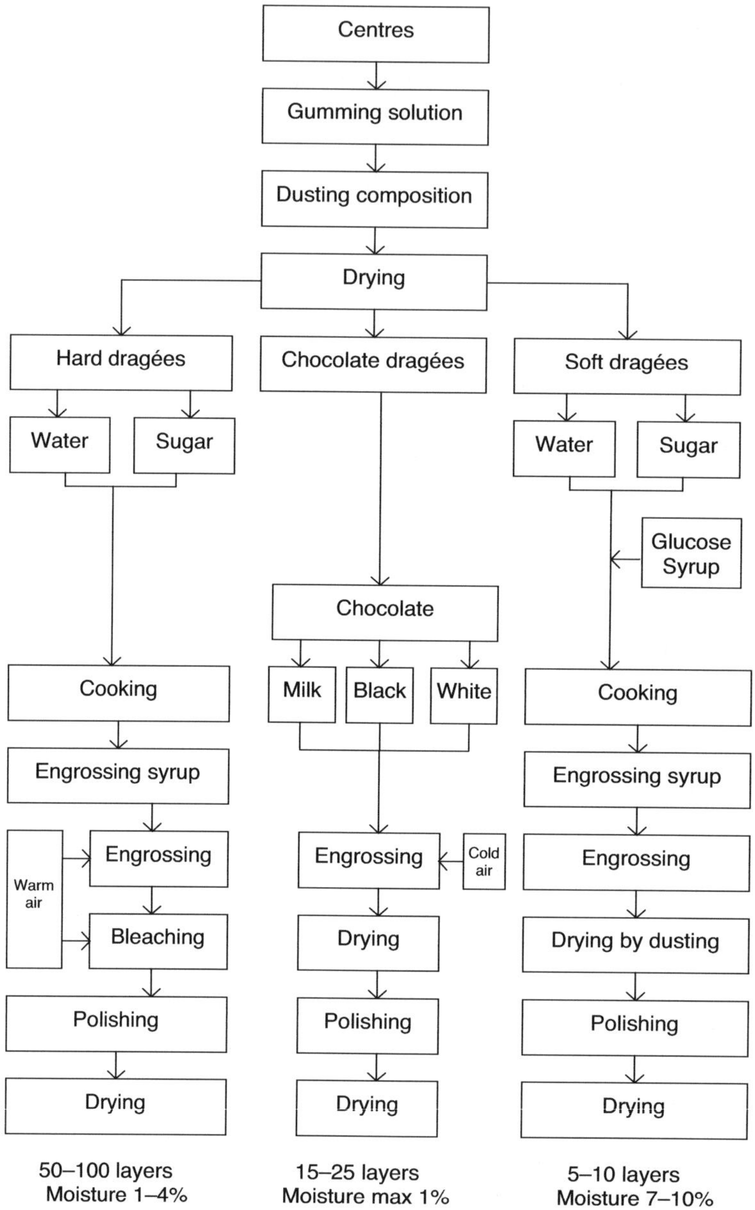

Fig. 2.3 Preparation of pan-coated products.

The most popular sugar-free product is polyol-coated chewing gum in which the gum base is used as a carrier for an active principle suitable for tooth health. Sorbitol, maltitol and xylitol can be used for hard coating. Each of these polyols has its own specific behaviour in terms of stability, hygroscopicity, cooling effect, crunchiness and sweetness. Xylitol and maltitol are the main polyols used for panning sugar-free chewing gum. Acacia gum is used in a hard-coating syrup containing about 65% polyol, 3% acacia gum, 1% titanium dioxide and 31% water. This is sprayed onto the surface of the chewing gum centre and then dried by air at 30°C and 40% humidity. One role of acacia gum is to decrease the crystallisation temperature of the polyol in order to apply the syrup at a lower temperature of 65–70°C so as not to damage the shape of the chewing gum centres. Other benefits of including acacia gum in the coating syrup are improving resistance of the polyol layers and increasing shelf life by decreasing hygroscopicity.

For salted products, such as dry roasted peanuts, acacia gum syrup is used as a glue to adhere salt and spices around the nut before roasting. In a rotating-pan coating cylinder, or in a tunnel equipped with helical brushes and nozzles, a 30% solution of acacia gum is used to coat the nuts. Afterwards, salt and spices are dusted over the wet and sticky surface and the nuts are roasted in an oven. The salt and spices remain well fixed around the nut after packaging. Film-forming properties of the gum prevent fat exudation, maintaining the original non-oxidised taste.

2.4.3 Emulsions

Acacia gum is used as an emulsifier and stabiliser in preparing oil-in-water emulsions. Acacia gum is not considered an actual emulsifier which contains a lipophilic and hydrophilic part in its molecule. Acacia gum is a water-soluble polysaccharide. However, it is possible to give it a hydrophilic–lipophilic balance value (Chun *et al.*, 1958). The protein contained in the AGP fraction of the molecule gives surface-active behaviour to the molecule and allows formation of a colloidal film around the oil droplets as shown in Fig. 2.4 (Randall *et al.*, 1998).

To guarantee a long shelf life of a concentrated or diluted emulsion (normally 1 year is required), it is necessary to ensure stability to avoid creaming, flocculation and coalescence (Dickinson and Galazka, 1991). Stabilisation of the emulsion is obtained by steric hindrance due to the high-molecular-weight fraction of the gum molecule and electric repulsion due to the uronic acids on each dispersed oil droplet.

Stability of the emulsion, expressed as the destabilisation velocity, V, follows Stokes' law. Different parameters are determined either by formulation or by the process of making the

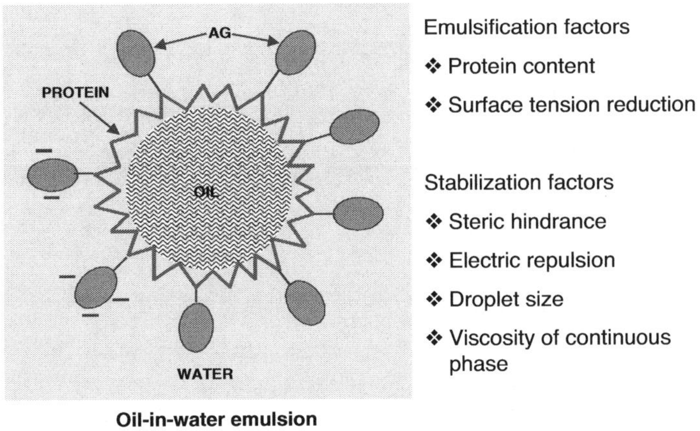

Fig. 2.4 Mechanism of emulsion stabilisation using acacia gum.

emulsion:

$$V = \frac{2gr^2(d_1 - d_2)}{9\eta}$$

where $(d_1 - d_2)$ is the difference in specific gravity between the dispersed and continuous phases. To get it as low as possible, oil with high specific gravity close to 1.0 will be used or oil with lower specific gravity, such as citrus oil with a value of $d_1 \sim 0.80$ will be weighted by a food-approved weighting agent, such as Estergum or sucrose acetate isobutyrate (Tan, 1998).

Here r is the radius of the oil droplet. The emulsion will be processed using high-shear mixing and homogenisation under pressure at 100–300 kg/cm^2 in order to reduce the size of the droplets to between 1 and 0.4 μm (Pandolfe, 1981). This size distribution avoids destabilisation by coalescence and provides a strong clouding effect after dilution (Fig. 2.5).

In the equation, η is the viscosity and has to be in the range of 30–100 mPa s to ensure efficient homogenisation and easy dispersion of the emulsion in syrup.

A comparison between gum from *A. senegal* and *A. seyal* shows that the amount of protein is double in *A. senegal*: 2% for *A. senegal* compared to 1% for *A. seyal*. Since there is more protein available in *A. senegal* than in *A. seyal*, it is obvious that if a stable concentrated or diluted emulsion is needed, gum from *A. senegal* is the one to use. An emulsion processed with *A. seyal* gum will give a slightly bigger oil droplet distribution and the stability will be limited to a few days.

Both acacia gums are very resistant to acidic media with no hydrolysis down to pH 2, which contributes to the guarantee of stability in acidic products, such as citrus and cola beverages.

The level of acacia gum used to stabilise emulsions depends upon the type, dosage and specific gravity of oil (Buffo and Reineccius, 2000). Generally, the gum level is between 12% and 20% of the total formula. Gum is dissolved in water at room temperature and then the oil

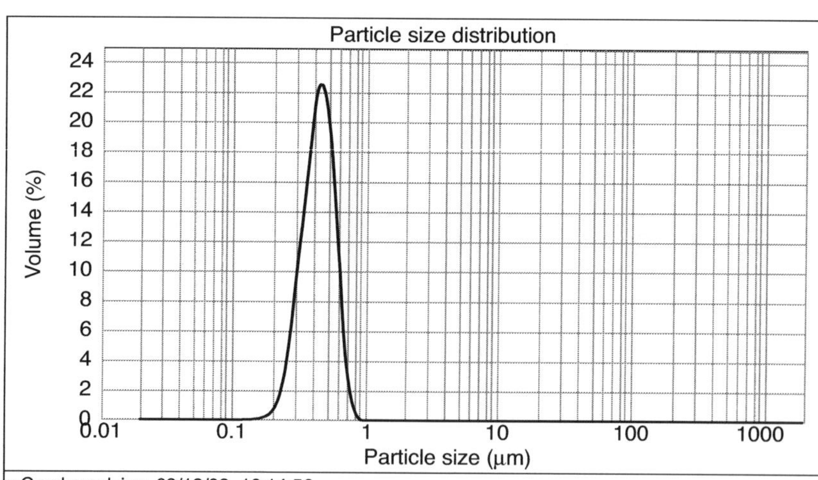

Fig. 2.5 Oil droplet size distribution for a homogenised emulsion obtained by laser granulometry.

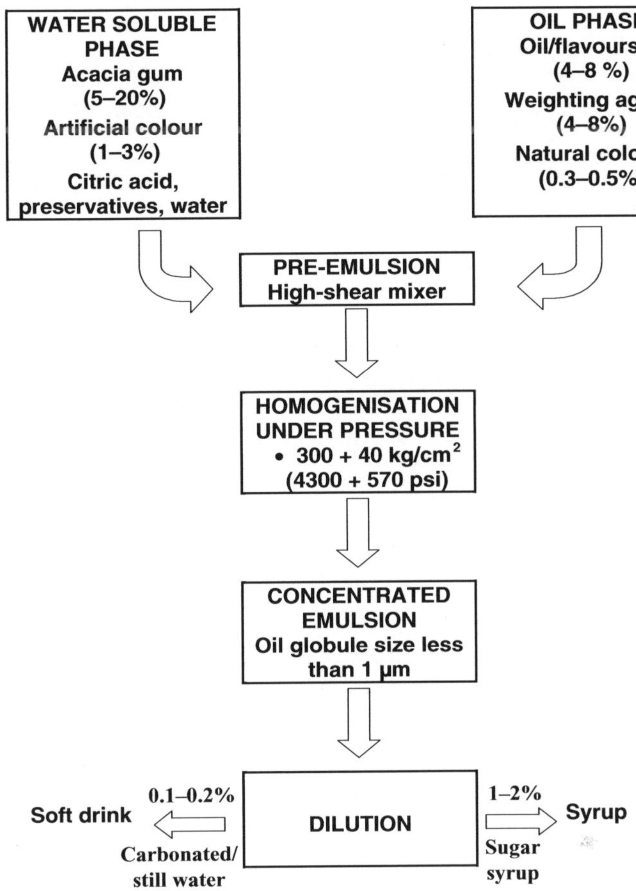

Fig. 2.6 Typical process for beverage emulsions.

phase is added under high-shear mixing to prepare the pre-emulsion with an oil droplet size around 5 μm. The pre-emulsion is homogenised under pressure to produce oil droplets below 1 μm. For pressure homogenisation, two passes are usually applied in a two-step system at pressures up to 300 kg/cm^2 (4300 psi) (Fig. 2.6).

Acacia gum is the main emulsifier used for preparing concentrated emulsions for soft drinks. All formulations are compatible with acacia gum including the following:

- Citrus oils, artificial flavours, cola oils, neutral vegetable oils, triglycerides, etc.
- Artificial colours, such as sunset yellow and tartrazine, or natural colours including carotene and oleoresins.

Concentrated emulsions for beverages, containing about 10–20% oil phase, are normally diluted at a ratio of either 10–20 g/L in high Brix syrup or 1–2 g/L in carbonated water to produce soft drinks (Thevenet, 2002). Such stabilised emulsions could also be diluted for flavouring alcoholic drinks, containing up to 20% alcohol, without any gum flocculation.

Forecasting emulsion stability is essential. Most methods used are based on oil droplet size distribution measurements by laser granulometry or microscopy applied to heated and non-heated emulsions and creaming observations made visually or by transmittance and back scattering. Of course, these two accelerated tests are supplemented by long-term observation of beverages.

Apart from beverage emulsions, acacia gum is used for producing water-dispersible natural colours in liquid form, such as carotene and oleoresin emulsions. For natural nutraceutical emulsions, acacia gum stabilises oil-soluble vitamins or unsaturated fatty acids using the same process as described previously.

2.4.4 Encapsulation

Encapsulation is a general term that includes many different techniques using different carriers including acacia gum (Risch, 1995). The main reasons for encapsulating an active principle are:

- protection against oxidation, water migration or internal reactions;
- production of a water-dispersible, free-flowing powder;
- production of a controlled release powder;
- reduction of hygroscopicity or reduction of dust pollution.

The encapsulation process is generally classified into two groups:

- *Matrix encapsulation.* The dispersion of oil droplets inside the matrix in which acacia gum may be used. Such a system allows 10–30% oil content in the carrier. The matrix system is normally produced by spray drying.
- *Membrane encapsulation.* The formation of membrane around a 'big' oil droplet from 20 to 500 μm in size. Such a system allows encapsulation of up to 90–95% oil.

Processes used to form the membrane system are mainly:

- complex coacervation where acacia gum could be involved, or
- extrusion where the membrane could be gelatine, agar or alginate.

The different technologies used for protecting sensitive products are:

- spray drying,
- spray coating,
- adsorption,
- coacervation,
- extrusion,
- hot-melt coating,
- molecular inclusion.

Each technology involves different equipment, gives a different quality of encapsulation and, of course, has different costs.

Acacia gum is used as a carrier for encapsulation mainly with two different technologies:

- complex coacervation (membrane encapsulation), and
- spray drying (matrix encapsulation).

Complex coacervation involves the reaction between two polymers, a polyanion and a polycation. Since *A. seyal* or *A. senegal* gums have uronic acids as constituents of their molecules, they are negatively charged with a zeta potential around −20 mV. Depending upon the isoelectric point, gelatine may be negatively charged above the isoelectric point or positively charged below the isoelectric point. Complex coacervation occurs when the oil dispersed in a dilute gelatine solution is mixed together with acacia gum solution and the pH reduced to 4, below the isoelectric point. The coacervate is formed around the oil droplet. It is then hardened by adding tannins or glutaraldehyde which reacts with the gelatine. Coacervates in which oil is entrapped can be used as a suspension in water or filtered and dried.

In addition to being the basic patented process for carbonless paper, this technology is used to encapsulate flavours used in food products where the heating or cooking treatment is very drastic, such as microwave cooking, retorting and products that are UHT treated.

Coacervation is also used for perfume encapsulation, for example in 'scratch and sniff' advertising or for contact glues.

Encapsulation using spray-drying technology is used frequently in the food industry because it is an efficient and economical technique and spray driers are available in many factories. This type of encapsulation is summarised in Fig. 2.7. Basically, the oil to be encapsulated is emulsified as small oil droplets below 1 μm using acacia gum as an emulsifier and carrier. The emulsification is performed by high-shear mixing and homogenisation. Then, the emulsion is dried by regular spray drying or by a multi-stage drying unit to produce powder with better water-dispersion properties. A typical formulation to produce powder containing 20% orange oil is as follows:

Formulation 2.1 Liquid emulsion formula

Ingredients	Quantity (%)
Orange oil	7
Acacia gum	28
Water	65

After homogenisation, this emulsion contains 35% dry substance and is spray dried to provide a powder having a composition as follows:

Ingredients	Quantity (%)
Encapsulated orange oil	20
Acacia gum carrier	80

The efficiency of the encapsulation and the quality of the encapsulated oil, expressed as percentage of encapsulated oil, percentage of surface oil, stability against oxidation and stability of oil suspension in an application, depends upon many parameters, especially:

- Concentration and type of flavour components
- Viscosity of the emulsion before drying

- Solid content in the input material
- Oil droplet size distribution in the emulsion
- Dryer inlet and outlet temperature

All these parameters have been studied and are the subject of many publications (Reineccius, 1988). Of course, the choice of the carrier or blend of carriers is one of the most important criteria. Acacia gum is used alone or in association with lower molecular weight polymers, such as maltodextrins with low dextrose equivalents. One purpose of using maltodextrins is to increase the level of dry substance (DS) of the emulsion before spray drying. A DS of 40% or more improves the quality of the encapsulation, as there is quicker formation of the network of the matrix, and also decreases the cost due to increased output of the dryer.

For encapsulation, acacia gum brings emulsification properties as well as film-forming behaviour. Acacia gum from *A. seyal* species has better film-forming properties, in terms of a barrier against oxidation, compared to *A. senegal*. It is worth noting that even though the emulsifying properties of *A. seyal* gum are inferior to those of *A. senegal*, they are sufficient to keep the emulsion stable for a few hours or days prior to spray drying as shown in

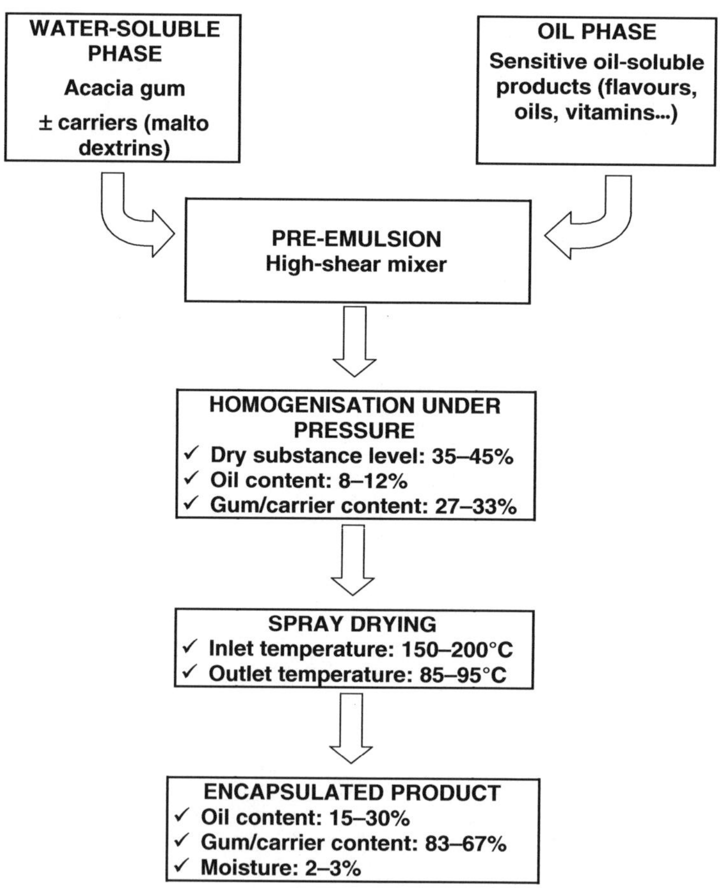

Fig. 2.7 Encapsulation by spray drying with acacia gum.

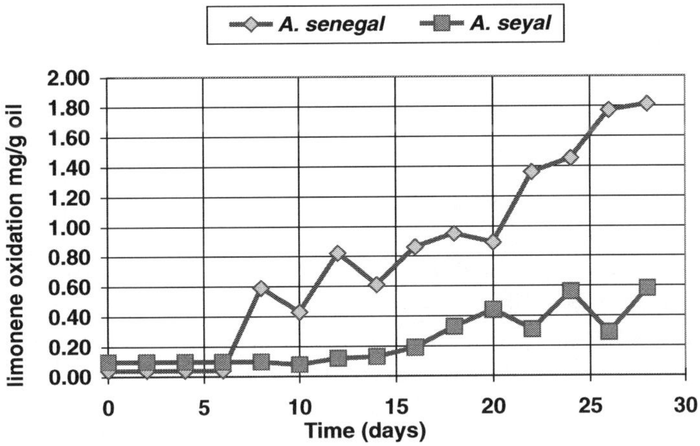

Fig. 2.8 Comparative stability against oxidation of orange oil encapsulated with different acacia gums (20% orange oil, storage at 37°C).

Fig. 2.8 (Thevenet, 1988). Citrus flavours, essential oils, neutral oils, oleoresins, natural colours, oil-soluble vitamins and nutraceuticals can be protected using this encapsulation technology.

Because of its water-binding properties, acacia gum is very often used in spray drying, together with maltodextrins, to reduce the hygroscopicity and to increase the yield of powders obtained from hygroscopic water-soluble products such as fruit juices, sugars or polyols.

2.4.5 Bakery

Acacia gum brings many benefits to bakery products in terms of processing, texture and shelf life due to its moisture regulation and film-forming properties.

In extruded products, such as breakfast cereals and snacks, the addition of a low level of acacia gum (2–5%) increases extruder output, due to a lubricant effect, allows better definition of the shape of the finished product and improves crispiness and crunchiness.

In cereal bars, besides nutritional benefits, the addition of 4–8% acacia gum controls the texture during storage by stabilising the moisture and effectively binding the dry components of the formula, such as fruits and cereals. Acacia gum is also used as a sugar substitute for reducing calorific value and improving the binding properties of the syrup.

Finally, shelf life is extended due to better moisture retention. Benefits are clearly shown in a wide range of bakery products, such as bread, sandwich loaf, croissant, buns, Melba toast and biscuits.

2.4.6 Wine stabilisation

Acacia gum is a traditional enological additive used to protect red and white wine against destabilisation. *A. senegal* and *A. seyal* are specified in the International Enological Codex (27-2000; N° INS 414).

Acacia gum has a chemical composition very similar to the natural colloidal substances contained in wine, the AGP fractions. The stabilising power of these inherent substances is well established. Acacia gum is normally added to the wine before the last filtration, at low levels of 10–30 g per 100 L, to prevent proteic, polyphenolic, copper, iron or tartaric off-notes and also to smooth the effects of tannin and to bring some mouthfeel.

2.5 HEALTH BENEFITS

Because of the increasing awareness of the connection between health and diet, a new concept of nutrition emerged in the early 1990s in Japan. This concept is now recognised worldwide as 'functional foods', that is foods that can improve physiological functions of the body or reduce the risk of specific diseases. The definition of functional foods, as well as their legal status, is being defined at the Codex level and by other regulatory bodies. This implies that every claim will have to be scrupulously scientifically demonstrated before it may be used on labels.

2.5.1 Dietary fibre

Dietary fibre is a general term covering a wide variety of substances which are neither digested nor absorbed in the small intestine. These substances include insoluble fibres, such as cellulose and hemicelluloses, and soluble fibres sub-classified as 'high viscosity' (guar, pectins...) and 'low viscosity' (acacia gum, fructo-oligosaccharides...). The health properties of dietary fibres depend upon their physicochemical properties. Benefits of dietary fibre include improved bowel functioning as well as a reduced colorectal risk and reduced risks of obesity and diabetes.

The promotion of acacia gum fibre is based on the results obtained from specific tests and clinical studies on a selected type of acacia gum with a traceable botanical origin, country and area of collection and defined chemical composition. Fibregum™, a range of selected acacia gum products from CNI Company, has a guaranteed dietary soluble fibre content of more than 90% on a dry basis as determined by the AOAC 985.29 and 991.43 methods. Compared to other low-viscosity soluble fibres, acacia gum fibres are highly appreciated for their resistance in acidic media. In finished products, the acacia gum fibre level and ability to improve mouthfeel and enhance flavour release are guaranteed for 1 year with a pH level down to 2.8. On packaging, fibre claims have to follow the recommendation of the Codex Alimentarius (see Table 2.2).

Table 2.2 Recommendations for fibre claims in Codex Alimentarius ALINORM 09/32/26 Appendix II.

High in fibre	≥6 g/100 g ≥3 g/100 kcal 20% DRV[a] per serving
Source of fibre	≥3 g/100 g ≥1.5 g/100 kcal or 10% DRV[a] per serving

[a] Daily Reference Value (DRV) and serving size to be determined at national level.

2.5.2 Prebiotics

Prebiotics were first defined in 1995 (Gibson and Roberfroid, 1995) as 'a non-digestible food ingredient that beneficially affects the host by selectively stimulating the growth and/or activity of one of a limited number of bacteria in the colon and thus improves host health'. The food component must resist digestion in the stomach and the small intestine, but it has to be fermented by the microflora colonising the gastrointestinal system. It has also to promote selectively the beneficial bacteria from the lactic acid bacteria group including bifidobacteria and lactobacilli.

More than 20 studies have been performed to explain the relationship between acacia gum, Fibregum™ and colonic microflora. The latest clinical studies confirmed that concentrations of bifidobacteria, lactobacilli and total lactic acid bacteria group were significantly increased by Fibregum™ at a dose of 10 g per day (Michel *et al.*, 1998; Cherbut *et al.*, 2003).

2.5.3 High gut tolerance

In contrast to other low-viscosity fibres, in clinical tests up to 50 g per day of Fibregum™ did not cause laxative side effects. This high gut tolerance may be explained by progressive fermentation along the gut. The high molecular weight, the high degree of branching and the compact structure make attack by bacteria difficult and slow. A slow fermentation lowers total gas production and avoids painful bloating and flatulence (Michel *et al.*, 1998; Cherbut *et al.*, 2003).

2.5.4 Hypoglycaemic effect

Studies performed on acacia gum showed that the intrinsic glycaemic index (GI) is almost nil as it is not digested in the small intestine (Sharma, 1985). Added to diets or finished products, studies demonstrated that a low level of Fibregum™ between 3% and 6% reduces GI. As a result, a bakery product, such as crispbread, may have its GI downgraded from high GI to medium GI. (High GI is >70 and medium GI is from 56 to 69 when measured using a validated scientific method developed at Sydney University's Glycaemic Index Research Service.)

Acacia gum is well known for its resistance to hydrolysis by salivary enzymes and local flora which explains why it is a non-cariogenic polysaccharide (Imfeld and Meance, 2003). Acacia gum is not digested in the small intestine and therefore does not constitute a source of metabolisable energy. However, like other fermentable dietary fibre, acacia gum is fermented by bacteria present in the large bowel, leading to the synthesis of short chain fatty acids that are metabolised by epithelial cells of the intestinal wall and other peripheral tissues, such as the liver. The calorific value of acacia gum is 2 kcal/g in Europe (Commission Directive 2008/100/EC of October 2008). Regulations in different regions consider different values, such as 1.7 kcal/g in the USA (Hills, 2008) and 1 kcal/g in Japan.

2.5.5 Applications of acacia gum for health benefits

Acacia gum is a natural vegetable product, a non-cariogenic soluble fibre with prebiotic and hypoglycaemic effects, having a low calorific value. It is used and labelled as an ingredient in numerous health foods (Meance, 2004a). Because of the low level of between 1% and 7% in the finished product, the addition of selected and guaranteed acacia gum does not change the texture of the product as it is a low-viscosity fibre. At such a level, apart from the nutritional aspects, acacia gum also acts as a film former, moisture stabiliser and mouthfeel enhancer (Meance, 2004b). Drinks, meal substitutes, cereal bars, bakery products, dairy products and confections already use acacia gum for its health benefits. After a careful audit of the origin and purification process by different associations, acacia gum is also certified as an organic material.

2.6 FUTURE DEVELOPMENTS

Acacia gum is a natural vegetable polysaccharide used in food and pharmaceutical applications for a variety of technological properties as a film former, emulsifier, texturising agent and stabiliser.

For the past few years, the development of nutraceutical products has been providing acacia gum with new applications as a stable and well-tolerated source of soluble fibre with proven prebiotic and hypoglycaemic effects. These new developments have just started. Acacia gum is one of the few fibres which have a complex structure that can link to other nutraceutical bases such as polyphenols or minerals.

Securing the supply of acacia gum has always been a major concern for the gum industry. The following steps have been taken to strengthen the credibility of this unique natural hydrocolloid:

- Diversification of the botanical sources of acacia gum from *A. senegal* and *A. seyal* species.
- Expanding production in the geographical harvesting zone by increasing plantation areas.
- Maintaining stable buffer stocks.

DISCLAIMER

The information contained in this chapter is correct to the best of our knowledge. The descriptions, recommendations or suggestions made are non-contractual and provide no guaranteed results. None of the statements are to be construed as violating any copyright or patent. They are intended only as a source of information.

References

Al-Assaf, S., Phillips, G.O. and Williams, P.A. (2005a) Studies on acacia exudate gums. Part I: the molecular weight of *Acacia senegal* gum exudate. *Food Hydrocolloids*, **19**, 647–660.

Al-Assaf, S., Phillips, G.O. and Williams, P.A. (2005b) Studies on Acacia exudate gums: part II. Molecular weight comparison of the *Vulgares* and *Gummiferae* series of Acacia gums. *Food Hydrocolloids*, **19**, 661–667.

Anderson, D.M.W. and Farquhar, J.G.K. (1974) The composition of eight Acacia gum exudates from the series *Gummiferae* and *Vulgares*. *Phytochemistry*, **18**, 609–610.

Anderson, D.M.W. and McDougall, F.J. (1987) The amino acid composition and quantitative sugar-amino acid relationships in sequential Smith-degradation products from gum arabic (*Acacia senegal* (L.) Willd.). *Food Additive and Contaminants*, **4**(2), 125–132.

Biswas, B. and Phillips, G.O. (2003) Computation of specific optical rotation from carbohydrate composition of exudate gums *Acacia senegal* and *Acacia seyal*. *Food Hydrocolloids*, **17**, 177–189.

Buffo, R. and Reineccius, G. (2000) Beverage emulsions and the utilization of gum acacia as emulsifier/stabilizer. *Perfumer and Flavorist*, **25**(July/August), 24–44.

Cherbut, C., Michel, C., Raison, V., Kravtchenko, T. and Méance, S. (2003) Acacia gum is a bifidogenic dietary fiber with high digestive tolerance in healthy humans. *Microbial Ecology in Health and Disease*, **15**, 43–50.

Chun, A.H.C., Joslin, R.S. and Martin, A.N. (1958) The hydrophile–lipophile balance of gums (part one). *Drug and Cosmetic Industry*, **82**(2), 164–260.

Connolly, S., Fenyo, J.C. and Vandevelde, M.C. (1987) Heterogeneity and homogeneity of an arabinogalactane-protein: *Acacia senegal* gum. *Food Hydrocolloids*, **1**(5/6), 477–480.

Connolly, S., Fenyo, J.C. and Vandevelde, M.C. (1988) Effect of a proteinase on the macromolecular distribution of *Acacia senegal* gum. *Carbohydrate Polymers*, **8**, 23–32.

Coppen, J.J.W. (1995) Exudate gums: gum arabic, gum talha and other acacia gums. In: J.J.W. Coppen (ed.), *Gums, Resins and Latexes of Plant Origin*. FAO, Rome, pp. 11–29.

Dickinson, E. and Galazka, V.B. (1991) Emulsifying behaviour of gum arabic. Part 1: effect of the nature of the oil phase on the emulsion droplet-size distribution. *Carbohydrate Polymers*, **14**, 373–383.

Dondain, G. and Phillips, G.O. (1999) The regulatory journey of gum arabic. *Foods and Food Ingredients Journal of Japan*, **179**, 38–56.

Edwards, W.P. (1995) Gums and gelling agents. In: E.B. Jackson (ed), *Sugar Confectionery Manufacture*. Blackie Academic & Professional, London, pp. 38–61.

Fincher, G.B., Stone, B.A. and Clarke, A.E. (1983) Arabinogalactan-proteins: structure, biosynthesis, and function. *Annual Review of Plant Physiology*, **34**, 47–70.

Flindt, C., Al-Assaf, S., Phillips, G.O. and Williams, P.A. (2005) Studies on acacia exudate gums. Part V. Structural features of *Acacia seyal*. *Food Hydrocolloids*, **19**, 687–701.

Gibson, G.R. and Roberfroid, M.B. (1995) Dietary modulation of the human colonic microbiota: introducing the concept of prebiotics. *Journal of Nutrition*, **125**(6), 1401–1412.

Hills, S. (2008) New gum arabic calorie count a boost to formulation. *Food Navigator*. Available at http://www.foodnavigator-usa.com/Product-Categories/Emulsifiers-stabilizers-hydrocolloids/New-gum-arabic-calorie-count-a-boost-to-formulation (accessed 5 August 2009).

Hassan, E.A., Al-Assaf, S., Phillips, G.O. and Williams, P.A. (2005) Studies on Acacia gums: Part III molecular weight characteristics of *Acacia seyal* var. *seyal* and *Acacia seyal* var. *fistula*. *Food Hydrocolloids*, **19**(4), 669–677.

Imeson, A. (1992) Exudate gums. In: A. Imeson (ed), *Thickening and Gelling Agents for Food*. Blackie Academic & Professional, London, pp. 66–97.

Imfeld, T. and Meance, S. (2003) Evaluation of the tooth safety of Acacia Gum FIBREGUM, at different concentrations in healthy humans, 1 January 2003, Personal Communication.

Islam, A.M., Phillips, G.O., Sljivo, A., Snowden, M.J. and Williams, P.A. (1997) A review of recent developments on the regulatory, structural and functional aspects of gum arabic. *Food Hydrocolloids*, **11**(4), 493–505.

Jurasek, P., Kosik, M., and Phillips, G.O. (1993) A chemometric study of the Acacia (gum arabic) and related natural gums. *Food Hydrocolloids*, **7**(1), 73–85.

Lynch, M.J. (1992) Panning: an overview. *The Manufacturing Confectioner*. MC Publishing, Chicago, pp. 59–64.

Meance, S. (2004a) Acacia gum (FIBREGUM), a very well tolerated specific natural prebiotic having a wide range of food applications – Part 1. *Agro Food Industry Hi-tech*, **15**, 24–28.

Meance, S. (2004b) Health effects and food applications of Acacia Gum, Fibregum – Part 2. *Agro Food Industry Hi-tech*, **15**, 32–35.

Meer, W. (1980) Gum arabic. In: R.L. Davidson (ed), *Handbook of Water-Soluble Gums and Resins*. McGraw-Hill, New York, pp. 8.1–8.24.

Michel, C., Kravtchenko, T., David, A., Gueneau, S., Kozlowski, F. and Cherbut, C. (1998) In vitro prebiotic effects of Acacia gums onto the human intestinal microbiota depends on both botanical origin and environmental pH. *Anaerobe*, **4**, 257–266.

Pandolfe, W.D. (1981) Effect of dispersed and continuous phase viscosity on droplet size of emulsions generated by homogenization. *Journal of Dispersion Science and Technology*, **2**(4), 459–474.

Picton, L., Bataille, I. and Muller, G. (2000) Analysis of a complex polysaccharide (gum arabic) by multi-angle laser light scattering coupled on-line to size exclusion chromatography and flow field flow fractionation. *Carbohydrate Research*, **42**, 23–31.

Randall, R.C., Phillips, G.O. and Williams, P.A. (1998) The role of the proteinaceous component on the emulsifying properties of gum arabic. *Food Hydrocolloids*, **2**(2), 131–140.

Reineccius, G. (1988) Spray-drying of food flavors. In: S.J. Risch and G. Reineccius (eds), *Flavor Encapsulation*, ACS Symposium Series 370. American Chemical Society, Washington, DC, pp. 55–66.

Risch, S.J. (1995) Encapsulation: overview of uses and techniques. In: S.J. Risch and G. Reineccius (eds), *Encapsulation and Controlled Release of Food Ingredients*, ACS Symposium Series 590. American Chemical Society, Washington, DC, pp. 2–7.

Sanchez, C., Renard, D., Robert, P., Schmitt, C. and Lefebvre, J. (2002) Structure and rheological properties of acacia gum dispersions. *Food Hydrocolloids*, **16**, 257–267.

Sharma, R.D. (1985) Hypoglycemic effect of gum acacia in healthy human subjects. *Nutrition Research*, **5**, 1437–1441.

Street, C.A. and Anderson, D.M.W. (1983) Refinement of structures previously proposed for gum arabic and other Acacia gum exudates. *Talanta*, **30**(11), 887–893.

Tan, C.-T. (1998) Beverage flavor emulsion – A form of emulsion liquid membrane microencapsulation. In: E.T. Contis, C.-T. Ho, C.J. Mussinan, T.H. Parliment, F. Shahidi and A.M. Spanier (eds), *Food Flavors: Formation, Analysis and Packaging Influences*, Elsevier Science, the Netherlands, pp. 29–42.

Thevenet, F. (1988) Acacia gums: stabilizers for flavor encapsulation. In: S.J. Risch and G. Reineccius (eds), *Flavor Encapsulation*, ACS Symposium Series 370. American Chemical Society, Washington, DC, pp. 37–44.

Thevenet, F. (2002) *Acacia Gums: Emulsifiers for Beverage Emulsions*. Third World Congress on Emulsions, Lyon, France.

Vassal, J. (1985) Gummiferous acacias and gum productivity some aspects of current research. *Bulletin IGSM*, **13**, 30–37.

3 Agar

Alan Imeson

ABSTRACT

Agar is extracted from red seaweed (*Rhodophyceae*) and has been used in foods for over 350 years. It is insoluble in cold water and hydrates when boiled. Cooling solutions below about 40°C produce very firm brittle gels which can be melted by heating above 85°C. Food applications include water dessert gels, aspics, confectionery jellies, canned meats, icings, piping gels and flan desserts.

3.1 INTRODUCTION

Agar is a seaweed hydrocolloid, or phycocolloid, with a long history of use as a gelling, thickening and stabilising food additive. Agar is considered to have been discovered in the mid-seventeenth century in Japan, 200 years before it was introduced to the West.

The Japanese name for this phycocolloid is 'Kanten' meaning 'cold weather', and in China it is 'Dongfen' or 'frozen powder', which both refer to the traditional methods of producing this material in winter (FAO, 1990). According to a Japanese legend, the original manufacturing method was discovered in the winter of 1658 when it is reported that a Japanese officer arrived at a little inn. The innkeeper, Minoya Tarozaemon, offered a traditional seaweed jelly for dinner which was prepared by cooking *Gelidium* seaweed with water. After dinner, the surplus jelly was thrown outside by the innkeeper. The jelly froze during the night and then thawed and dried in the sun, leaving a dry, white residue. When this was boiled in water and cooled, it produced a clearer jelly than the original and, thus, the method for extracting and purifying agar was accidentally discovered (Armisen, 1995).

The use of agar in foods is widespread throughout the Far East, including Japan, China, Taiwan, Korea, the Philippines and Indonesia, and it has contributed significantly to the food customs in these countries for 350 years.

In fact, the word 'agar' is Malayan where it is used in the double form, 'agar-agar', to describe jelly (FAO, 1990). Chinese workers in Malaya imported Japanese 'Kanten' for their own use and also called it agar-agar. Later, European traders learnt about the use of this Japanese phycocolloid for making fruit jellies and, when it was introduced into Europe, the Malayan term became attached to the Japanese seaweed extract.

In Taiwanese, the name for agar translates as 'vegetable swiftlet' referring to the similarity of the gel texture to the swiftlet nest used to produce birds' nest soup. The introduction of agar into Europe dates back to 1859, almost 200 years after its initial discovery, when Payen presented agar to the Academy of Sciences in Paris, together with swiftlet nests, brought from China as food products (Payen, 1859).

Agar is widely used for solid culture media in microbiology. Although this application is credited to Robert Koch, it is now accepted that his wife, Angelina, first suggested the use of this material (Guiry, 2009). She had learnt of the use of agar-agar in jams and jellies from a Dutch immigrant returning from Java. This gelling agent melted at higher temperatures than gelatine and enabled the jellies to tolerate the hot and humid climate conditions. Subsequently, Koch and his assistant Hesse perfected the extraction of agar for use in microbiological plates leading to Koch's discovery of the tuberculosis bacillus and the promotion of agar for microbiological media in general (Koch, 1882).

Agar production in Japan increased up to the end of the nineteenth century when exports to the West began. When the US Food and Drug Administration (FDA) started classifying food additives, agar was immediately accorded the status of 'generally recognised as safe' (GRAS) (FDA, 1972) and authorised the use of agar-agar in different food categories with specified maximum use levels.

A review of the toxicological status of agar by the FAO/WHO noted that 'it is consumed traditionally as a food though not as a nutrient' and estimated that the acceptable daily intake in man should be 'not limited' (FAO/WHO, 1974).

3.2 RAW MATERIALS

Agar is a family of linear galactan polysaccharides obtained from the cellular walls of red seaweeds, *Rhodophyceae*. These seaweeds are widely distributed around the world and come from 20 countries; the most important sources are the coasts of Japan, Spain, Portugal, Morocco, Senegal, Chile, Mexico, the southern USA, India, the Philippines and Madagascar (McHugh, 1991).

Gelidium and *Gracilaria* are the main species utilised for commercial agar production. *Gelidium* species were the original materials used in Japan, but shortages in World War II led to the discovery that *Gracilaria* species were suitable if the seaweed was treated with alkali during processing (FAO, 1990).

The highest gel strength agar is generally obtained from *Gelidium* seaweeds. These plants have a cylindrical stalk at the base anchored to the rocks by a holdfast and open out into fronds about 2 mm thick. The plants are usually found on westerly facing coasts on rocks situated just below low-tide level (the littoral zone) where they are collected by divers or manually harvested at the very low tides around the equinox. The main countries supplying material include Spain, Portugal, Morocco, Japan, South Korea, China, Chile and South Africa. These small, slow-growing plants are only 10–35 cm high, and attempts to cultivate them have not been commercially successful so the raw material is obtained from natural seaweed beds.

Each year, around 35 000–40 000 tonnes dry weight of *Gracilaria* seaweed are processed, amounting to about two-thirds of the annual quantity used for agar production (see Table 3.1). Most *Gracilaria* is obtained from seaweeds growing naturally. To meet the increasing demand for food-grade agar, additional seaweed sources continue to be investigated. Over recent years, potential new supplies of *Gracilaria* have been identified from *Gracilaria verrucosa* in Turkey (Yenigül, 1993), *Gracilaria dura* from the Thau lagoon in the Mediterranean (Marinho-Soriano and Bourret, 2005) and *Gracilaria tenuistipitata* in the Philippines (Montaño *et al.*, 1999).

Table 3.1 Agarophyte seaweeds harvested in 2001 (tonnes dry weight).

Gracilaria	
Europe	200
Africa	300
Americas	25 000
Asia Pacific	11 500
Subtotal	37 000
Gelidium	
Europe	6 600
Africa	7 200
Americas	500
Asia Pacific	4 300
Subtotal	18 600
Pterocladia	
Europe	50
Total	55 650

Source: McHugh (2003).

Gracilaria seaweeds have been successfully cultivated on a commercial scale, particularly in Southeast Asia and South America. A number of techniques are used to cultivate *Gracilaria* including bottom culture, raft culture, stake-rope culture and pond culture with this latter method being divided into monoculture or polyculture with shrimp and other species (FAO, 1990). Each year, South China and Taiwan produce about 3000 and 1000 tonnes of dried weed, respectively. The cultivation of *Gracilaria* has also been very successful in Chile and both wild and cultivated material is available from Argentina, Japan, South Africa, Indonesia, the Philippines, South Korea and India.

Other species used as minor sources of raw material are *Pterocladia*, a small seaweed from the Azores and New Zealand, and *Gelidiella* from India, Egypt and Madagascar.

Annually, about 55 000 tonnes of dried seaweed are used to produce 7500 tonnes of agar. Chile, Spain and Japan produce 60% of the total agar output. The agar industry has grown slowly over recent years, recording growth of 1–2% over the past 30 years (McHugh, 2002).

Agar-containing seaweeds contain the polysaccharides as an amorphous matrix within the framework in the cell walls. This framework, which comprises less than 10% of the dry weight of the seaweed, is formed from linear, neutral polymers, with cellulose being the most common. Xylans and mannans may also be present in the framework.

Agar is the principal matrix polysaccharide and it is believed to be secreted by the Golgi apparatus in the cells. Scanning electron micrographs of *Gelidium sesquipedale* have clearly shown the very thick cell walls of the interior cells, as well as rhizoids, which are much longer, smaller diameter cells which run longitudinally through the seaweed (Vignon *et al.*, 1994). The rhizoid walls are much thicker than the other cells and it is considered that the greatest amounts of agar are found here. A more detailed study of the rhizoids is expected to give some information on the enzymatic processes involved in the biosynthesis of agar.

3.3 PRODUCTION

The production of agar was initiated in Japan in the middle of the seventeenth century. In the traditional production process, agarophyte extracts, called Tokoroten in Japan, are poured into trays and left to gel by cooling. Depending on the volume of the gel, it can be cut into 'strip' or 'square' form before using the natural freezing process to produce what are commonly known as 'natural agars' (Armisen, 1997). Firm brittle gels are not freeze–thaw stable as the rigid gel network is physically disrupted by the growth of ice crystals and so the freeze–thaw process enables the gel to be dewatered and concentrated prior to drying.

Before World War II, the commercial production of agar in Japan was based on numerous small factories using the traditional production process. These small family concerns were labour intensive, the process was not mechanised, and product quality was inconsistent and unstandardised. It has only been relatively recently that large industrial plants have been utilised.

In the USA, agar production started in 1922 in Glendale, California, employing industrial freezers to freeze and dewater the gel (Matsuoka, 1921). The process was developed further with support from the US government which wanted the USA to be self-sufficient in agar for its use in microbiological media.

During World War II, the shortage of agar was an incentive for countries to utilise *G. sesquipedale*, which is very similar to *Gelidium pacificum* used in Japan. Agar production was initiated in Portugal and Spain (Armisen and Galatas, 1987) and, subsequently, the commercially valuable agar industry in Iberia was established by the merger of several small manufacturers of food-grade agar (Hispanagar, 2009).

Originally, *Gracilaria* seaweeds were considered unsuitable for agar production as the extracts gave very soft gels. However, in the 1950s, it was found that pre-treatment with alkali was able to modify the extract and produce firm-gelling material that was suitable for food use (McHugh, 2002). Also at this time, the use of syneresis was introduced in Japan for agar produced from *Gracilaria* and, subsequently, adapted and mechanised by Iberian manufacturers for use with *Gelidium*, *Pterocladia* and other seaweeds.

The extraction process utilises syneresis, a phenomenon in which the free interstitial water is expelled from the gel network. This may be achieved as a result of the polymer chains associating during storage or by mechanical pressure. As described in Chapter 4, syneresis in alginate gels is caused naturally by an excess of calcium ions which progressively cross-link adjacent alginate chains, and the increased number of junction zones cause the network to tighten and express free water. With agar, syneresis is initiated by the application of external pressure: it is the most effective method to concentrate the polysaccharides in the gel with the least energy consumption. This technique also improves the purity of the extract as the interstitial water, which contains impurities such as mineral salts and other soluble materials, is expressed in larger volumes by pressure than in the natural freeze–thaw process.

The production process for agar is shown in Fig. 3.1. Bales of washed, dried seaweed are shipped to the manufacturing plant marked with details of the species and date and place of collection. *Gelidium* seaweeds from France, Iberia and North Africa can be stored for many months, and even several years, if they are well dried, without losing the properties needed for agar production, such as high yield and gel strength (R. Armisen, personal communication). However, the agar quality deteriorates if the seaweed is humid. Recent studies have shown that agar extracted from *Gracilaria* seaweeds stored up to 3 months has physical and textural properties suitable for food use. However, longer periods of storage are not recommended as gel strength greatly decreases, most probably as a result of agarase hydrolysis and depolymerisation (Romero *et al.*, 2008). Thus, the raw materials are closely controlled to ensure that they give extracts with the required product characteristics.

The production process can be separated into key stages of pre-treatment, extraction, filtration, concentration, dehydration and grinding (Armisen and Galatas, 1987). Pre-treatment

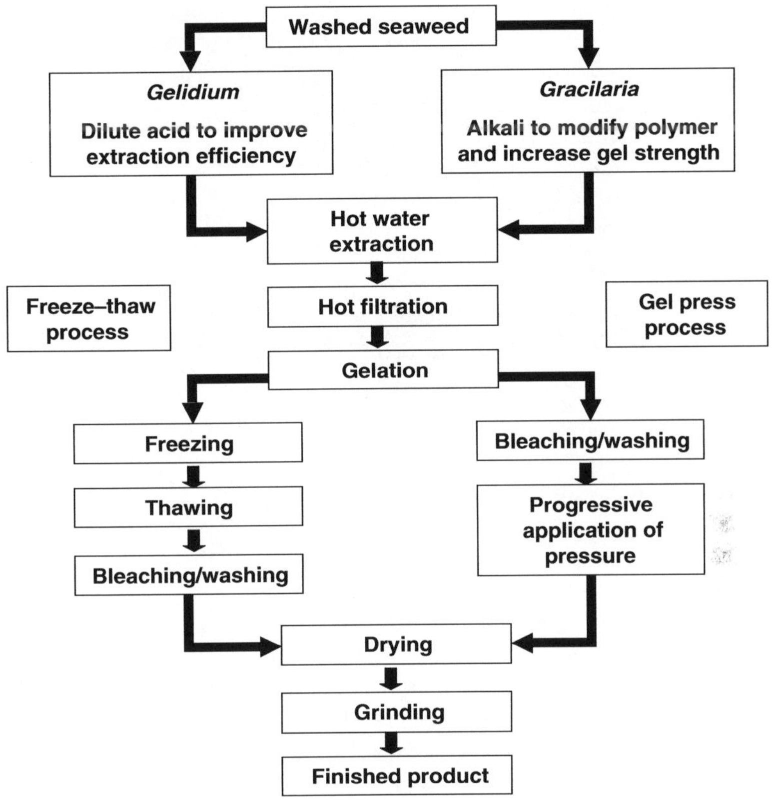

Fig. 3.1 Manufacturing processes for agar.

for *Gracilaria* seaweeds involves washing followed by alkali treatment in 2–5% sodium hydroxide at 85–90°C for 1 h to convert sulphate groups to 3,6-anhydrogalactose. Without this pre-treatment, the gel strength of the extract is unsuitable for most food applications. *Gelidium* weeds are more resistant to extraction and may be pre-treated with acid to improve extraction efficiency. Agar is extracted as a dilute solution of about 1–2% in water by treating chopped weed with hot water under pressure for 2–4 h at 105–110°C for *Gelidium* spp. and 95–100°C for *Gracilaria* materials. The extract is filtered hot to remove insoluble material and then cooled to form a gel.

In the freeze–thaw process, the gel is slowly frozen in order to develop large ice crystals. As the gel is rigid and inflexible, it cannot accommodate the large ice crystals so the polysaccharide is concentrated into bundles, forming a sponge-like structure. When the block is thawed, the water drains away. Pressure may be used to increase the volume of expressed water, concentrating the agar to 10–12% (McHugh, 2003).

The gel press process uses syneresis to dewater the gel. Mechanical pressure of 5–10 kg/cm^2 is applied progressively to the gel (Setexam, 2009). The polymer chains associate and free water is expressed, increasing the agar concentration to about 20%. The gel press process is more energy efficient than the freeze–thaw process and it gives a more concentrated cake which can be dried more efficiently.

After concentrating, the agar strips or flakes are dried with hot air and ground to the appropriate mesh size, usually 80–100 mesh (100–150 μm).

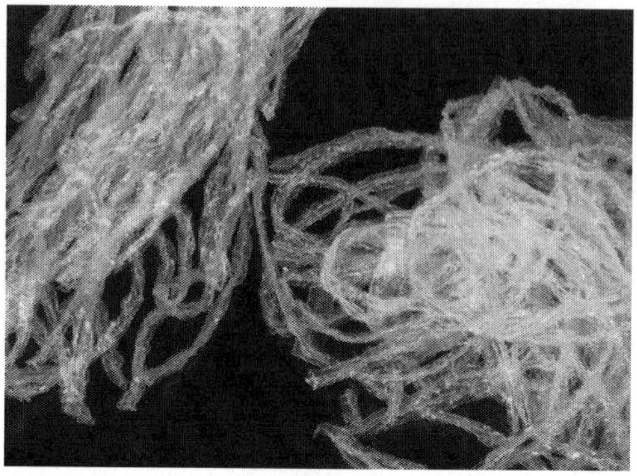

Fig. 3.2 Strip agar. (Reproduced with permission of Wing Yip Store, www.wingyipstore.co.uk.) For a colour version of this figure, please see Plate 1 of the colour plate section.

'Natural' agars, which are sold commercially in strip (Fig. 3.2) or square form, are produced mainly in the Far East, principally in Japan and Korea, and used in traditional oriental cooking. In the West, these agars are sold in small quantities through specialist shops and health food stores.

The gelling capacity of strip agar is very weak, being in the range of 150–300 g/cm^2 for a 1.5% gel measured by the Nikan-Sui method. The production of 'natural agar' has continually decreased over the past 40 years, as traditional oriental cooking is substituting agar which comes in powder or tablet forms. The annual production of 'natural agar' is around 100 tonnes, amounting to around 1.5% of the world production.

Some foods are sensitive to high temperatures, so the need to fully hydrate agar by boiling may limit its use in some gel applications. As a consequence, readily-soluble agar has been developed by further processing of dried agar powder or dehydrated agar gel granules obtained from the gel press or freeze–thaw processes. The agar is mixed with water and other materials such as sugar, locust bean gum, guar gum or other dispersants and passed through a single- or twin-screw extruder. When the co-processed material is dispersed in water above the gel point, the solutes dissolve and the agar network is able to associate with water molecules and hydrate. The improved dispersion and hydration properties allow the gelling agent to be used more easily in confectionery, bakery, dairy and water gel applications (Labbar *et al.*, 1996).

3.4 COMPOSITION AND STRUCTURE

Agar has been recognised as a polysaccharide since its introduction into Europe in the middle of the nineteenth century. Initially, it was assigned a linear galactan configuration, but later small quantities of sulphate were identified within the macromolecule.

The sulphate content of agar is below 4.5% which is very low compared to that of carrageenans, another group of polysaccharides which are obtained from red seaweeds, *Rhodophyceae*. Typically, the sulphate content of agar is 1.5–2.5%, in contrast to the ester sulphate content of kappa, iota and lambda carrageenans of 22, 32 and 37%, respectively (Blakemore and Harpell, 2009).

Different species of *Gracilaria* contain agarophytes with different levels of sulphate, and alkali treatment during extraction converts any L-galactose-6-sulphate groups to

Table 3.2 The chemical composition and gelling parameters of agar from Chinese *Gracilaria* spp.

Gracilaria species	Gel strength (g/cm^2)	Gelling temperature (°C)	Melting temperature (°C)	Hysteresis (°C)	3,6-anhydro-L-galactose (%)	Sulphate (%)
G. asiatica	64	29.3	80.8	51.5	25.7	8.3
	89	35.5	81.8	46.3	27.0	5.4
	187	35.3	84.5	49.2	32.5	4.5
G. tenuistipitata	36	29.3	76.5	47.2	23.9	10.1
	180	39.0	86.8	47.8	32.0	3.6
G. blodgettii	16	29.8	75.5	45.7	22.0	9.0
	258	42.0	94.5	52.5	28.2	4.2
G. hainanensis	58	40.8	91.8	51.0	26.9	6.1
	113	39.8	90.0	50.2	30.4	6.0
G. sjeostedtii	59	28.5	84.8	56.3	28.8	7.3
	153	29.5	88.8	59.3	31.7	5.3
G. corda	15	29.4	76.2	46.8	21.9	7.1
G. eucheumoides	12	34.5	57.0	22.5	20.7	8.8

3,6-anhydro-L-galactose, modifying the chain conformation and producing a concomitant increase in gel strength of the extract. Alkali treatment also affects the gelling and melting temperatures and, hence, gel hysteresis, as shown in Table 3.2 (FAO, 1990).

Furcellaran, a polysaccharide obtained from *Furcellaria* species of red seaweed, has, rather misleadingly, been called 'Danish agar' as a consequence of the original source of the material and its firm-gelling properties (FAO, 1990). However, furcellaran contains 16–20% sulphate content and is structurally similar to kappa carrageenan. 'Russian agar' sold in southern Russia and Ukraine is another extract that gives firm, brittle gels, although it is suggested that this is carrageenan that is obtained from *Ahnfeltia* and *Coccotylus* species (Guiry, 2009).

3.4.1 Agarose

Araki separated two components in agar from *Gelidium amansii*, a major fraction of a neutral, firmly gelling polysaccharide called agarose and minor fraction of a weakly gelling, charged polymer called agaropectin (Araki, 1937). Agarose is a linear polymer composed of (1–3)-linked agarobiose units of β-D-galactopyranose (1–4)-linked to 3,6-anhydro-α-L-galactopyranose, shown in Fig. 3.3 (Araki, 1956; Chaplin, 2009).

The determination of molecular weight by ultracentrifugation, molecular exclusion chromatography and intrinsic viscosity measurements assigns a molecular weight to agarose of at least 120 000 daltons, equivalent to a linear chain length of 400 units of agarose or 800 exoses (Armisen and Galatas, 1987).

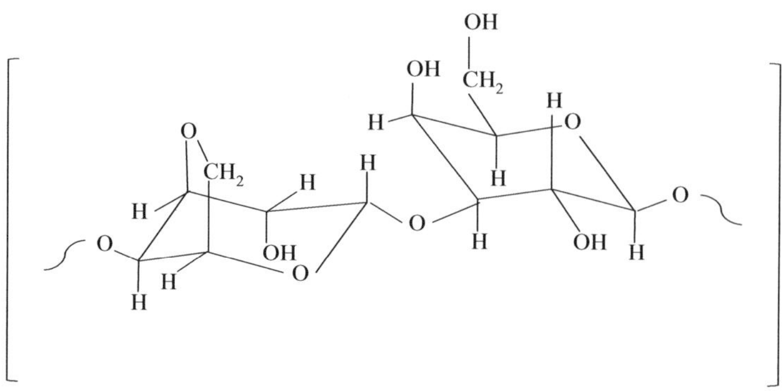

- (β1–4)-3,6-anhydro-L-galactose- (α1-3)-D-galactose-

Fig. 3.3 Agarose: the structural unit of agar.

3.4.2 Agaropectin

Like agarose, agaropectin is based on the same structural unit of agarobiose but different side groups include up to 8% sulphate groups and methyl and pyruvic acid acetyl groups (Lahaye and Rochas, 1991). The substituents are typical of the genus and species of the agarophytes used to produce agar, and other variations may occur within the plant growth cycle, the season of collection or from some other ecological factors. The presence of these side groups prevents this polymer adopting a regular structure so that it does not contribute significantly to gel formation.

Agaropectins have a molecular weight of around 12 600 daltons, equivalent to 75–100 residues. However, they may form structures with increased molecular weight as a result of the formation of calcium bridges between sulphate groups on adjacent agaropectin molecules. These bridges may be broken by adding a sequestering agent, such as sodium hexametaphosphate or sodium tripolyphosphate, which reduces the apparent viscosity.

3.5 FUNCTIONAL PROPERTIES

3.5.1 Gelation

The ability to form reversible gels by simply cooling hot, aqueous solutions is the most important property of agar. This gel-forming ability has led to the large number of practical applications where agar is used as a food additive or in other applications in microbiology, biochemistry or molecular biology, as well as in industrial applications.

Agar is insoluble in cold water but hydrates to form random coils in boiling water. Gelation depends exclusively on the formation of hydrogen bonds, where the random coils associate to form single helices (Foord and Atkins, 1989; Jimenez-Barbero *et al.*, 1989) and double helices (Rees and Welsh, 1977). These left-handed threefold helices are stabilised by the presence of water molecules bound inside the double helical cavity (Labropoulos *et al.*, 2002) and exterior hydroxyl groups allow aggregation of up to 10 000 of these helices to form microdomains of spherical microgels (Boral *et al.*, 2008). Following the phase separation process (Matsuo *et al.*, 2002), the microgels aggregate to form gels on cooling below 30–40°C. This transition

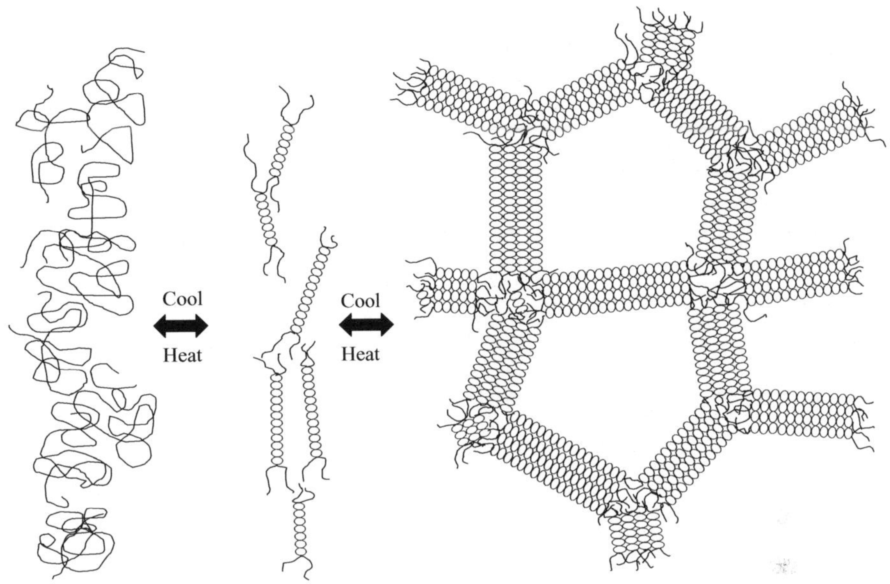

Fig. 3.4 Agar gel formation.

occurs at higher temperatures with increasing agar concentration and decreasing cooling rate. For example, the gelling temperature for *Gelidium* agar increases from 32 to 38°C and melting temperature rises from 86 to 89°C as the concentration increases from 0.5 to 2.0% (Armisen, 1997). Figure 3.4 illustrates the gelation process of agar: random coils associate on cooling to form helices followed by further aggregation of the helices to give the aggregated structures of agar gels.

Agar forms gels at very low concentrations, with a concentration threshold for gelation around 0.2%. Traditionally, gel strengths are measured by the Nikan-Sui method, also known as the Kobe test, which is based on measuring the load that causes a standard gel to break in 20 s using a cylindrical piston with an area of 1 cm^2. Gel strengths are measured on gels produced by boiling a solution containing 1.5% agar, cooling and maturing at 20°C for 15 h. Food-grade agar, in which the agarose fraction is almost exclusively responsible for the gel strength, typically has a gel strength around 800 g/cm^2.

Another very important property of agar is gel hysteresis, the difference between the gelling and melting temperatures. For *Gelidium* species, the gel point ranges from 28–31°C and for *Gracilaria* between 29–42°C, with higher gel points being observed in extracts with higher levels of methyl ester content (FAO, 1990). The melting temperatures range from 80–90°C for *Gelidium* to 76–92°C for *Gracilaria*. The hysteresis value for *Gelidium* and low methyl ester *Gracilaria* agars is around 50–60°C, but this falls as the methyl ester content increases. The difference between gelling and melting temperatures is much larger than that for many other gelling agents, such as kappa carrageenan which has a hysteresis value of 15–20°C, and this agar property is utilised in many applications in food and microbiology.

Unlike some other gelling agents, agar is readily incorporated into food formulations as it does not require cations to gel. Furthermore, this means that any variation in the cations contributed by the other components used in the recipe, such as dairy and fruit products which may contribute calcium, does not affect the agar gel performance.

The agar gelation process is totally reversible. The gel melts on heating and resets on cooling. This cycle can be repeated many times without any significant change in the mechanical properties of the gel, provided the agar is not used in very low pH conditions below 4, or used with oxidising agents.

Agar gels comprise a macromolecular network of agarose molecules associated through hydrogen bonds. Consequently, the presence of proton scavengers such as potassium iodide, sodium thiocyanate, urea, guanidine, etc., will block the gelling process by impeding the formation of hydrogen bonds and hence prevent agarose gel formation.

3.5.2 Synergy and incompatibility

Agar comprises a neutral polymer chain which exhibits limited reactivity to other materials. In contrast to carrageenan, agar does not show any synergy with protein due to the lower content of sulphate ester groups on the polymer chain. The limited synergies and antagonisms shown by agar are described below.

3.5.2.1 *Synergy with carob or locust bean gum*

Locust bean gum (LBG) is a galactomannan with a non-random distribution of galactose side chains along the mannan backbone. Up to ten contiguous units of mannan may be unsubstituted and these may align and hydrogen bond to other polysaccharide structures (Wielinga, 2009). Gel strength data for gels containing 1.5% total gum content of *Gelidium* agar and LBG are shown in Fig. 3.5 (Armisen and Galatas, 1987).

Gel strength increases to a maximum with 10% LBG in the mixture and falls to the original agar gel strength with 20% LBG in the mixture. The increase in gel strength is only about 8% and is much more modest than the significant increases obtained with combinations of kappa carrageenan and LBG. The inclusion of LBG has a marked effect on gel texture and produces more elastic, cohesive gels than obtained with *Gelidium* agar alone.

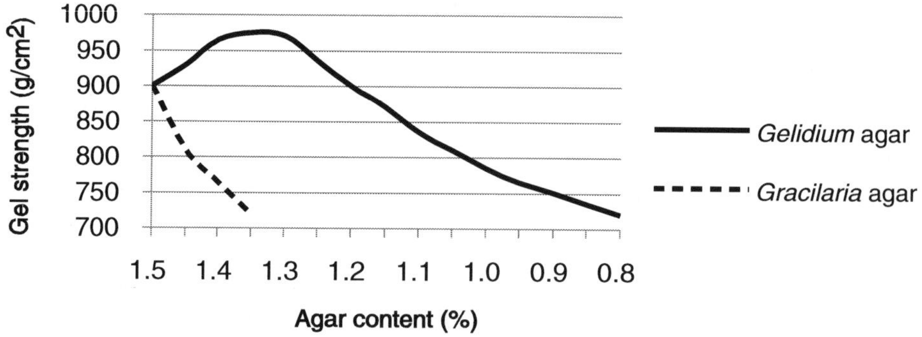

Fig. 3.5 Gel strength of agar–locust bean gum mixtures (1.5% total gum).

In contrast, combinations of *Gracilaria* agar and LBG do not exhibit any synergy. With this combination, the substitution of agar by LBG simply reduces the gelling component leading to a progressive loss of gel strength as the agar content is reduced (Fig. 3.5).

The differences in synergy are attributed to the content and distribution of ester sulphate groups of the different agars. Compared to *Gracilaria* agar, the low level of sulphate and 3.6 anhydrogalactose in *Gelidium* agar produces a more regular conformation and pitch to the agar helix and this is able to associate with LBG, modifying the agar gel characteristics.

3.5.2.2 *High sugar concentrations*

Some high gel strength, low sulphate content, *Gelidium* agars and some very high gel strength (high molecular weight) *Gracilaria* agars may exhibit enhanced gel strength in systems containing above 60% sugar. This synergy could be caused by the different pitch of the helices which depend on the quantity and distribution of the sulphate ester groups (Armisen, 1995).

3.5.2.3 *Tannic acid*

Tannic acid or pentadigaloyl glucose can be found in fruits such as quince, some apple varieties and plums. This is a proton scavenger and the presence of significant quantities of tannic acid can markedly inhibit the gelling process, although this effect can be overcome by adding low levels of glycerol (Armisen, 1995).

In non-food applications, the presence of other proton scavengers, such as potassium iodide, guanidine, urea and sodium thiocyanate, can inhibit agar gelation.

3.6 APPLICATIONS

3.6.1 Legislation

When the FDA classified food additives, an extensive bibliographic study was made into the long and extensive use of agar and it was immediately accorded the status of 'generally recognised as safe' (GRAS) (FDA, 1972). The use of agar was permitted in different food categories with specified maximum use levels (FDA, 1984):

Baked goods and baking mixes	0.8%
Confections and frostings	2.0%
Soft candies	1.2%
All other food categories	0.25%

A review of the toxicological status of agar by FAO/WHO (1974) noted: 'It is consumed traditionally as a food though not as a nutrient. The effect on weight gain observed in rats was probably due to the lack of utilization of agar, or to its laxative effects, or both.' Thus it was estimated that the acceptable daily intake in man should be 'not limited'.

In Europe, agar is defined as 'the hydrophilic colloidal polysaccharide consisting mainly of D-galactose units. On about every tenth D-galactopyranose unit one of the hydroxyl groups is esterified with sulphuric acid which is neutralised by calcium, magnesium, potassium or sodium. It is extracted from certain natural strains of marine algae of the families *Gelidiaceae* and *Sphaerococcaceae* and related red algae of the class *Rhodophyceae*' (EU, 1998). It is approved for use in food *quantum satis*, the level required to achieve the required technological effect in a product, and is assigned the number E 406. This compares with furcellaran, an extract obtained from *Furcellaria* species of red seaweed that has been misleadingly called 'Danish agar'. Furcellaran is a sulphated polysaccharide and is classified as carrageenan, E 407.

3.6.2 Food applications

The most important advantages for agar in different food applications derive from the characteristic firm texture and heat tolerance of the gels, stability in acidic conditions and limited reactivity to other food components. The key properties are listed below (Armisen, 1991, 1995):

- The large gelling capacity enables agar to be used at very low concentrations. The threshold concentration for gelation is 0.2% and levels in food products are typically between 0.5% and 2.0%.
- The large gel hysteresis, the difference between setting and melting temperatures, is much greater than that of any other reversible gelling agent, so that liquid solutions may be held at 40°C before setting and, once gelled, products remain stable up to 80°C.
- Agar forms gels over a wide pH range. The neutral polymer chain confers good resistance to acid hydrolysis at the normal pH values found in foods, such as in fruit products. However, agar can be hydrolysed by acid at high temperatures, so for pH values below 5, it is recommended that the pH of the food is lowered just before cooling to form the gel.
- No counter ions are needed for gelation and, hence, there is no characteristic metallic taste in the final products as found with alginates or carrageenan, which makes agar a very good product for delicately flavoured foods. Also, variations in the levels of cations in the food product do not produce a change in the textural properties of the gel.
- In cultured foods, such as yogurt, agar does not inhibit the growth of inoculated bacteria.
- It has good compatibility with other polysaccharides and proteins at normal use levels. This enables agar to give consistent gelled dairy desserts and avoid textural variations which can result from differences in milk quality.
- It does not require a minimum sugar solids level for gelation and so it can be used for low-sugar jams and jellies where reduced sugar levels and/or high intensity sweeteners may be employed. In some cases, high sugar concentrations assist the association of the agar network and gel strength increases.
- Agar now has 350 years of safe use in food. It has a high soluble fibre content that is not metabolised in humans and so it does not add calories to foods.

This combination of benefits results in agar maintaining its position as the hydrocolloid of choice in a number of specific food applications, particularly where firm, short-textured gels with good heat stability and good moisture stabilisation are required (Hispanagar, 2009):

- Water gels, such as water dessert jellies, vegetable, meat and fish aspics and artificial caviar.
- Confectionery including sweets and candies, fruit jellies, nougat, candy fillings, piping gels, jellies and jams.
- Bakery products, such as icings and glazes for pastries, cakes and doughnuts.

Table 3.3 Agar applications in different regions of the world.

Application	Regions		
	Asia	**USA/Europe**	**Latin America**
Ice cream	+	+	+
Milk shake	+	+	?
Sherbet	+	+	+
Custard pudding	+	+	+
Cakes	+	+	+
Pie filling	−	+	+
Flat icing	+	+	+
Meringue	−	+	−
Cookies	+	+	+
Candy (agar jelly)	+	+	+
Fruit jelly dessert	++	+	+
Jams and jellies	+	+	+
Processed cheese	−	+	?
Fermented dairy products	+	++	+
Wine clarification	?	+	+
Gelled meats	++	++	?
Dulce de batata	−	−	++
Mitsumame	++	−	−
Red bean jelly	++	−	−
Tokoroten noodles	++	+	+?
Agar jelly beverages	+	−	−

−, little or no usage; +, moderate usage; ++, frequent usage.
Reproduced from Modliszewski (1990), with permission.

- Dairy products including flans, puddings, custards, creams, flavoured milks and ice cream.
- Fermented products, for example yogurt and sour cream.
- Canned meat and fish products such as meat in jelly and fish pate.
- Soups and sauces such as béchamel sauce.
- Fining agent for clarifying wine, juice and vinegar.
- Health foods to contribute bulk and fibre.

Despite the longevity of its use in food, it is noteworthy that the food applications of agar are confined to specific geographical regions of the world or for preparing culinary dishes traditional to certain cultures. Outside these areas, there is little development of new applications due to food preferences being linked to customs and traditions. Thus, although agar was introduced into Europe and the USA over a century ago, there has been limited incorporation into new food products. Table 3.3 lists the different applications for agar in foods in relation to different cultural areas in the world (Modliszewski, 1990).

The preparation of various traditional oriental dishes with strip, square and sheet forms of natural agar is diminishing as they are being reformulated with powdered material. The physical formats of strip, square and sheet agars enable them to be added easily to formulations without lump formation. Powdered agar requires the use of suitable dispersion techniques to avoid lump formation and ensure full hydration of the material to maximise efficacy. The techniques are the same as those used for other gelling agents, such as carrageenan and gellan gum. As agar is insoluble in cold water, the powder may be simply dispersed into cold water and agitated as the temperature is raised. To speed up product preparation, the powder may be added directly to hot water. In this case, it should be pre-dispersed with five times its weight of sugar or other powder to physically separate the particles and avoid lumping. Higher levels of sugar may impede gum hydration, so, for example for confectionery jellies, it is recommended that bulk of the sugar be added after the agar is fully hydrated. Once the agar has been dispersed without lumps, the solution should be boiled to maximise hydration. Agar can be hydrolysed when heated in solution at pH values above 8.0 and below 5.5. Alkaline conditions are not often found in food systems, but for acidic products, it is recommended that the acid be added after the agar has been hydrated and cooled.

3.6.2.1 *Water dessert gels*

This is the most traditional use of agar, and a typical recipe is shown in Formulation 3.1. The powder is added to water and must be boiled to ensure the complete hydration of the gelling agent.

Formulation 3.1 Water dessert jelly

Ingredients	Quantity (%)
Sugar	12.00–15.00
Agar	0.80–0.85
Colour	as required
Flavour	as required
Water or fruit juice	up to 100.00

Preparation is simple. The agar is boiled in water or fruit juice until the solids dissolve and then sugar and/or sweetener, flavouring and colouring are added as required. The liquid is poured into moulds and left to set. The liquid may also be mixed with chopped or diced fruit or berries as a dessert or it can be used as a jelly filling or topping in a multilayer product such as cake. It is possible to use agar for desserts using pineapple or other fruits which contain enzymes that degrade protein and prevent gelatine being used for such desserts. Gel hysteresis avoids any softening or melting when the product is stored at ambient temperature before serving. As agar gels do not melt in the mouth, the concentration is adjusted to ensure that the gel breaks down easily to release flavour when eaten.

Gel hysteresis is the key attribute used in the preparation of the Japanese dessert Mitsumame. This is made from agar gel cubes containing sugar, flavour and colour which are mixed with fruit pieces in syrup and canned. Hysteresis enables the cans to be sterilised without the agar cubes melting into the syrup.

A savoury gel may be made by boiling agar in stock and pouring the cooled liquid over pre-cooked vegetables, meat or fish to give a firm aspic that can be used to seal the surface of a product to prevent moisture loss during storage.

In canned meats, the agar may be distributed by dusting over the meat and tumbling before filling into cans and processing at 120°C for 45–60 min. The agar sets on cooling and gives a moist succulent product that can be sliced easily. Typical use levels are around 0.8–1.0%. For softer-textured products, such as pate, lower levels of agar may be used.

A novel gelling application for agar is to prepare imitation caviar. A mixture of fish extract, colour and flavour is mixed into a hot agar solution. Droplets are extruded into a setting bath of cold sunflower oil to set the beads. After hardening the beads, closely mimic the appearance of caviar and they are sold as cost-effective alternatives to natural caviar in the southern Russia and Ukraine.

3.6.2.2 *Confectionery jellies*

A brittle gel that breaks easily when eaten can be made using 1.0–1.3% agar in the following high sugar solids recipe (Formulation 3.2).

Formulation 3.2 Gum sweets

Ingredients	Quantity (%)
Water	53.80
Sugar	44.85
Tartaric acid	0.23
Agar powder	1.12
Flavour and colour	as required
Total	100.00

The sweets are prepared by dissolving agar mixed with about five times its weight of sugar into the water. The mix is brought to the boil and heated continuously whilst the balance of sugar is added. Finally, acid, colour and flavour are added and the mix is deposited in moulds or formed as cylinders and cut to size before coating with icing sugar.

3.6.2.3 *Icings*

Icings are a high-solids product in which sugar crystals are suspended in a saturated sugar solution. They are applied to various bakery products to enhance the appearance and flavour and to extend shelf life. Agar is a well-established stabiliser for icings as the hot coating quickly sets when applied to the product and it strongly binds water to prevent surface drying and cracking (Formulation 3.3).

Formulation 3.3 Doughnut icing

Ingredients	Quantity (%)
Powder sugar	69.08
Water	16.55
Granulated sugar	6.00
Glucose syrup	4.00
Cocoa powder	3.00
Salt	0.35
Calcium sulphate	0.27
Chocolate flavour	0.25
Vanilla flavour	0.20
Agar powder	0.20
Potassium sorbate	0.10
Total	100.00

Agar is hydrated by boiling it in the water with the sugar and glucose syrup. The remaining dry ingredients are added and stirring continued for 30 min at 65°C (140–150°F). The icing is then cooled to 49–54°C (120–130°F) and applied hot to the doughnuts where it quickly sets. An icing suitable for wrapping bakery products may be made by including 1.0% starch in the recipe above.

3.6.2.4 *Dairy desserts*

Hot-filled flan desserts have a similar firm texture to those made with carrageenan and furcellaran. Because the agar molecule is uncharged, it does not interact with proteins and, consequently, the use level required to form a milk gel is similar to that for water desserts. However, unlike carrageenan, any variation in milk protein quality is not reflected by changes in the texture of the food product (Formulation 3.4).

Formulation 3.4 Flan desserts

Ingredients	Quantity (%)
Granulated sugar	10.00
Agar powder	0.30–0.50
Starch	0.00–2.00
Colour and flavour	as required
Milk	up to 100.00

Multilayered products, such as coffee, chocolate and vanilla gels, may be made applying additional layers after the first layer has gelled: hysteresis prevents the gel melting and ensures that the layers do not merge at the boundary. Custards and sauces may be thickened by combining agar with starch.

3.6.2.5 *Health foods*

Agar is a satiating material as it swells in the stomach to increase bulk and give a feeling of fullness. This is the basis of a current fashion diet in Japan, the Kanten diet, which is also receiving publicity in the USA. Agar also moderates glucose and fat uptake from foods and reduces cholesterol levels. It acts as a prebiotic as the fibre can be fermented by bacteria in the colon. Consumption from food products is generally 1–2 g/day, but for pharmacological use, levels in excess of 6 g/day are used for a mild laxative effect and it can act as an intestinal regulator.

3.6.3 Non-food applications

Although more than 90% of the world production of agar is used in food applications, significant commercial volumes are used in biotechnology. The most significant application is solid culture media for microbiology. The unique combination of properties of some agars has made it the principal gel former in this area. Bacteriological agars are mainly produced from *Gelidium* and, in lower quantities, from *Pterocladia*. Agars obtained from other agarophytes and alternative gelling agents have failed to match this combination of properties (Armisen, 1991, 1995):

- The gel point at 36 ± 1.5°C does not damage nutrients, vitamins, antibiotics, cells and other components added to the liquid agar prior to gelation.
- The large gel hysteresis, with a melting point at 87 ± 1.5°C, enables gels to be incubated up to 50°C for the study of thermophilic organisms.
- The polymer chains associate through hydrogen bonds at neutral pH so that gels can be formed without cations that could affect microbial growth.
- Agar is not metabolised or degraded by the hydrolytic enzymes in bacteria, yeast and fungi.
- The polymer does not react with the components used in media formulation and contains no charged groups that may affect nutrient migration or contain any substances that may inhibit or moderate cell growth.
- It is stable to heat treatment and withstands the long periods of autoclaving needed to sterilise the growth media prior to inoculation.
- It is easy to inoculate and mix liquids prior to gelation or to apply cultures to gelled plates.
- The gel has good clarity so it is easy to identify and count colonies.

This combination of properties underpins the use of agar as the key gelling agent for solid media in microbiology and similar application areas.

In plant breeding, agar is used to produce solid substrates for cloning plants and to propagate high-value plants, such as orchids. Agar is particularly useful as it does not affect the activity of plant growth hormones, such as auxins and gibberellins.

For breeding insects, such as *Drosophila* species used in genetic studies, and the non-seasonal production of silk worms, agar provides the support for food consumed by larvae. Other non-food uses include impression gels to make exact replicas of dental, sculpture and archaeological pieces. In these applications, high concentrations of up to 8% agar are used in the gel and dehydration is avoided by including glycerol in the formulation. Finally, gelled electrolytes are used to make leak-proof batteries.

3.7 FUTURE DEVELOPMENTS

Agar has a well-known combination of properties that lead to its use where firm brittle gels are desired, such as water dessert gels, confectionery jellies, icing, aspics, canned meats, dairy flans, sauces and bakery fillings. It has a long history of safe use in food over 350 years in Asia and over 100 years in the West. Growth outside the traditional areas of application is likely to come from the recognition that agar provides functional benefits related to health and nutrition. It forms gels that are slow to transit the stomach increasing satiety by increasing the feeling of fullness, and the increased viscosity moderates glucose and fat uptake and controls cholesterol.

Several new functional foods have been launched in Japan, Taiwan and China using the gel properties of agar and this trend is likely to continue.

Acknowledgements

I would like to thank Professor Rafael Armisen and Catherine Side for their continued interest and constructive comments in the preparation of this chapter, and for permission to use some figures and formulations.

Dedication

I would like to dedicate this chapter to my daughter Katie, who provided the inspiration to complete this final piece of this very lengthy project.

References

Araki, C. (1937) Acetylation of agar like substance of *Gelidium amansii*. *Journal of the Chemical Society of Japan*, **58**, 1338–1350.

Araki, C. (1956) Structure of agarose constituent of agar-agar. *Bulletin of the Chemical Society of Japan*, **29**, 43–44.

Armisen, R. (1991) Agar and agarose biotechnological applications. *Hydrobiologia*, **221**, 157–166.

Armisen, R. (1995) World-wide use and importance of *Gracilaria*. *Journal of Applied Phycology*, **7**, 231–243.

Armisen, R. (1997) Agar. In: A. Imeson (ed), *Thickening and Gelling Agents for Food*, 2nd edn. Blackie Academic and Professional, London, pp. 1–21.

Armisen, R. and Galatas, F. (1987) Production, properties and uses of agar. In: D.J. Mc Hugh (ed), *Production and Utilization of Products from Commercial Seaweeds*. Fisheries *technical paper 288*, FAO, Rome, pp. 1–57. Available at http://www.fao.org/docrep/X5822E/x5822e03.htm (accessed 5 August 2009).

Blakemore, W.R. and Harpell, A.R. (2009) Carrageenan. In: A. Imeson (ed), *Food Stabilisers, Thickeners and Gelling Agents*. Blackwell Publishing, Oxford, pp. 73–94.

Boral, S., Saxena, A. and Bohidar, H.B. (2008) Universal growth of microdomains and gelation transition in Agar hydrogels. *Journal of Physical Chemistry B*, **112**(12), 3625–3632.

Chaplin M. (2009) Agar in water structure and science. Available at http://www.lsbu.ac.uk/water/hyagar.html (accessed 5 August 2009).

EU (1998) E 406 Agar, *Commission Directive* 98/86/EC, OJ, L334, 9.12.98, pp. 8–9. Available at http://eur-lex.europa.eu/LexUriServ/LexUriServ.do?uri=OJ:L:1998:334:0001:0063:EN:PDF (accessed 5 August 2009).

FAO (1990) Training manual on *Gracilaria* culture and seaweed processing in China, Regional seafarming development and demonstration project. *FAO Corporate Document Repository*, project report No. 6, pp 107. Available at http://www.fao.org/docrep/field/003/AB730E/AB730E03.htm (accessed 5 August 2009).

FAO/WHO (1974) Agar (WHO Food Additives Series 5) in Seventeenth Report of the Joint FAO/WHO Expert Committee on Food Additives, *World Health Organization technical report series*, 1974, No. 539, pp. 10–11.

FDA (1972) Agar Agar, GRAS (generally recognized as safe) Food Ingredients. Food and Drug Administration, PB-221225 NTIS. US Department of Commerce, Washington, DC, 33 pp.

FDA (1984) Agar-agar. Code of Federal Regulations Title **21:** *Direct Food Substances affirmed as Generally Recognized as Safe*, **21**, §184.1115.

Foord, S.A. and Atkins, E.D.T. (1989) New x-ray diffraction result from agarose: extended single helix structure and implications for gelation mechanism. *Biopolymers*, **28**, 1345–1365.

Guiry, M. (2009) Agars. The Seaweed Site, National University of Ireland, Galway. Available at http://www.seaweed.ie/uses_general/agars.html (accessed 5 August 2009).

Hispanagar (2009) Products: Agar-agar. Available at http://www.hispanagar.com/agar.htm (accessed 5 August 2009).

Jimenez-Barbero, J., Boufar-Roupe, C., Rochas, C. and Perez, S. (1989). Modeling studies of solvent effects on the conformational stability of agarobiose and neoagarobiose and their relationship to agarose. *International Journal of Macromolecules*, **11**, 265–272.

Koch, R. (1882) Die Aetiologie Der Tuberculose. *Berliner Klinische Wochenschrift*, **15**, 221–230.

Labbar, T., Labbar, R. and Riad, A. (1996) Production process of quick soluble agar. US Patent 5496936 and European Patent 0598141.

Labropoulos, K.C., Niesz, D.E., Danforth, S.C. and Kevrekidis, P.G. (2002) Dynamic rheology of agar gels: theory and experiment. Part II. gelation behavior of agar sols and fitting of a theoretical rheological model. *Carbohydrate Polymer* **50**(4), 407–415.

Lahaye, M. and Rochas, C. (1991) Chemical structure and physico-chemical properties of agar. *Hydrobiologia*, **221**, 137–148.

Marinho-Soriano, E. and Bourret, E. (2005) Polysaccharides from the red seaweed *Gracilaria dura* (Gracilariales, Rhodophyta). *Bioresource Technology*, **96**(3), 379–382.

Matsuo, M., Tanaka, T. and Ma, L. (2002) Gelation mechanism of agarose and κ-carrageenan solutions estimated in terms of concentration fluctuation. *Polymer*, **43**, 5299–5309.

Matsuoka, C. (1921) Process for making agar-agar. US Patent 1399359.

McHugh, D.J. (1991) Worldwide distribution of commercial resources of seaweeds including *Gelidium*. *Hydrobiologia*, **221**(1), 19–29.

McHugh, D.J. (2002) Prospects for seaweed production in developing countries, *FAO Fisheries Circulars*, No. 968, pp. 28.

McHugh, D.J. (2003) A guide to the seaweed industry, *FAO Fisheries Technical Paper* **441**, pp. 12–22. Available at http://www.fao.org/docrep/006/Y4765E/y4765e00.HTM (accessed 5 August 2009).

Modliszewski, J. (1990) Food Uses of *Gelidium* Extracted Agars. International Workshop on *Gelidium*, Santander, Spain, 3–8 September 1990.

Montaño, N.E., Villanueva, R.D. and Romero, J.B. (1999) Chemical characteristics and gelling properties of agar from two Philippine *Gracilaria* spp. (Gracilariales, Rhodophyta). *Journal of Applied Phycology*, **11**(1), 27–34.

Payen, M. (1859) Sur la gelose et les nids de salangane. *Comptes Rendus De L Academie Des Sciences Paris*, **15**, 521–532.

Rees, D.A. and Welsh, E.J. (1977) Secondary and tertiary structure of polysaccharides in solution and gels. *Angewandte Chemie International Edition in English*, **15**, 214–224.

Romero, J.B., Villanueva, R.D. and Montaño, M.N.E. (2008) Stability of agar in the seaweed *Gracilaria eucheumatoides* (Gracilariales, Rhodophyta) during postharvest storage. *Bioresource Technology*, **99**(17), 8151–8155.

Setexam (2009) The products, processing. Available at http://www.setexam.com/products_moteur.htm (accessed 5 August 2009).

Vignon, M.R., Rochas, C., Chanzy, H., Vuong, R. and Tekely, P. (1994) *Gelidium sesquipedale* (Gelidiales, Rhodophyta) II An ultrastructural and morphological study. *Botanica Marina*, **37**, 331–340.

Wielinga, W.C. (2009) Seed gums. In: A. Imeson (ed), *Food Stabilisers, Thickeners and Gelling Agents*. Blackwell Publishing, Oxford, pp. 275–292.

Yenigül, M. (1993) Seasonal changes in the chemical and gelling characteristics of agar from *Gracilaria verrucosa* collected in Turkey. *Hydrobiologia*, **260–261**(1), 627–631.

4 Alginates

Trond Helgerud, Olav Gåserød, Therese Fjæreide,
Peder O. Andersen, Christian K. Larsen

ABSTRACT

Alginates are derived from various species of brown seaweed found off the coasts of the North Atlantic, South America and Asia. They are produced as a range of salts, but sodium alginate is predominantly used in foods. Sodium alginate hydrates in cold or hot water to give viscous solutions. The controlled interaction between sodium alginate and calcium salts gives cold-setting gels that are shear irreversible and heat stable. Control is affected using citrate or phosphate sequestrants, or by processing at temperatures above about 70°C and cooling. Typical food applications include reformed foods such as onion rings and olive fillings, cold-setting bakery cream fillings and heat-stable bakery and fruit fillings.

4.1 INTRODUCTION

Alginate finds its use in a wide range of food applications. It is derived from brown seaweed and provides a unique combination of properties including cold solubility, cold-setting gels, non-melting gels (heat/temperature independent) and freeze–thaw stable gels. This chapter gives an introduction on how the nature of brown algae influences the properties of the extracted alginates and how this functionality can be utilised in foods.

Alginate was first described by Stanford (1881), and the name is now used as the general term for the range of alginic acid salts that can be employed in foods. Alginate is used for thickening, stabilising, gelling and film-forming purposes, and this chapter provides a range of examples. There is a wide range of applications outside the food world, and alginate is a very exciting hydrocolloid widely explored by scientific groups. A vast number of scientific papers describe in depth the details of alginate chemistry and properties.

4.2 PRODUCTION

4.2.1 Raw materials

Alginate occurs in the cell walls and intercellular spaces of brown algae. The alginate molecules provide both flexibility and strength to the plants and these properties are adapted as necessary for growth conditions in the sea. A diverse range of alginate applications have

Plate 1 Strip agar. (Reproduced with permission of Wing Yip Store, http://www.wingyipstore.co.uk.)

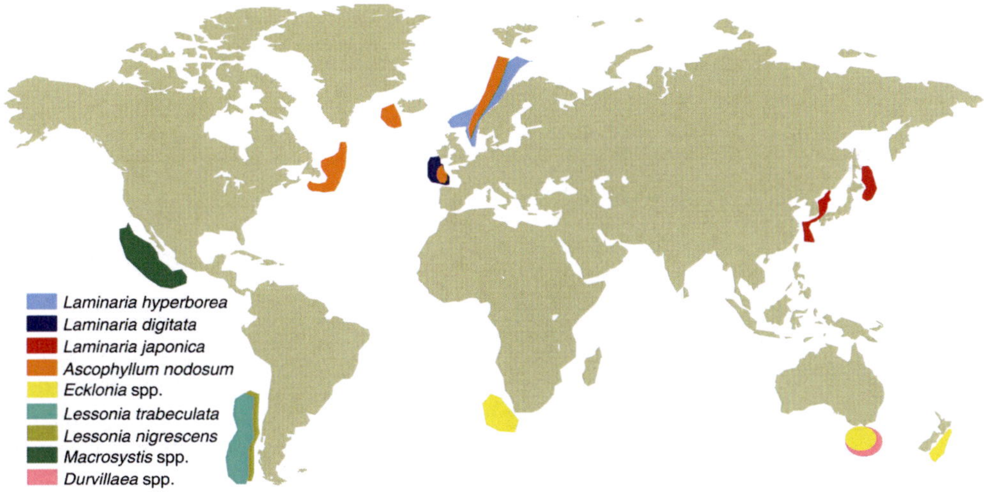

Plate 2 Industrially utilised brown seaweeds. (Reproduced with kind permission of FMC Corporation.)

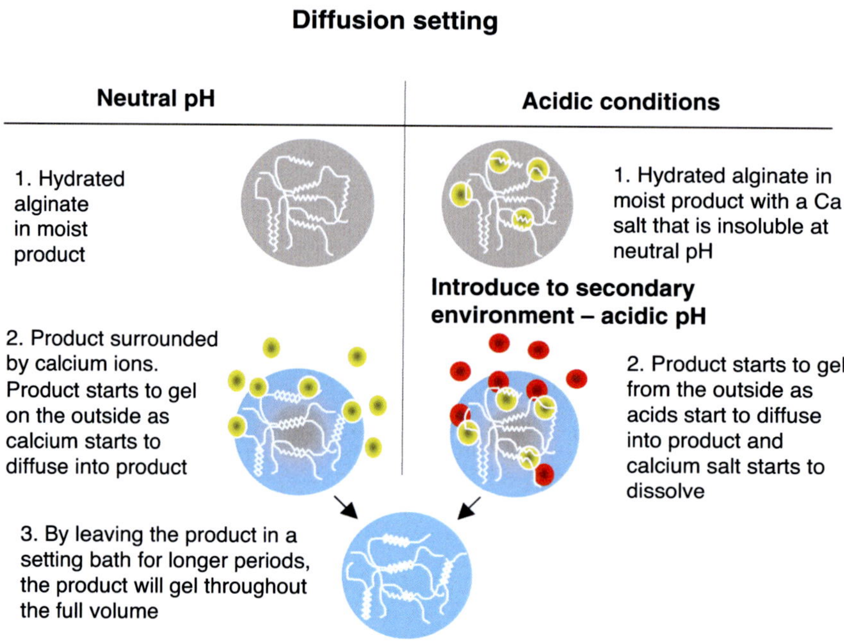

Plate 3 Alginate gelation by diffusion setting. (Reproduced with kind permission of FMC Corporation.)

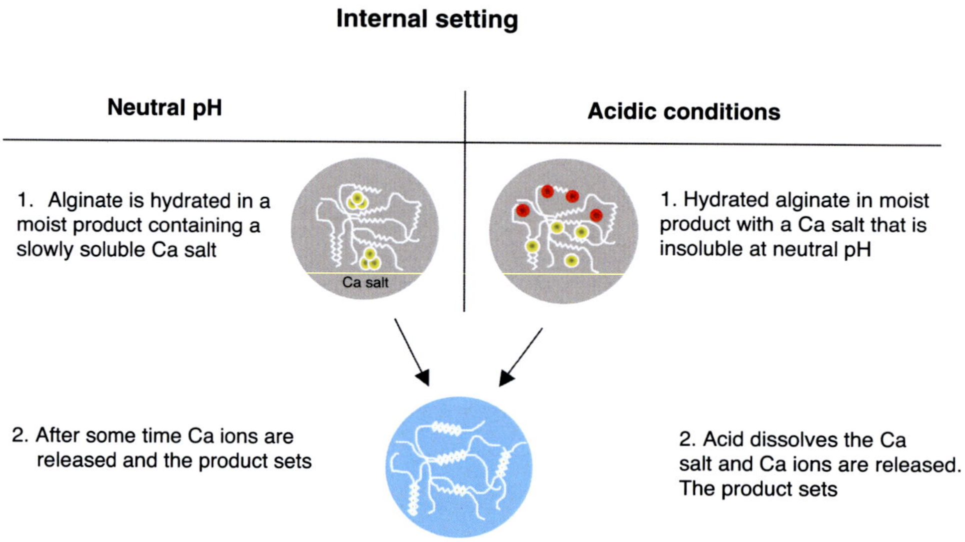

Plate 4 Gelling by internal setting. (Reproduced with kind permission of FMC Corporation.)

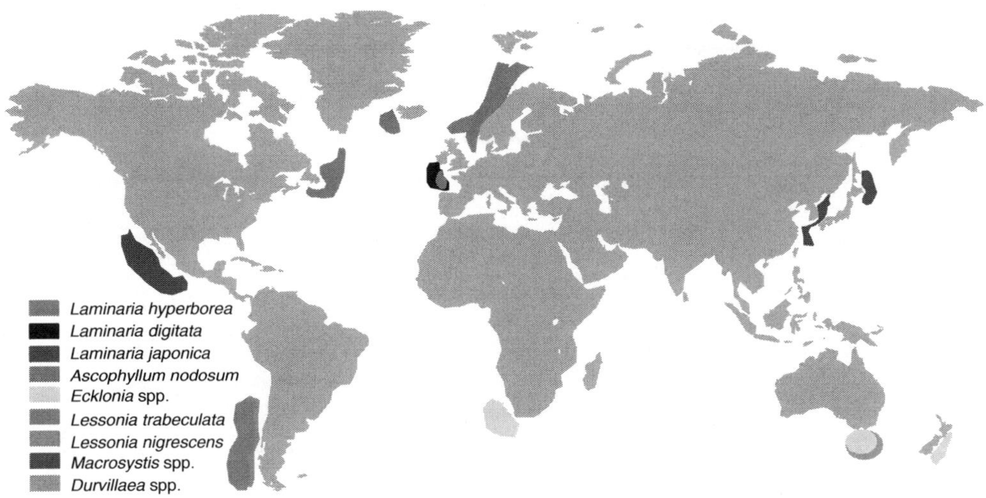

Fig. 4.1 Industrially utilised brown seaweeds. (Reproduced with kind permission of FMC Corporation.) For a colour version of this figure, please see Plate 2 of the colour plate section.

been developed during more than 60 years of commercial utilisation. In food applications, alginate provides texturising properties such as thickening, stabilising and gelling.

Brown algae require clean water with temperatures between 4 and 18°C. As photosynthetic organisms, they are restricted to locations with appropriate light conditions, from the tidal zone to a depth of 50 m, depending on the species.

The locations of brown algae industrially utilised for alginate production are shown in Fig. 4.1. The brown algae most widely used for the industrial production of alginate include *Laminaria hyperborea*, *Laminaria digitata*, *Laminaria japonica*, *Ascophyllum nodosum*, *Ecklonia maxima*, *Macrocystis pyrifera*, *Durvillea antarctica*, *Lessonia nigrescens* and *Lessonia trabeculata*. The plants are generally harvested from the sea or collected from the shore.

Cultivation methods are also used, and an example of this is the cultivation of *Laminaria japonica* in China. It has been found difficult, and thereby too costly, to cultivate the other species mentioned, as these seaweeds cannot be grown simply from cuttings of mature plants. The regeneration would, in these cases, involve the production of spores from mature plants from which new plants could grow (McHugh, 2003).

The alginate application, whether as thickener, stabiliser, gel former or film-forming agent, generally determines which seaweed is used as the source of alginate in order to achieve optimum performance. The global annual production of alginate is estimated at approximately 38 000 tonnes.

4.2.2 Manufacturing

Alginic acid is extracted from brown seaweed through a long and slow process. This is the free acid form of alginate and the water-insoluble intermediate in the commercial manufacture of alginates. Alginic acid has limited stability as chains are broken by auto-catalysed acid hydrolysis. In order to make stable, water-soluble alginate products, the alginic acid is

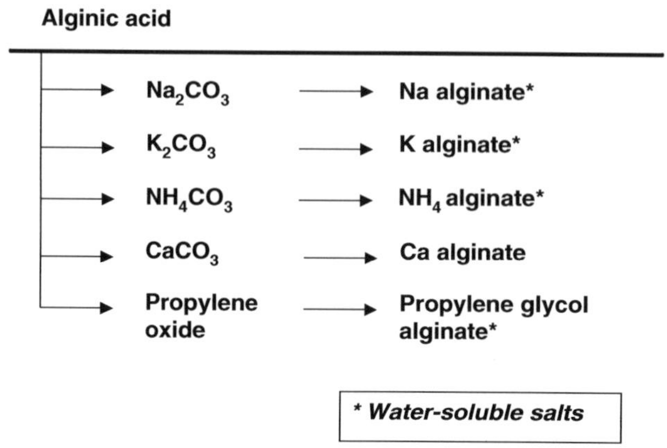

Fig. 4.2 Alginate salts produced from alginic acid for food use. (Reproduced with kind permission of FMC Corporation.)

transformed into the range of commercial alginate salts by the incorporation of different inorganic salts as shown in Fig. 4.2.

4.2.3 Regulatory status

Alginates have been used in foods for a long time and approved for use in a variety of applications and in a number of countries. Specifications for alginic acid, sodium, potassium, calcium and ammonium alginate salts and propylene glycol alginate (PGA) are listed in the US Food Chemicals Codex and are also considered generally recognised as safe (GRAS) in accordance with US Food and Drug Regulations (CFR 21). In Europe, alginic acid, sodium, potassium, calcium and ammonium alginate salts and PGA are listed as a range of E-numbers from E400 to E405. The same products are also listed as Food Additives by FAO/WHO Joint Expert Committee on Food Additives (JECFA).

4.3 CHEMICAL COMPOSITION

4.3.1 General

Alginates are natural, high-molecular-weight polymers. They are salts of alginic acid with a degree of polymerisation, after commercial processing, usually in the range 50–3000, corresponding to molecular weights of approximately 10–600 kDa. Alginic acid is a copolymer of the building blocks β-D-mannuronic acid (M) and its C-5 epimer, α-L-guluronic acid (G), linked together to form a linear polysaccharide with (1,4)-glycosidic bonds. Molecular structures are presented in Fig. 4.3.

4.3.2 Configuration

It has been shown that the monomeric M- and G-residues in alginates are joined together in sections consisting of homopolymeric M-blocks (-MMMMM-) and G-blocks (-GGGGG-) or heteropolymeric blocks of alternating M and G (MGMGMG). In the polymer chain, the monomers will tend to find their most energetically favourable structure. For G–G, it

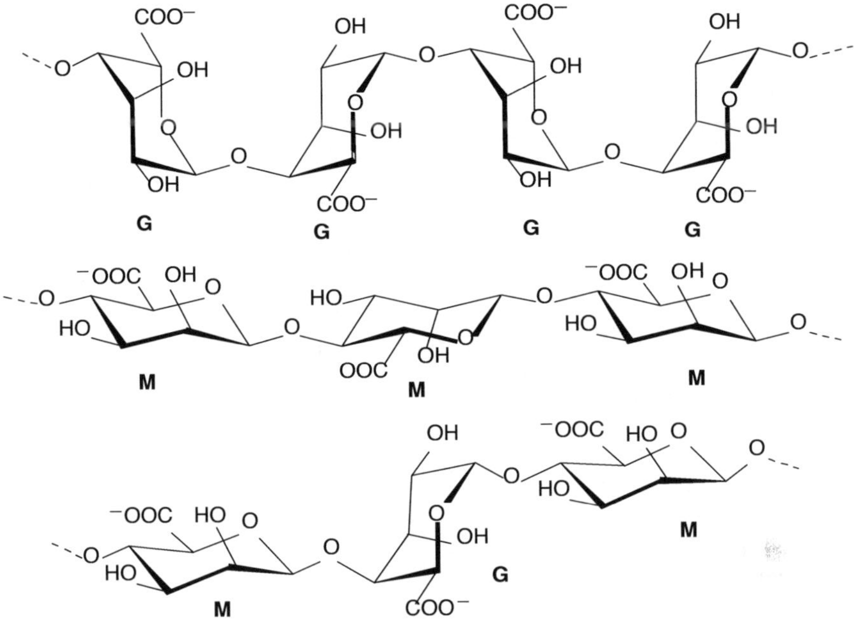

Fig. 4.3 Block structures in alginate. From top: G-blocks, M-blocks and MG-blocks (alternating M and G). (Reproduced with kind permission of FMC Corporation.)

is the $^{1}C_{4}$ chair form linked together with an α-(1,4)-glycosidic bond. For M–M, it is the $^{4}C_{1}$ chair form linked together with a β-(1,4)-glycosidic bond. The epimeric configuration of the carboxylic acid groups determines how sugar units are connected. Thus there is an equatorial–equatorial glycosidic bond in M–M, an axial–axial glycosidic bond in G–G and an equatorial–axial glycosidic bond in M–G. The charge and bulk of the carboxylic acid groups give these different conformations differing degrees of extension and flexibility; it has been found that M–G is the most flexible conformation and G–G the most rigid arrangement (flexibility MG > MM > GG) (Smidsrød, 2001; Smidsrød *et al.*, 1973). After shielding the charge, for example in 0.1 M sodium chloride solution, it was found that the conformations have similar extension, suggesting that stiffness is mainly an electrostatic effect rather than a steric effect (Vold *et al.*, 2006).

4.3.3 Alginate biosynthesis

The block structure distribution of an alginate is determined by the biosynthesis of the polymer in seaweed and its subsequent genetic and environmental control. The first stages of the pathway for alginate biosynthesis in brown algae yield polymannuronate, the homopolymer of mannuronic acid (M). A series of alginate epimerases, both membrane bound and extracellular, then act on the polymannuronate chain and progress along the chain effecting the epimerisation from M to G in certain regions of the polymer in various patterns (Larsen, 1981). Although most of the studies on biosynthesis have been done on alginate-producing bacteria, the metabolic pathways are, in major terms, similar in seaweed, including epimerase activity (Skjåk-Bræk, 1992). The epimerase type and activity is the key to the final alginate structure. Seven bacterial epimerases have been cloned and isolated from the bacteria *Azotobacter vinelandii*, all introducing different structure into the alginate chain ranging from

Table 4.1 Typical composition of alginates obtained from different seaweeds.

Alginate source	F_G	F_{GG}	F_{MM}	$F_{GM,MG}$
Laminaria hyperborea (stem)	0.68	0.56	0.2	0.12
Laminaria hyperborea (leaf)	0.55	0.38	0.28	0.17
Laminaria digitata	0.41	0.25	0.43	0.16
Laminaria japonica	0.35	0.18	0.48	0.17
Ascophyllum nodosum (old tissue)	0.36	0.16	0.44	0.2
Macrocystis pyrifera	0.39	0.16	0.38	0.23
Lessonia nigrescens	0.38	0.19	0.43	0.19
Durvillea antarctica	0.29	0.15	0.57	0.14
Lessonia trabeculata (stem)	0.62	0.47	0.23	0.15

Source: NMR Data from Moe *et al.* (1995), except *Lessonia trabeculata* (stem), data extracted from Venegas *et al.* (1993), where FG is calculated from GG-block fraction plus ½ MG-fraction, F_{GG} and F_{MM} equals, respectively, the reported GG- and MM-block fractions and $F_{MG,GM}$ is ½ the MG-fraction.

alternating MG sequences to long G-blocks (Ertesvåg *et al.*, 1999). Genes with similarities to the bacterial epimerase genes have also been found in brown seaweed (Nyvall *et al.*, 2003).

It has been proposed that the algae may respond to external conditions and regulate the alginate composition by biosynthesis. For instance, it has been shown that leaves of sporophytes grown in sheltered areas yield a significantly higher ratio of GG-blocks after transplantation to a more exposed environment (Venegas *et al.*, 1993).

4.3.4 Block structure analysis

Evaluation of the block structure of alginate can be performed in several ways, for example by chemical extraction of GG-, MM- and MG-blocks (Haug *et al.*, 1967, 1974), by an enzymatic method (Østgaard, 1993) or, more commonly, by NMR analysis (Grasdalen, 1983). Using ^{1}H NMR, detailed ratios of F_G, F_M, F_{GG}, F_{MM}, $F_{MG,GM}$, F_{GGG}, F_{GGM} and F_{MGM} can be obtained. Average values of the sample are obtained but are relatively consistent for various commercial grades of alginates, and can be used to distinguish alginates from different raw materials and determine gel strength potential. Table 4.1 lists some of the values determined by NMR, except the values for *Lessonia trabeculata*, which were determined by chemical extraction (Venegas *et al.*, 1993).

4.4 FUNCTIONAL PROPERTIES

4.4.1 Dissolution of alginate

In order to prepare solutions of alginate effectively, it is necessary to completely disperse and dissolve the alginate particles in the system. This may be done by the careful addition of material during high shear mixing or by pre-blending the alginate with other food ingredients before addition to the aqueous system. Typically, the alginate powder may be dry blended with particles, such as sugar, or it may be suspended in a hydrophobic solvent, for example vegetable oil. The dissolution of alginate is dependent on particle size as smaller particles dissolve faster (Fig. 4.4). However, smaller particles are more likely to agglomerate and form lumps, increasing the importance of using good dispersion and high shear with these products (Larsen *et al.*, 2003). In addition, it has been found that the dry matter content,

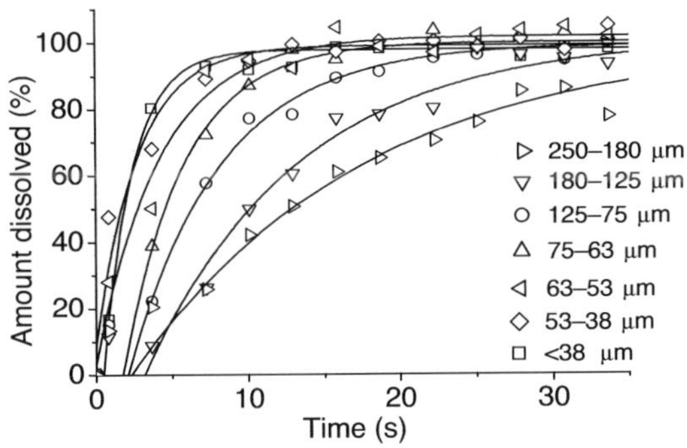

Fig. 4.4 Dissolution of alginate powder from different size fractions. (Reproduced with kind permission of FMC Corporation.)

ionic levels, such as water hardness, and temperature of the dissolution medium affect the dissolution rate of alginate.

4.4.2 Viscosity

The intrinsic viscosity of an alginate solution is determined by the length of the alginate molecules involved. When sodium alginate is dissolved in pure water, the intrinsic viscosity can be directly correlated with molecular weight through the Mark Houwink Sakurada (MHS) equation. Methods for doing this and values for alginates from *Laminaria hyperborea* (stem) and *Macrocystis pyrifera* have been determined by Martinsen *et al.* (1991). Molecular weight distribution can be determined by techniques combining chromatography and light-scattering detectors such as size exclusion chromatography, SEC-MALLS (Vold *et al.*, 2006).

The viscosity, as measured by a rheometer or viscometer, also depends on the composition of the aqueous system. The addition of other salts, high levels of sugars or polyols and non-solvents, such as alcohols, also highly influence the final viscosity of the food. The introduction of a small amount of calcium ions to an alginate solution would, for example, give a steep rise in solution viscosity due to partial and non-permanent cross-linking.

A viscous solution of alginate exhibits shear-thinning properties. This is a consequence of the long polymer chains and the stiffness of the hydrated molecules. Alginate is a linear and highly charged polysaccharide, properties that provide rigidity to the molecule. At low shear rates, such as during storage or under very low-speed stirring, the alginate molecules will be directed more or less randomly. By increasing the shear rate, the molecules will start to direct themselves in a more-parallel way. As a result, the apparent viscosity will decrease when shear rate is increased past the primary Newtonian area of the system. The shear-thinning effect depends on the amount of salts present, as ions shield the charge of the alginate molecule and, hence, reduce the rigidity of the molecule.

Appropriate viscosity in a certain application can be obtained by selecting from a wide range of commercially available grades of alginates. High- and low-molecular-weight grades will provide slightly different textures, and the alginate concentration can be adjusted to fulfil special textural and rheological requirements. Figure 4.5 shows the viscosity range for some alginate grades at different concentrations. To formulate against a target product viscosity, one could use either more of a lower viscosity grade alginate or less of a high viscosity grade

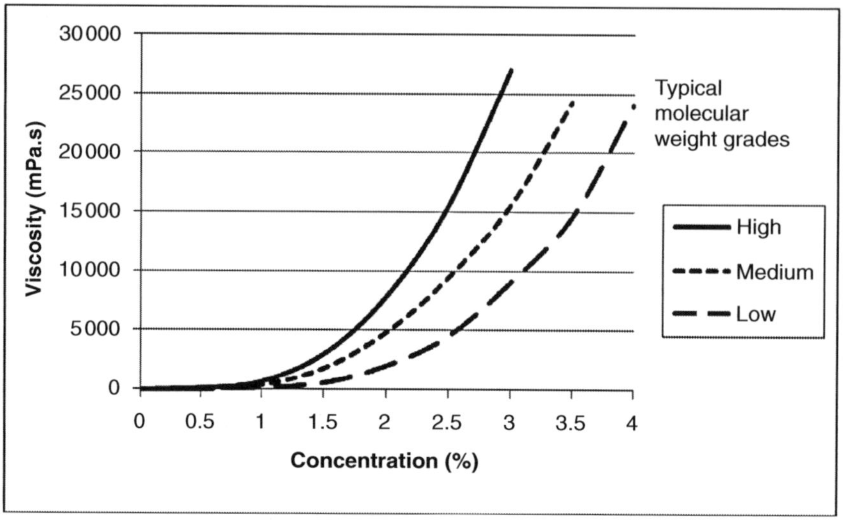

Fig. 4.5 Alginate viscosity dependency on concentration and molecular weight. (Reproduced with kind permission of FMC Corporation.)

alginate. Lower molecular weight alginates will exhibit a longer shelf life because of the lower effect of chain cleavage on apparent viscosity.

4.4.3 Gelation

The most important property resulting from the block structure of alginates is the ability to form gels. The main advantage of alginate as a gel former is its ability to form heat-stable gels which can set at room temperatures. In food applications, it is primarily gel formation with calcium ions which is of interest. However, at slightly acid pH, such as in fruit and jam applications, the alginate gel will be an acid-type gel or a mixed calcium/acid gel. This is described below in more detail.

In order to react with calcium to form a gel, alginate has to contain a certain proportion of guluronic acid, and the guluronic acid monomers must occur in blocks. The junction zone of the alginate gel network is formed when a G-block in one alginate molecule is physically linked to a G-block in another alginate molecule through chain–Ca^{2+}–chain interactions. The M-blocks and the MG-blocks will not participate in the junction zones but form so-called elastic segments in the gel network.

The interaction between alginate and calcium ions is commonly visualised through the 'egg-box model', where the calcium ions fit into the structural void in the alginate chain, like eggs in an egg box (Grant *et al.*, 1973), as shown in Fig. 4.6. Also it has been found that the egg-box model can be explained from energy considerations when looking at Lennard-Jones potentials, Van der Waals forces and electrostatic interactions (Braccini and Pérez, 2001). It is argued that the anti-parallel chain pairing between two guluronate blocks is the most likely association structure in the gel state, and the shape of this structure is similar to the commonly mentioned egg-box structure. More recently, small angle X-ray scattering analysis by Stokke *et al.* (1997) showed that the cross-sectional thickness of the junction zones expand when the calcium concentration and/or the content of G-blocks in the alginate

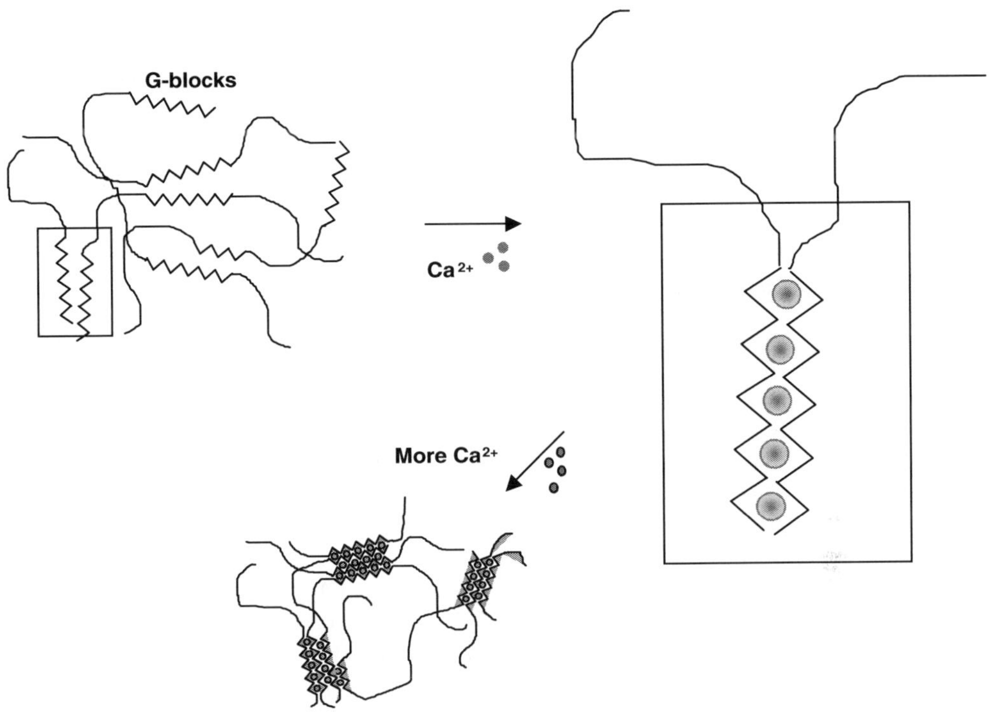

Fig. 4.6 Alginate gelation with calcium ions in the 'egg-box model'. (Reproduced with kind permission of FMC Corporation.)

increases. The explanation is that more than two G-blocks are joined in the lateral junction zones which imply a modification of the original egg-box model (Fig. 4.6) and a strong indication of the presence of multilayer junction zones.

The gel-forming capacity and the resulting gel strength obtained from alginate is very closely related to the number and length of junction zones, that is the amount of G-blocks and the average G-block length. High G content and long G-blocks give alginates with the highest calcium reactivity and the strongest gel-forming potential. The molecular weight of the chain and level of available calcium ions are also of importance. In the lower range of molecular weight, up to about 100–150 kDa, the gel strength of saturated calcium alginate gels will increase with molecular weight. Above 100–150 kDa, the gel strength increases only slightly at saturated calcium levels (Smidsrød and Haug, 1972; Draget *et al.*, 1993).

From a chemical point of view, the formation of a calcium alginate gel should be considered as an ion-exchange process. The counterion of a water-soluble alginate is exchanged with calcium giving the sol-gel transition. The counterion, such as sodium, has a lower affinity for the alginate chain compared to calcium. However, it is observed that sodium has a higher affinity for alginate than potassium (Draget *et al.*, 1998). This may have practical implications for preparing gels with low levels of calcium, but the effect is minor when using relatively high calcium levels. Because of the difference in affinity for sodium and potassium, calcium alginate is formed slightly faster from potassium alginate compared to sodium alginate. Potassium alginate may have advantages in applications with high sodium content, for example when finished gels are stored in a high concentration salt brine, because the reverse exchange of sodium with calcium is reduced and a subsequent loss of gel strength can be minimised.

An acid alginate gel is formed in a similar way through the involvement of junction zones. G-blocks contribute to the greatest extent to gel formation and gel strength (Draget *et al.*, 1994, 2003), but M-blocks are also able to support the formation of weak cross-links between chains. Generally, alternating MG-blocks will interfere with the formation of intermolecular cross-linking.

4.4.4 Syneresis

After gelation, the water molecules are entrapped by the alginate gel network but are still free to migrate by diffusion. The gel retains water through hydrogen bonds, but if the gel network contracts, some water will be squeezed out. This effect is called syneresis and is commonly seen in many biopolymer gel systems. For alginate gels, syneresis depends on parameters such as M:G profile, calcium concentration, setting mechanism and molecular weight. Controlling these factors is important to avoid undesirable syneresis.

Gels prepared from lower molecular weight alginates show less syneresis than gels of higher molecular weight material. This is probably due to a lower number of intact elastic segments between junction zones, resulting in a lower capability of the network to reorganise and contract during the continued gelling process. Alginates containing high levels of alternating MG structures, and having more flexible elastic segments in the gel network, exhibit higher degrees of syneresis (Draget *et al.*, 2001). Most important for syneresis is the calcium to alginate ratio in the system. In a balanced formulation, with just enough calcium to saturate all the G-blocks of the alginate, syneresis will usually be negligible. Where there is excess calcium present, syneresis may be higher. Internal gelation tends to give gels with less syneresis compared to externally set alginate gels, basically because the ratio of calcium to alginate is easier to control.

4.4.5 Gel strength measurement

Gels prepared from alginates of various origin and M:G composition may be characterised by their gel strength. No industry standard for gel strength measurement exists, but gel strength is commonly measured by preparing 0.7–1.0% alginate gels and then measuring the force required to turn a probe in the gel or to compress the gel. Food Industries Research Association (FIRA) values refer to gel strength measurements using an FIRA jelly tester to measure homogeneous alginate gels set by internal-setting mechanisms (Toft, 1982). The force required to compress a gel can be measured with instruments such as the Instron® or Texture Analyzer.

4.5 GEL FORMATION TECHNIQUES

4.5.1 General

Alginates will form gels with acids and with all multivalent cations except magnesium. In the food industry, calcium is by far the most widely used gelling cation for alginate systems. Gel formation can be controlled through controlling the release of calcium or acid into an alginate solution. Both acid and calcium alginate gels are thermo-irreversible and will form over a wide temperature range. A thermo-reversible gel can be made under acidic conditions

Table 4.2 Calcium salts frequently used in combination with alginate.

Calcium salt	Gelling mechanism	Molecular weight	% Calcium	Cold solubility	E-Number
Calcium acetate (monohydrate) Formula: $Ca(CH_2COO)_2 \cdot H_2O$	Diffusion, bath or spray	176.18	22.75	27%	E263
Calcium Chloride Formula: $CaCl_2$		110.99	36.11	40%	E509
Calcium Lactate Formula: $Ca(C_3H_5O_3)_2 \cdot 5H_2O$		308.30	13.00	5%	E327
Calcium sulphate anhydrous Formula: $CaSO_4$	Internal, neutral gelation	136.15	29.44	0.20%	E516
Calcium Sulphate dihydrate Formula: $CaSO_4 \cdot 2H_2O$		172.16	23.28	0.27%	E516
Calcium Carbonate Formula: $CaCO_3$	Internal, acid gelation	100.09	40.04	0.0015%	E170
Dicalcium phosphate anhydrous Formula: $CaHPO_4$		136.10	29.00	$<$0.02%	E341b
Dicalcium Phosphate dihydrate Formula: $CaHPO_4 \cdot 2H_2O$ (faster Ca^{2+} release than $CaHPO_4$)		172.10	23.29	0.02%	E341b
Tricalcium citrate Formula: $Ca_3(C_6H_5O_7) \cdot 4H_2O$		570.50	21.08	0.09%	E333
Tricalcium Phosphate Formula: $Ca_3(PO_4)_2$ (slower than $CaHPO_4$ for Ca^{2+} release)		405.15	29.63	0.002–0.003%	E341c

below pH 4.0 and, preferably, around pH 3.4, by using a combination of alginate and high methyl-esterified pectin (Toft, 1982). However, the most important property of alginate as a gel former is its ability to make heat-stable gels in cold systems.

As a general rule, the alginate needs to be properly hydrated before contact with calcium in order to fully utilise the gelling potential. The gelling mechanisms applied in alginate systems can be divided into categories of diffusion setting, internal setting and setting by cooling. A list of calcium sources is given in Table 4.2 and of sequestrants in Table 4.3.

4.5.2 Diffusion setting, neutral pH

When utilising diffusion setting, a food-mix-containing hydrated alginate is brought into contact with dissolved calcium, either by dipping the food item in a calcium bath or by spraying it with a calcium salt dissolved in water. Calcium chloride is the most commonly used source, but any readily soluble calcium salt may be used. The calcium ions diffuse into the food matrix containing alginate, forming a calcium alginate gel when the calcium ions react with the alginate. Cross-linking of the alginate will start at the interface and then proceed inwards as the alginate is saturated with gelling ions on the surface.

This process is especially suitable for relatively thin or small dimension materials, such as pimiento strips, or to provide a thin coating on a product surface, as with onion rings. The diffusion rate can be increased by raising temperature, by raising the concentration of

Table 4.3 Sequestrants used to control the gelation of alginate with calcium.

Functions	Ortho (1 phosphorus)	Pyro (2 phosphorus)	Tripoly (3 phosphorus)	Hexameta (more than 3 phosphorus)
Buffer action	+++++	+++	+	0
Sequestration capacity	0	+++	+++	++++
Dispersion	0	++	++	++++
Solubility	+++++	++	+++	++++
Common sequestrant used	Sodium orthophosphate	Tetrasodium pyrophosphate (TSPP)	Sodium tripolyphosphate (STPP)	Sodium hexametaphosphate (Calgon, SHMP)
E-number	E339	E450a	E450b	E452

+ represents the degree of efficacy, i.e. +++++ is more effective than ++++ which is more effective than +++, etc.

calcium in the setting bath or spray and by using a strongly calcium-reactive alginate, that is an alginate with a high proportion of G-blocks.

4.5.3 Diffusion setting, acidic pH

In this system, an acid-soluble calcium salt that is insoluble at neutral pH is mixed with the alginate. When an acid comes into contact with the surface of the mix, the calcium salt solubilises as protons diffuse into the alginate–calcium salt matrix and the pH is reduced. The released calcium ions will then react with the alginate and form the gel. Figure 4.7 shows the gelling mechanism during diffusion setting under neutral and acidic conditions. Examples of suitable calcium salts are calcium carbonate or dicalcium phosphate.

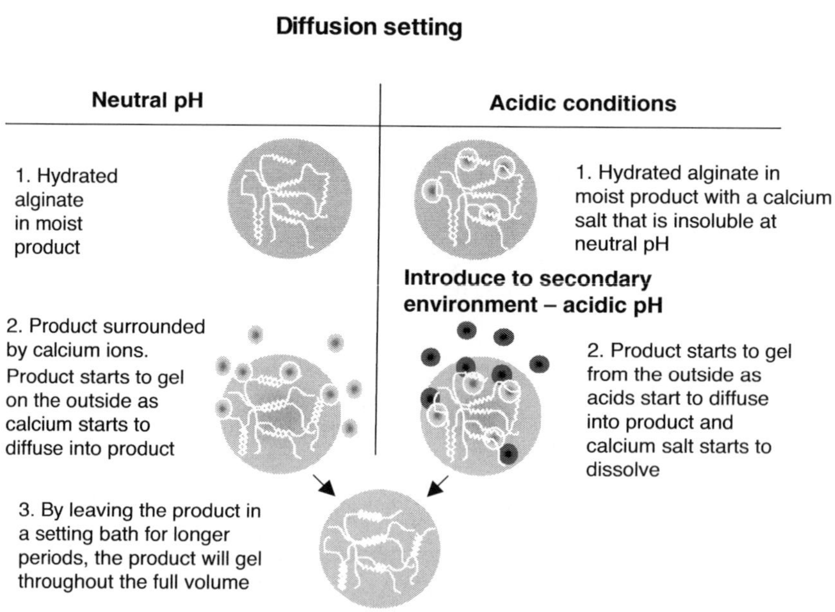

Fig. 4.7 Alginate gelation by diffusion setting. (Reproduced with kind permission of FMC Corporation.) For a colour version of this figure, please see Plate 3 of the colour plate section.

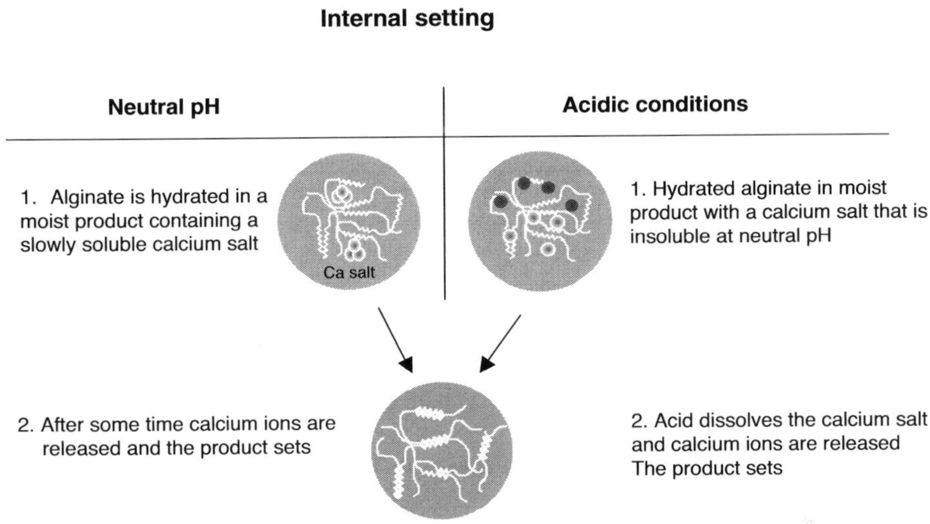

Fig. 4.8 Gelling by internal setting. (Reproduced with kind permission of FMC Corporation.) For a colour version of this figure, please see Plate 4 of the colour plate section.

4.5.4 Internal setting, neutral pH

In this process, calcium is released homogenously within the product under controlled conditions. It employs a combination of alginate, a slowly soluble calcium salt and a suitable calcium sequestrant, such as a phosphate or citrate. Calcium sulphate is commonly used because of its low cost and convenient solubility. The sequestrant will bind free calcium and prevent pre-gelation of the alginate during the time required to mix the food with the alginate and process it into its desired shape (Fig. 4.8). The amount of sequestrant to be used is dependent on the delay required to process the food: longer processing times will require higher concentrations of sequestrant.

4.5.5 Internal setting, acid pH

The internal-setting process may be performed at both neutral and acidic pH. Calcium carbonate and some calcium phosphates are widely used as calcium sources because their solubility, and the subsequent release of calcium ions, is pH dependent. The food is mixed with alginate and calcium source at neutral pH, and calcium release is then obtained by acidification with citric acid, adipic acid or another suitable acid. Glucono-delta-lactone is sometimes used as an acidifier because of the slow and controlled acidification provided.

4.5.6 Combined setting

The diffusion and internal-setting systems may be combined, thus providing a rapidly gelled outer surface or coating before the product sets throughout by means of internal gel setting through slow calcium release. The coating is often desired to make the product mechanically robust for further processing, before internal setting has provided a firm structure.

4.5.7 Setting by cooling

In this process, the alginate, calcium salt and a sequestrant are dissolved in a hot liquid. The elevated temperature prevents gelation because the alginate chains are in thermal motion, which prevents association of the chains through cross-linking. Setting will start when the solution is cooled and will then result in a heat-stable calcium alginate gel. Gels set in this way show less syneresis than gels made by diffusion or regular internal setting (Mancini and McHugh, 2000). By careful formulation, gels can be made to set over the range 0–50°C, but this process is limited to relatively soft textures.

4.5.8 Alginate–pectin gels

When used alone, HM pectins are only able to form gels at high sugar solid levels within a narrow pH range, and low methyl-esterified (LM) pectins form gels at low sugar levels with divalent cations, such as calcium. By introducing sodium alginate, it is possible to make alginate–HM pectin gels at low solid levels and over a wider pH range (Thom *et al.*, 1982; Toft, 1982; Toft *et al.*, 1986).

Fruits naturally rich in pectin, such as apples, form gels when a sodium alginate solution is added after cooking. Firm gels can be formed when combining HM pectin and sodium alginate solution high in guluronic acid. A 1:1 ratio has been reported to provide the highest gel strength in these systems (Thom *et al.*, 1982; Toft, 1982). A denser polymer network is seen for a combined gel of HM pectin and alginate compared to LM pectin gels (Walkenstrøm *et al.*, 2003). There is a synergy between alginate and pectin, as a mix of the two generally results in a firmer gel than the same amount of each individual biopolymer. In contrast to thermally stable calcium alginate gels, alginate–pectin gels are thermo-reversible. The alginate–pectin synergy is one of very few interactions for alginate with other hydrocolloids and, so far, the only one of commercial value.

4.6 APPLICATIONS

4.6.1 General

When an alginate gel is to be formed at pH values above 3.5, the pK_a value of alginic acid, alginate and a cross-linking agent such as calcium have to be present. In order to obtain an optimal alginate gel, the gelling process has to be controlled to avoid release of calcium until the alginate is fully hydrated, and all food processing completed, before gelation commences. This is mostly done using a sequestrant to ensure that the availability of calcium is controlled. The chemical and physical changes involved are rather complex, and knowledge of alginate properties and calcium release with the chosen combination of sequestrant and calcium source is required. In fact, the alginate grade and the calcium-sequestering system, including calcium source and sequestering agent(s), must be matched with the process developed to manufacture the food. However, successful product development of a restructured food product will usually involve trials and optimisation work, especially during scale-up. Additionally, all major alginate suppliers are able to offer technical guidance and support to develop alginate gel systems. They may also provide a complete blend of alginate, sequestrant and calcium salt to ensure that consistent results are obtained.

4.6.2 Restructured food

Almost any food material that can be pulped can also be restructured, recombined or reformed, providing several benefits for the food manufacturer. Restructuring allows the utilisation of cheap raw materials, for example fruit or vegetable pulp, meat pieces and fish mass (see Table 4.5). It may allow the use of a standard automated process making use of consistent mechanical properties, for example handling resistance, to give a standard product in terms of shape, reproducibility, heat resistance and freeze–thaw stability. Diffusion setting and internal setting mechanisms can be applied separately, or in combination, when restructuring foods.

The amount of alginate needed is usually between 1% and 2% of the final product weight. An example recipe for restructured onion rings is given in Formulation 4.1.

Formulation 4.1 Restructured onion rings

Ingredient	Amount (%)
Water	70.50
Onion powder (16 mesh grade)	15.90
Wheat flour	12.64
Alginate (fine mesh, high viscosity, high gel strength)	0.82
Salt (NaCl)	0.14
Total	100.00

Setting bath: calcium chloride, $CaCl_2 \cdot 2H_2O$, 5% solution in water (2.5 L setting bath per 1000 g onion mass).

Method

1. Mix the dried onion with 90% of the water and hydrate for approximately 1 h.
2. When the onion is hydrated, stir the onion mass into the rest of the water.
3. Thoroughly blend alginate, flour and salt and stir into the onion mixture until it is homogeneous.
4. Extrude the onion mixture through an annulus and transfer to the setting bath where the onion is allowed to gel (Fig. 4.9).

4.6.3 Bakery creams and fruit fillings

Several applications for alginate exist in bakery products. One important application which demonstrates alginate gelling properties is the thickening of cold-prepared bakery creams. Alginate will ensure that the cream is bake stable as well as freeze–thaw stable. A blend containing sodium alginate, calcium salt and sequestrant is used to give internal setting in neutral conditions. This mixture is blended with milk powder, sugar, starch and any other dry ingredients of the bakery cream. When preparing the cream, tap water is added to the

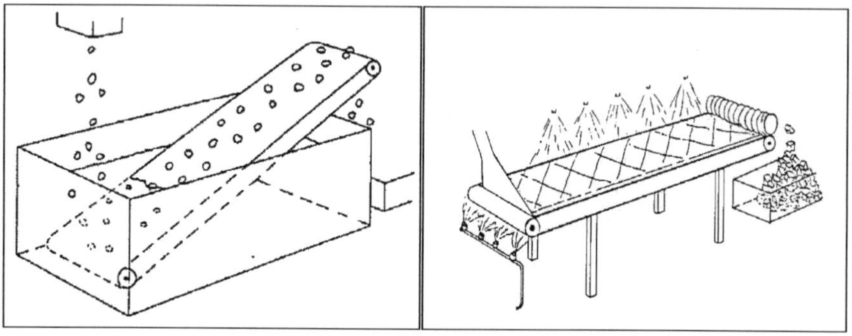

Fig. 4.9 Alginate gel setting with calcium bath and calcium spray. (Reproduced with kind permission of FMC Corporation.)

dry blend and mixed well. The resulting smooth, viscous cream is then applied to the cake. After some minutes, the slightly gelled cream is heat stable and will keep its shape during the baking process.

Alginate is frequently used in fruit fillings, giving the filling bake stability to prevent it from leaking out of the pastry during baking. These products can be either cold or hot processed as desired. By creating a viscous or gelled filling, the alginate reduces moisture transfer from the fruit filling into the pie pastry. The pectin and calcium naturally present in the fruit will contribute to gel formation as synergistic gelation between alginate and pectin is often obtained.

Thickening of batter is another alginate application. Alginate provides a short flow which gives good deposition control and pickup which is easier to handle on production machinery.

Finally, the water-binding properties of alginate will provide an additional moisture-retaining benefit in pastries.

4.6.4 Dessert jellies and mousse

Instant water dessert jelly is a good example of internal gel setting with alginate in acid conditions. The product is presented as a dry powder mix with alginate, calcium source, sequestrant, acidity regulator (buffer) and flavour components that can be dissolved in cold water to form a dessert jelly within a few minutes. The dry mix must be carefully formulated in order to give a blend that is capable of creating a jelly with the desired texture and setting rate. Calcium and other divalent ions present in tap water (hard water) may cause premature gelling in this type of application. This may present a challenge for products marketed over a wide geographical area in which water hardness varies.

Instant and ready-to-eat chocolate mousses may be prepared based on alginate. Alginate is added to stabilise the whipped mousse foam and to give structure to the product. In an instant product, the cold-gelling property of alginate is utilised, whereas for the ready-to-eat version, it is possible to form a gel without the use of sequestrants by utilising the self-sequestering capability of milk, described in more detail in Section 4.8.

4.7 THICKENING AND STABILISING

4.7.1 Thickening

As described previously, alginate develops viscosity when hydrated in aqueous solutions and gels upon the addition of calcium. In practice, it is difficult to make a sharp distinction between thickening and gelling functionality when other ingredients in the product are in play. The three most important parameters are the molecular weight and M:G profile of the alginate and the amount of calcium available. A very weak gel may appear as a thick solution and thickening may often be the result of limited alginate cross-linking by calcium.

Alginate is used in cheese sauces to provide increased viscosity. Equally important, it gives the cheese sauce the desired cling needed to stay on food items such as pasta. Alginate also reduces surface skin formation. A highly viscous, high-M alginate is often used to get the desired thickening effect while avoiding too strong an interaction with calcium, present in other ingredients in the sauce, which could subsequently lead to gelation.

The thickening properties of alginate can be used in gravy in two contrasting ways: as a temporary suspending agent or as a delayed thickening agent. An example of the first process is the use of alginates to prevent meat from settling out in gravy prior to filling into cans. When the filled cans are retorted, the heat treatment reduces the thickening effect of alginate, which is no longer required, by thermal degradation of the polymer chains. The opposite effect can be used to facilitate heat exchange in cans during retorting. The effect is obtained by using a mixture of alginate and a slowly soluble calcium salt. The combination of slow solubility and heat delays gel formation until the heat treatment has taken place. Rapid heat transfer in the low viscosity sauce then allows a shorter heating cycle than with a pre-gelled product.

4.7.2 Stabilising

Water-soluble alginates can act as stabilisers in systems consisting of particles or droplets dispersed in water. Such systems include oil-in-water emulsions, for example ice cream and salad dressings, and solids suspended in water, such as fruit juice. The alginate acts as a stabiliser to prevent separation by increasing the viscosity of the aqueous phase and by producing charged films at the interfaces, so that the individual particles or droplets tend to repel each other avoiding coalescence and phase separation.

Ice cream was the first application for alginate in the food industry. Alginate addition reduces the size of the ice crystals and provides body and a smooth texture. Alginate prevents syneresis and delays the meltdown of the ice cream. In ice cream, as in many other dairy systems where calcium ions are naturally present, the alginate is usually combined with a low level of a sodium phosphate, used as a sequestering agent to prevent premature gelling.

Food emulsions, including certain sauces and salad dressings, can be thickened and stabilised to prevent phase separation by including alginate in the aqueous phase. In neutral systems, a sodium alginate would be used, whereas in low pH systems PGA may be better suited, as described below.

Alginate is also used in low-fat spreads to provide the desired texture and to stabilise the emulsion.

Low-fat spreads are water-in-oil emulsions. Alginate is used to stabilise and weakly gel the water droplets finely distributed in the oil phase. A soft gel based on a relatively high-M alginate is required to achieve a texture that mimics full-fat spreads.

4.7.3 Thickening/stabilising with PGA

In low-pH foods and beverages, with pH values below or around 3.5, the pK_a of alginic acid, regular alginate will not be an effective thickener or stabiliser. As the pH decreases, alginate will be partly protonated and lose negative charge. Below the pH corresponding to the pK_a value, alginic acid will eventually start to precipitate. PGA, in contrast, possesses valuable functionality, even in low-pH conditions, since a portion of the carboxylic acid groups are esterified and cannot be protonated.

PGA is manufactured by reacting alginic acid with propylene oxide. The presence of the lipophilic propylene glycol ester groups provides PGA with good emulsifying properties and makes it more acid tolerant and less calcium reactive than sodium alginate. The remaining non-esterified acid groups retain some negative charge, even down to pH 2.75. These may participate in weak, but significant, cross-linking with calcium and proteins, which at this low pH, below the isoelectric point, usually carry a net positive charge. These PGA properties are utilised in the stabilisation of milk proteins under acidic conditions, such as in low-pH dairy beverages, for example drinking yogurt, and for pulp stabilisation in acidic beverages in general. PGA is appreciated as a particularly effective thickener and stabiliser in low-pH salad dressings and mayonnaise. PGA is also used to stabilise the foam head in certain types of beer through interaction with the beer proteins.

4.8 DAIRY PRODUCTS

Alginate is used as a stabiliser, thickener and a gelling agent in dairy products including thickened and canned cream, chocolate mousse, yogurt, bakery creams, milk shake, ice cream and cheese. Since milk naturally contains calcium, the system needs to be sequestered in order to ensure proper hydration of the alginate otherwise a gelled layer will form on the surface of unhydrated alginate particles, a phenomenon known as 'fish eyes'. Proper hydration can be achieved by adding a sequestrant, for example phosphate or citrate, with the alginate to pick up the free calcium before it can interact with alginate. The most commonly used sequestrant in milk systems is tetrasodium pyrophosphate. Another way to sequester the calcium is to heat the milk; milk proteins bind the calcium more tightly as the temperature increases (Tranchant, 2000). This will, in combination with the increasing thermal movements of the molecules, prevent the alignment of the alginate molecules and calcium ions and prevent junction formation. Above approximately 70°C, the calcium in milk will bind harder to the micelles and alginate can hydrate without interacting with the calcium. Gelation will occur upon cooling, and a simple alginate milk gel can be obtained without the use of other additives.

4.9 FILM FORMATION

A film is formed when a thin layer of alginate gel or solution is dried. This type of film provides several benefits in the food industry, for example reduced water loss, barrier properties for diffusion control and shape control. An alginate film can be used in pastries to prevent moisture from the filling passing into the rest of the cake; it can be used in cake icings to prevent adhesion to the wrapping and, simultaneously, to act as an anti-cracking agent.

Alginate films can be prepared as water-soluble or water-insoluble materials. Traditionally, insoluble calcium alginate films have been prepared in situ by first spraying a sodium alginate solution onto the food or dipping it into a sodium alginate bath. Then calcium chloride solution is applied either as a spray or through a bath. This type of film has been used to protect frozen fish from oxidation and loss of water by stabilising the coating layer of ice glaze and making it more impermeable to oxygen. The technique can also be used to bind herbs and spices to the surface of, for example, meat pieces. Several benefits including reduced moisture loss, reduced shrinkage and less off-odour have been reported (Williams *et al.*, 1978). Sensory attributes are also improved by applying an alginate film to pork patties (Wanstedt *et al.*, 1981). More recently, alginate has been used as a film former in casings for sausages made from vegetable materials.

Another more recent use of sodium alginate is within the field of rapidly dissolving films, such as breath strips. In this case, the film carries strong flavour agents, for example menthol, or it may even contain pharmaceutical actives. The use of a plasticiser such as glycerine is usually required to obtain flexible films.

4.10 ENCAPSULATION

Over recent years, alginates have increasingly found use in numerous encapsulation applications, such as encapsulation of probiotics, flavours and functional food oils. In these applications, the mild setting conditions of alginates are of a particular benefit, as sensitive products may be encapsulated without deterioration.

There exist a number of possible encapsulations techniques to be used within the food area, and commonly used techniques for alginates are extrusion, emulsification and coating (Gibbs *et al.*, 1999; Chan *et al.*, 2002). In the encapsulation of probiotics, it is found that alginates effectively protect bacteria from acid exposure in the stomach (Kailasapathy, 2002; Chandramouli *et al.*, 2004; Chan and Zhang, 2005). Encapsulated bacteria can be added to dairy products such as yogurt, cheese, facilitating higher stability and bacterial productivity (Krasaekoopt *et al.*, 2003).

Encapsulation of flavouring agents, such as flavour oils, can be done by extruding or dripping an emulsion of alginate and the oil into a calcium salt solution (King, 1988). Yet another technique is to use a double-tube nozzle in which the material to be encapsulated is extruded through the inner tube, while the alginate solution is extruded through the outer tube (see Fig. 4.10). The encapsulation of functional food oils, such as fish oils, wheat germ oil, and evening primrose oils, is also successfully performed in order to improve properties such as taste, stability or oxidation (Chan *et al.*, 2000).

4.11 OTHER APPLICATIONS

4.11.1 Nutritional benefits

Significant research during recent years, both academic and by industry, points towards the beneficial effects on human health of alginates in the diet. These effects include a reduction

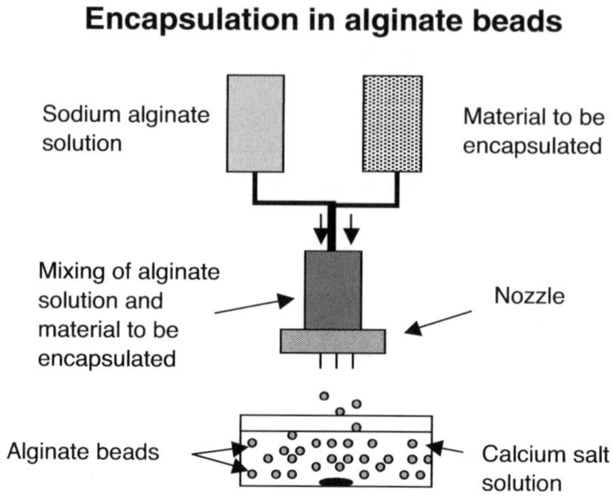

Fig. 4.10 Equipment for encapsulation of a liquid or suspension with alginate. (Reproduced with kind permission of FMC Corporation.)

of intestinal absorption rates, reduction of colonic luminal toxicity, alteration of colonic microflora and improvements in the colonic mucosal barrier (Brownlee *et al.*, 2005).

As alginate is a soluble dietary fibre, it provides the effect of increasing the intestinal viscosity, which may reduce blood levels of glucose and cholesterol (Sandberg *et al.*, 1994; Wolf *et al.*, 2002). However, viscous foods have low palatability. Incorporating calcium alginate powder into bread or soy beverages allows high concentrations of alginate to be added to foods without any significant effect on palatability. The viscosity develops after the calcium alginate is exposed to the stomach acid and neutralised in the intestine (Gåserød *et al.*, 2008).

The effect of alginate on satiety and food intake has been demonstrated with beverages containing high guluronic acid alginate with insoluble calcium salts. These beverages can form intragastric lumps when ingested to increase the feeling of fullness and decrease hunger (Hoad *et al.*, 2004). Further studies have shown that incorporating calcium-gelled alginate–pectin fibres into beverages can increase satiety and reduce energy intake, particularly in individuals who control energy intake less rigidly (Pelkman *et al.*, 2007). Such gels or fibres may have benefits in weight control plans and may help in the prevention and/or treatment of being overweight or obese.

A number of other potential health benefits specific to alginates are indicated in the scientific literature. Low-molecular-weight alginates have prebiotic properties and appear to beneficially alter the colonic microflora by increasing the bifidobacteria population and also reduce bacterial toxin levels in the colonic lumen (Terada *et al.*, 1995). Alginates rich in polymannuronic acid have gastrointestinal and systemic health properties not seen with other dietary fibres which may further their use as functional food ingredients (Brownlee *et al.*, 2005).

4.11.2 Extended shelf life

Alginate has been shown to extend shelf life for bakery goods including bread, muffins, cup cakes and soft cookies. Alginate, being a very hydrophilic polymer, binds and retains water, keeping the product moist and soft over an extended time. It also improves dough viscosity and is easy to dry blend with other ingredients. It is believed that alginate creates a network that prevents the retrogradation of starch and, in this way, contributes to slowing the ageing process of bakery products (Butt *et al.*, 2001).

Alginate has also found use in gluten-free bakery products, due to its ability to bind and retain water, hence providing an acceptable shelf life to the products. Gluten-free pre-mixes for bread, crisp bread and chocolate cakes are examples of products containing alginate.

4.11.3 Gourmet food applications

The use of alginate in gourmet foods is a small but growing market. Some natural products, for example fish eggs and caviar, can be imitated by alginate capsules (Ueda, 1986). Alginate beads can also be used as garniture in restaurant dishes and to prepare novel dishes (Adria and Adria, 2008). Artificial bird's nest and shark fins using alginate as the gelling and structuring agent have been made in certain countries in Asia over many years.

4.12 SUMMARY

The global consumption of alginate is still growing. The future general demand for more effective utilisation of the world's food resources will certainly continue to increase the volume of alginate consumed for food purposes (Tables 4.4 and 4.5). While some markets move in the direction of using less-processed raw materials, the need for convenience will trigger new developments in the area of restructured food. Food technology process development will be

Table 4.4 Alginate properties utilised in food products.

Gel forming	Pet food Restructured fruit and vegetables Restructured fish and meat Puddings and desserts Cold prepared bakery creams Fruit preparations, fruit fillings Mousse Encapsulation, bead formation
Thickening/water binding	Tomato ketchup, tomato sauce Soups, sauces, cheese sauce Milk shakes Thickened cream
Stabilising	Ice cream Mayonnaise Whipped cream Low fat spread Salad dressing (PGA) Fruit juice (PGA) Beers, lagers (PGA) Baked goods
Film forming	Glazes for frozen meat and fish Film coatings for fresh meats Coatings for cakes and cookies Vegetable coating

Table 4.5 Alginate application areas.

Restructured foods	Onion rings
	Pimiento fillings for olives
	Fruit fillings and dried fruit pieces
	Fish, meat and poultry
	Anchovy olive fillings
	Apple pieces for pie fillings
	Cocktail berries
	Meat chunks for pet food
	Shrimp-like fish products
	Fish patties
Pet food	Alginate-gelled chunks
Bakery products	Bake-stable filling creams
	Bake-stable fruit fillings
	Extended shelf-life baked goods
Ice cream	Ice cream
	Sorbet (PGA)
Jams and marmalades	Fruit preparations for yogurts and ice cream
	Low sugar jam
	High-quality marmalade with high fruit content
Dressing and sauces (emulsions)	Salad dressings (PGA)
	Mayonnaise (PGA)
	Ketchup, tomato sauce
	Low-calorie margarine/spreads
Desserts and dairy products	Instant desserts
	Jelly, puddings
	Mousse
	Thickened cream
Toppings, sauces and ripple syrups	Cheese sauce
	Ice cream ripple syrup
Drinks	Beers, lagers (PGA)
	Juice, squash (PGA)
	Liqueurs (PGA)
Biotechnology	Immobilisation of bacteria and yeast
Feed	Fish feed

targeted towards semi-continuous and continuous production and away from batch processing. Alginates will be used in many of these developments because gel formation can take place at chilled and ambient temperatures and the sophisticated processes for restructuring products using alginate gel technology are highly developed.

In parallel with further development and use of alginate in traditional alginate applications, there is a lot of development work invested in new applications of alginate. New products based on alginate can be foreseen for the encapsulation of probiotics and flavours, confectionary films and gourmet applications (molecular food design).

References

Adria, A. and Adria, F. (2008) elBulli Texturas, Sferificación. Recipes. Available at http://www.albertyferranadria.com/esp/texturas-sferificacion.html (accessed 5 August 2009).

Braccini, I. and Pérez, S. (2001) Molecular basis of Ca^{2+}-induced gelation in alginates and pectins: The egg-box model revisited. *Biomacromolecules*, **2**, 1089–1096.

Brownlee, I.A., Allen, A., Pearson, J.P., Dettmar, P.W., Havler, M.E., Atherton, M.R. and Onsøyen, E. (2005) Alginate as a source of dietary fiber. *Critical Reviews in Food Science and Nutrition*, **45**, 497–510.

Butt, M.S., Anjum, F.M., Samad, A., Kausar, T. and Mukhtar, M.T. (2001) Effect of different gums on the quality and shelf life of bread. *International Journal of Agriculture and Biology*, **3**(4), 482–483.

Chan, E.S. and Zhang, Z. (2005) Bioencapsulation by compression coating of probiotic bacteria for their protection in an acidic medium. *Process Biochemistry*, **40**(10), 3346–3351.

Chan, L.W., Lee, H.Y. and Heng, P.W.S. (2002) Production of alginate microspheres by internal gelation using an emulsification method. *International Journal of Pharmaceutics*, **242**(1–2), 259–262.

Chan, L.W., Lim, L.T. and Heng, P.W. (2000) Microencapsulation of oils using sodium alginate. *Journal of Microencapsulation*, **17**(6), 757–766.

Chandramouli, V., Kailasapathy, K., Peiris, P. and Jones, M. (2004) An improved method of microencapsulation and its evaluation to protect *Lactobacillus* spp. in simulated gastric conditions. *Journal of Microbiological Methods*, **56**(1), 27–35.

Draget, K.I., Gåserød, O., Aune, I., Andersen, P.O., Storbakken, B., Stokke, B.T. and Smidsrød, O. (2001) Effects of molecular weight and elastic segment flexibility on syneresis in Ca-alginate gels. *Food Hydrocolloids*, **15**(4–6), 485–490.

Draget, K.I., Skjåk-Bræk, G. and Smidsrød, O. (1994) Alginic acid gels: the effect of alginate chemical composition and molecular weight. *Carbohydrate Polymer*, **25**(1), 31–38.

Draget, K.I., Simensen, M.K., Onsøyen, E. and Smidsrød, O. (1993) Gel strength of Ca-limited gels made in situ. *Hydrobiologia*, **260/261**, 563–569.

Draget, K.I., Steinsvåg, K., Onsøyen, E. and Smidsrød, O. (1998) Na- and K-alginate effect on Ca^{2+} gelation. *Carbohydrate Polymer*, **35**, 1–6.

Draget, K.I., Stokke, B.T., Yuchi, Y., Urakawa, H. and Kajiwara, K. (2003) Small-angle X-ray scattering and rheological characterization of alginate gels. 3. Alginic acid gels. *Biomacromolecules*, **4**(6), 1661–1668.

Ertesvåg, H., Høidal, H.K., Schjerven, H., Svanem, B.I. and Valla, S. (1999) Mannuronan C-5-epimerases and their application for in vitro and in vivo design of new alginates useful in biotechnology. *Metabolic Engineering*, **1**(3), 262–269.

Gåserød, O., Haraldsen, H. and Lynch, G. (2008) Calcium alginate as a gastro-activated dietary fibre. In: P.J. Williams and G.O. Phillips (eds), *Gums & Stabilisers for the Food Industry 14*. The Royal Society of Chemistry, Cambridge, pp. 349–357.

Gåserød, O., Larsen, C.K., Bendiksen, M. and Smidsrød, O. (2002) A study of dissolution rates of alginates using the rheometer. *FMC BioPolymer poster*.

Gibbs, B.F., Kermasha, S., Alli, I. and Mulligan, C.N. (1999) Encapsulation in the food industry: a review. *International Journal of Food Sciences and Nutrition*, **50**(3), 213–224.

Grant, G.T., Morris, E.R., Rees, D.A., Smith, P.J.C. and Thom, D. (1973) Biological interactions between polysaccharides and divalent cations: the egg-box model. *FEBS Letters*, **32**(1), 195–198.

Grasdalen, H. (1983) High-field, 1-H-N.M.R. spectroscopy of alginate: sequential structure and linkage conformations. *Carbohydrate Research*, **118**, 255–260.

Haug, A., Larsen, B. and Smidsrød, O. (1967) Studies on the sequence of uronic acid residues in alginic acid. *Acta Chemica Scandinavica*, **21**, 691–704.

Haug, A., Larsen, B. and Smidsrød, O. (1974) Uronic acid sequence in alginate from different sources. *Carbohydrate Research*, **32**, 217–225.

Hoad, C.L., Rayment, P., Spiller, R.C., Marciani, L., Alonso, B.C., Traynor, C., Mela, D.J., Peters, H.P.F. and Gowland, P. (2004) In vivo imaging of intragastric gelation and its effect on satiety in humans. *Journal of Nutrition*, **134**, 2293–2300.

Kailasapathy, K. (2002) Microencapsulation of probiotic bacteria: technology and potential applications. *Current Issues in Intestinal Microbiology*, **3**(2), 39–48.

King, A.H. (1988) Flavor encapsulation with alginates. *ACS Symposium Series*, **370**, 122–125.

Krasaekoopt, W., Bhandari, B. and Deeth, H. (2003) Evaluation of encapsulation techniques of probiotics for yoghurt. *International Dairy Journal*, **13**(1), 3–13.

Larsen, B. (1981) Biosynthesis of alginate. *Proceedings of the International Seaweed Symposium*, **10**, 7–34.

Larsen, C.K., Gåserød, O. and Smidsrød, O. (2003) A novel method for measuring hydration and dissolution kinetics of alginate powders. *Carbohydrate Polymers*, **51**(2), 125–237.

Mancini, F. and McHugh, T.H. (2000) Fruit–alginate interactions in novel restructured products. *Nahrung*, **44**(3), 152–157.

Martinsen, A., Skjåk-Bræk, G., Smidsrød, O., Zanetti, F. and Paoletti, S. (1991) Comparison of different methods for determination of molecular weight and molecular weight distribution of alginates. *Carbohydrate Polymers*, **15**, 171–193.

McHugh, D.J. (2003) A guide to the seaweed industry. FAO Fisheries Technical Paper **441**, pp. 12–22. Available at http://www.fao.org/docrep/006/Y4765E/y4765e00.HTM (accessed 5 August 2009).

Moe, S.T., Draget, K.I., Skjåk-Bræk, G. and Smidsrød, O. (1995) Alginates. In: A.M. Stephen (ed), *Food Polysaccharides and Their Applications*. Marcel Dekker Inc., New York, pp. 245–286.

Nyvall, P., Corre, E., Boisset, C., Barbeyron, T., Rousvoal, S., Scornet, D., Kloareg, B. and Boyen, C. (2003) Characterization of mannuronan C-5-epimerase genes from the brown alga Laminaria digitata. *Plant Physiology* **133**(2), 726–735.

Østgaard, K. (1993) Determination of alginate composition by a simple enzymatic assay. *Hydrobiologia*, **260/261**, 513–520.

Pelkman, C.J., Navia, J.L., Miller, A.E. and Pohle, R.J. (2007) Novel calcium-gelled, alginate-pectin beverage reduced energy intake in nondieting overweight and obese women: interactions with dietary restraint status. *American Journal of Clinical Nutrition*, **86**(6), 1595–1602.

Sandberg, A.S., Andersson, H., Bosaeus, I., Carlsson, N.G., Hasselblad, K. and Harrod, M. (1994) Alginate, small bowel sterol excretion, and absorption of nutrients in ileostomy subjects. *American Journal of Clinical Nutrition*, **60**(5), 751–756.

Skjåk-Bræk, G. (1992) Alginates: biosyntheses and some structure–function relationship relevant to biomedical and biotechnological applications. *Biochemistry of Plant Polysaccharides*, **20**, 27–33.

Smidsrød, O. (2001) Effects of molecular weight and elastic segment flexibility on syneresis in Ca-alginate gels. *Food Hydrocolloids*, **15**, 485–490.

Smidsrød, O., Glover, R.M. and Whittington, S.G. (1973) The relative extension of alginates having different chemical composition. *Carbohydrate Research*, **27**, 107.

Smidsrød, O. and Haug, A. (1972) Properties of poly (1,4-hexauronates) in the gel state. *Acta Chemica Scandinavica*, **26**(1), 79–88.

Stanford, E.C.C (1881) Improvements in the manufacture of useful products from seaweeds. British Patent 142.

Stokke, B.T., Draget, K.I., Yuguchi, Y., Urakawa, H. and Kajiwara, K. (1997) Small-angle X-ray scattering and rheological characterization of alginate gels. *Macromolecules Symposium*, **120**, 91–101.

Terada, A., Hara, H. and Mitsuoka, T. (1995) Effect of dietary alginate on the faecal microbiota and faecal metabolic activity in humans. *Microbiology Ecology Health and Disease* **8**, 259

Thom, D., Dea, I.C.M., Morris, E.R. and Powell, D.A. (1982) Interchain associations of alginate and pectins. *Progress in Food and Nutrition Science*, **6**, 97–108.

Toft, K. (1982) Interactions between pectins and alginates. *Progress in Food and Nutrition Science*, **6**, 89–96.

Toft, K., Grasdalen, H. and Smidsrød, O. (1986) Synergistic gelation of alginates and pectins. In: M.L. Fishman and J.J. Jen (eds), *Chemistry and Function of Pectins*. ACS Symposium Series 310, American Chemical Society, Washington, DC, pp. 117.

Tranchant, C.C. (2000) Coagulation behaviour of differently acidified and renneted milk and the effects of pre-treatment of milk. *Ph.D. Dissertation*, University of Guelph, pp. 381.

Ueda, T. (1986) Method for preparing fish-egg-like edible products. US Patent 4,702,921.

Venegas, M., Matsuhiro, B. and Edding, M.E. (1993) Alginate composition of Lessonia trabeculata (phaeophyta: Laminariales) growing in exposed and sheltered habitats. *Botanica Marina*, **36**, 47–51.

Vold, I.M, Kristiansen, K.A. and Christensen, B.E. (2006) A study of the chain stiffness and extension of alginates, in vitro epimerized alginates, and periodate-oxidized alginates using size-exclusion chromatography combined with light scattering and viscosity detectors. *Biomacromolecules*, **7**(7), 2136–2146.

Walkenstrøm P., Kidman, S., Hermansson, A-M., Rasmussen, P.B. and Hoegh, L. (2003) Microstructure and rheological behavior of alginate/pectin mixed gels. *Food Hydrocolloids*, **17**, 593–603.

Wanstedt, K.G., Seideman, S.C., Donnely, L.S. and Quenzer, N.M. (1981) Sensory attributes of precooked, calcium alginate-coated pork patties. *Journal of Food Protection*, **44**, 732–735.

Williams, S.K., Oblinger, J.L. and West, R.L. (1978) Evaluation of a calcium alginate film for use on beef cuts. *Journal of Food Sciences*, **43**, 292–296.

Wolf, B.W., Lai, C.S., Kipnes, M.S., Ataya, D.G., Wheeler, K.B., Zinker, B.A., Garleb, K.A. and Firkins, J.L. (2002) Glycemic and insulinemic responses of nondiabetic healthy adult subjects to an experimental acid-induced viscosity complex incorporated into a glucose beverage. *Nutrition*, **18**, 621.

5 Carrageenan

William R. Blakemore and Alan R. Harpell

ABSTRACT

Carrageenan is obtained from red seaweeds harvested around the coasts of the North Atlantic, South America and the Far East. Different species provide a number of carrageenan extracts that have a backbone of galactose but differ in the proportion and location of ester sulphate groups and the proportion of 3,6-anhydrogalactose. Kappa carrageenan and furcellaran form thermally reversible firm, brittle gels. Iota carrageenan gives soft, elastic gels. Kappa carrageenan interacts synergistically with polymannans, such as locust bean gum and konjac, to give strong cohesive gels. Blends of kappa with iota carrageenan or polymannans are used to give a range of gel textures used in injected meats, canned meats, water dessert gels and glazes. Hybrid (kappa/iota) carrageenan is particularly suitable for firm, creamy textures in dairy desserts. Lambda carrageenan is non-gelling but thickens instant drinks and dairy desserts. A specific interaction between kappa carrageenan and kappa casein is widely used to stabilise dairy products including milk beverages, ice cream mixes and processed cheese products.

5.1 INTRODUCTION

For centuries, red seaweeds (*Rhodophyceae*) have been harvested and used as foods in the Far East and Europe. There are many different species of red seaweeds but they all contain naturally occurring polysaccharides that fill the voids within the cellulose structure of the plant. This family of natural polysaccharides includes carrageenan, furcellaran and agar.

Carrageenan extracted from seaweed is not assimilated by the human body, providing only fibre with no nutritional value, but it does provide unique functional characteristics that can be used to gel, thicken and stabilise food products and food systems. Multiple commercial red seaweed species provide a sub-family of carrageenan extracts, with differences in composition and molecular conformation, which lead to a wide spectrum of rheological profiles, gel properties and textures, molecular charge densities and interactions with other gums and proteins.

Carrageenans have a backbone of galactose but differ in the proportion and location of ester sulphate groups and the proportion of 3,6-anhydrogalactose. There are three primary types. Kappa carrageenan and iota carrageenan form thermally reversible gels, which range in texture from firm and brittle to soft and elastic. Lambda carrageenan is non-gelling. Kappa carrageenan interacts synergistically with other gums to modify further the gel texture, for example with polymannans such as locust bean gum and konjac. A specific interaction between kappa carrageenan and kappa casein is widely used to stabilise dairy products.

5.2 RAW MATERIALS

The main species of *Rhodophyceae* (red seaweeds) used in the commercial production of carrageenan include *Kappaphycus alvarezii* ('Cottonii') and *Eucheuma denticulatum* ('Spinosum'), these warm-water cultivated species producing kappa carrageenan and iota carrageenan, respectively. These are spiny, bushy plants, about 50 cm high, which grow on reefs and in shallow lagoons primarily in the Philippines, Indonesia and East Africa. As these species are farmed in tropical waters, and plants mature in 8–12 weeks, they are available year round and production and harvesting can be planned in line with demand.

In contrast, cold-water species are harvested annually during the summer months. Annual supplies are variable and limited, and proper demand forecasting is essential. *Chondrus crispus* is the most familiar of these red seaweeds and is found as a small bushy plant, only about 10 cm in height, widely distributed around the coasts of the North Atlantic. Carrageenan extracted from this species comprises both kappa and lambda types, although it has been shown that these do not occur within the same plant but in individual plants, haploid and diploid respectively, which grow together (McCandless *et al.*, 1973). *Gigartina* species are large plants up to 5 m in length which are collected from the coastal waters off Chile and Peru to give kappa and lambda carrageenans, again from individual plants. *Furcellaria* species are found in the cold waters around Northern Europe and Asia and yield extracts, furcellarans, similar to kappa and lambda carrageenans.

As with other natural agricultural products, seaweeds need careful post-harvest treatment, specifically to be dried as rapidly as possible to target moisture contents, before removal of impurities such as sand and stones, baling and shipping to the manufacturing location. The key step in raw material handling is rapid drying to a safe moisture content and then maintaining this moisture level until processing. Target moistures are in the range 18–35%, depending on seaweed type and natural salt content. Manufacturing plants located near the harvesting site may utilise fresh wet seaweed to avoid the costly drying and subsequent rehydration stages. Working with wet seaweed requires additional care as natural microbial and enzymatic depolymerisation of the carrageenan can occur and destroy the valued functionalities.

5.3 MANUFACTURING

Different manufacturing processes are used to make carrageenan extract and semi-refined carrageenan, also known as processed *Eucheuma* seaweed (PES). The primary process difference between 'extract' and 'PES' relates to whether the carrageenan is solubilised or not. The extract process solubilises the carrageenan and removes the solids, whereas the PES process leaves the carrageenan within the seaweed cellulosic structural matrix. The two processes are shown in Fig. 5.1.

At the manufacturing site, the baled seaweeds are tested for quality and yield, and lots are selected in line with the desired extract end functionalities. Proper selection of the raw materials and an understanding of the influence of process parameters on the properties of the final carrageenan are vital to the production of high-quality and consistent end products.

After the seaweeds are identified and chosen to make a particular extract, they are washed to remove sand and stones before applying appropriate amounts of various alkalis to swell the seaweed and extract the carrageenan. The specific alkali, for example sodium, potassium or calcium hydroxide, is selected depending on the carrageenan salt to be produced. As

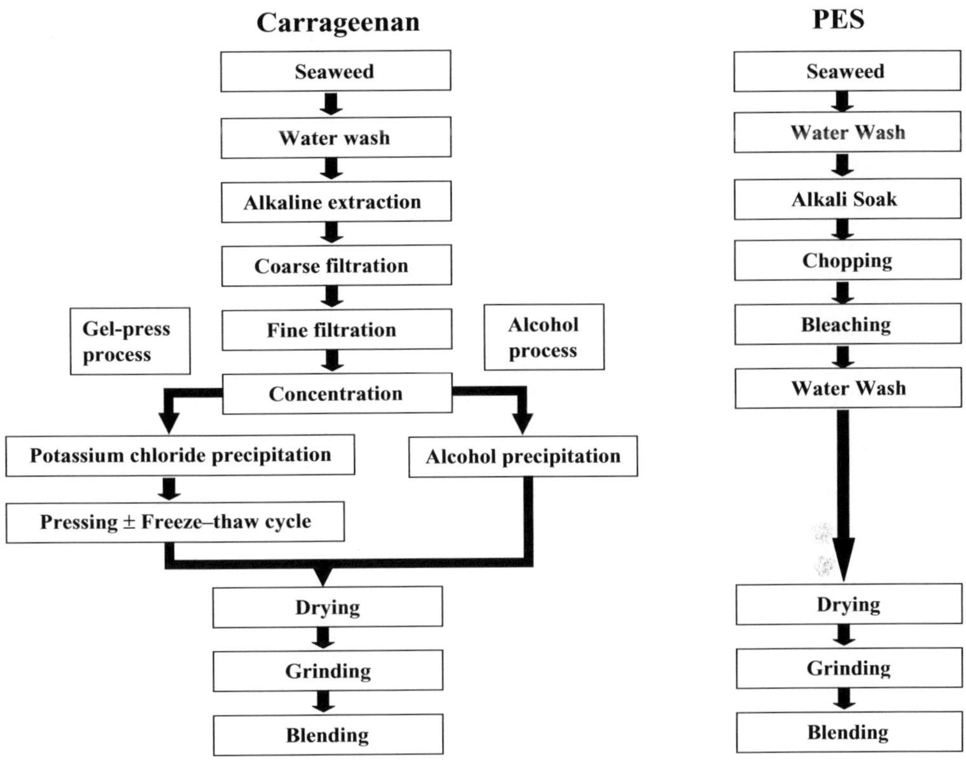

Fig. 5.1 Manufacturing processes for carrageenan and processed *Eucheuma* seaweed (PES). (Reproduced with kind permission from FMC Corporation.)

discussed later, this has important consequences for the properties of the resultant extract, including dispersion, hydration, thickening and gel formation. Prolonged treatment with alkali promotes an internal molecular rearrangement that modifies the polysaccharide backbone. The anhydride bridge that is formed by this reaction effectively reduces the number of bends or 'kinks' in the molecular chain so that inter-chain associations are optimised and strengthened. As a consequence, kappa carrageenan gels have higher rupture strength, deform less before breaking, and are more brittle.

After extraction and structural modification, the dilute carrageenan extracts are filtered and clarified by high-speed centrifugation and concentrated by a range of methods. The concentrated solutions are then precipitated with isopropyl alcohol to give a fibrous mass that is squeezed to remove the alcoholic liquor and dried.

An alternative recovery process utilises the specific selectivity of kappa carrageenan for potassium salts to form a gel. As kappa carrageenan solution is extruded into a concentrated solution of potassium chloride, a fibrous gel mass is formed. The precipitated gel mass exhibits syneresis, the exudation of free water as molecular structures aggregate and tighten. The mass is further dewatered under pressure to make 'gel press' carrageenan. The precipitated carrageenan may be frozen and thawed to assist this dewatering step as this action increases the tightening of the molecular structure and increases the resulting syneresis. The pressed fibres are then dried and ground to the appropriate particle size.

Each manufacturer carefully controls the raw materials and process parameters to produce a variety of extracts with well-defined properties. Individual extracts are characterised by their thickening and gelling properties. Finished products are made by blending one or more extracts, with or without standardising agents, in order to maintain consistent quality from lot to lot, to provide the specific properties needed to meet the agreed customer requirements and to function in target applications.

PES, also known as Philippines natural grade (PNG), semi-refined carrageenan (SRC), alternatively refined carrageenan (ARC) and alkali-modified flour (AMF), is processed by direct alkaline modification of the carrageenan structure within the structural matrix of cellulose in the seaweed. This is a more economical process as it avoids extracting carrageenan into the dilute solution necessary for solids removal and the resulting expensive concentration steps for the effective recovery of the carrageenan extract.

The PES process is compared to the traditional extraction process in Fig. 5.1. Seaweed selection and washing is the same for both processes. At the modification step, the *Eucheuma* seaweed is soaked in potassium hydroxide solution in situ before chopping and bleaching to enhance the colour of the finished powder. The key to this type of processing is to have excess potassium cations to prevent carrageenan solubility, this being achieved by potassium hydroxide or in some cases by sodium hydroxide/potassium chloride combinations. This process step is adapted from the historical alkali treatment of *Gracilaria* and it is common to reuse or recycle the alkali. After washing, the drying, grinding and blending steps are the same as for the extract carrageenan process. The microbiological specifications for PES used in foods are identical to those of carrageenan. The requirements for AMF are less strict as this product is targeted for petfood and non-food applications.

5.4 REGULATION

Carrageenan has a long history of use in food. Reviews of the extensive carrageenan literature by various regulatory agencies and scientific experts have resulted in carrageenan and PES being permitted globally for use in food, as shown in Fig. 5.2.

In the European Union, carrageenan and PES are approved food additives and assigned the numbers E407 and E407a, respectively, in the list of permitted emulsifiers, stabilisers, thickening and gelling agents for use *quantum satis*, the level required to achieve a given technological benefit. Previously, furcellaran was given a separate classification of E408 under EEC food legislation. However, a reassessment of carrageenan and furcellaran recognised the structural and functional similarity of the two materials and reclassified furcellaran as E407.

In the USA, the US Food and Drug Administration makes no distinction between carrageenan and PES and both are regulated as carrageenan.

'Carrageenan' must not be confused with 'poligeenan', also identified in literature as 'degraded carrageenan'. For use as permitted food additives, carrageenan and PES have strict limitations on viscosity measurement to clearly distinguish them from poligeenan (IARC, 1983; Cohen and Ito, 2002; JECFA, 2002; SCF, 2003). The major food compendial monographs include a viscosity specification of 'not less than 5 cps for 1.5% solution at 75°C'. This lower limit of 5 cps is approximately equivalent to an average molecular weight (Mw) of 100 000 daltons. Commercial carrageenan and PES normally have Mw in the range 200 000–800 000 daltons. Viscosity has proven to be a suitable tool for monitoring the molecular weight of carrageenan over the decades.

US Food and Drug Administration (FDA)

Chondrus extract – generally recognised as safe (GRAS) when used as a stabiliser – 21 CFR 182.7255
Carrageenan – food additive regulation – 21 CFR 172.620

Food Chemicals Codex

Specifications – 3rd edition National Academy Press, 1981 (and 2nd supplement to the 3rd edition, 1986)

European Community (EC)

Specifications – Directive 2004/45/EC of 16 April 2004
Food Additive Citations – Directives 95/2/EC (carrageenan) and 96/85/EC (PES)

Joint FAO/WHO Expert Committee on Food Additives (JECFA)*

Evaluation – WHO Food Additives Series 48 prepared by the 57th meeting of the Joint FAO/WHO Expert Committee on Food Additives (JECFA), published by WHO, 2002.
Specifications – FAO Food and Nutrition Paper 52, Addendum 9 (carrageenan and PES)
*Food and Agriculture Organisation/World Health Organisation

Fig. 5.2 Carrageenan/processed *Euchema* seaweed (PES) regulatory references. (Reproduced with kind permission from FMC Corporation.)

In contrast to carrageenan, poligeenan has a Mw of about 10 000–20 000 daltons. It is prepared by deliberate acid-hydrolysis of the starting material carrageenan at high temperatures for extended periods of time. Poligeenan is used in clinical treatments, for example in barium sulphate radiological contrast solutions and stomach ulcer therapy. Poligeenan is completely non-functional in food applications. Poligeenan at high dosage levels has shown some specific adverse colonic effects in specific animals, in particular guinea pigs, and when administered in drinking water rather than in food. Unfortunately, many researchers over the years have missed the clear distinction between 'carrageenan' and 'poligeenan/degraded carrageenan' and have incorrectly attributed the adverse toxicological effects of poligeenan to carrageenan, resulting in unfounded positions and false criticisms of carrageenan's record of safe use in food.

Recently, the possible degradation of carrageenan during food processing and digestion was researched and reviewed. A study on the fate of carrageenan subjected to a range of food-processing conditions revealed that normal food processes do not significantly increase the proportion of low-molecular-weight material (Marrs, 1998). The presence of associated cations prevents carrageenan hydrolysis during digestion.

As recently as 2003, the Scientific Committee for Food (SCF) concluded that 'there is no evidence of any adverse effects in humans from exposure to food grade carrageenan, or that exposure to degraded carrageenan from the use of food-grade carrageenan is occurring.' Nevertheless, the SCF proposed the establishment of a Mw limit of not more than 5% below 50 000 daltons, 'if feasible'. The intent was to ensure that the presence of degraded carrageenan was kept to a minimum. Subsequent to this review, another 90-day feeding study was undertaken (Weiner *et al.*, 2007). Rats were fed a specific type of kappa carrageenan containing an average 7% less than 50 000 daltons and a viscosity of 8 cps. Even with this amount of low-molecular-weight material, there was no evidence of erosions, ulcerations, inflammation, regeneration, hyperplasia or any other abnormalities of the gastrointestinal tract. Even though a validated analytical method for accurately measuring the lower Mw tail of carrageenan has yet to be developed, the European Commission adopted and published the SCF's proposed specification in Directive 2004/45/EC.

5.5 STRUCTURE

Carrageenan is a high-molecular-weight linear hydrophilic polysaccharide comprising repeating disaccharide units of galactose and 3,6-anhydrogalactose (3,6 AG), both sulphated and non-sulphated, joined by alternating α-(1,3) and β-(1,4) glycosidic links. An almost continuous spectrum of carrageenans exists, but the work of Rees and coworkers (Anderson *et al.*, 1965) was able to distinguish and attribute definite chemical structures to a small number of idealised polysaccharides, these disaccharide repeating structures defining the primary carrageenans as shown in Fig. 5.3. The main carrageenan types, kappa, iota, and lambda, can be prepared in pure form by selective extraction techniques from specific seaweeds and plants within those species. Mu and nu carrageenans are the precursor structures in carrageenan which are converted to kappa and iota carrageenans, respectively, by alkaline modification.

In the natural state in live seaweed, unmodified kappa and iota carrageenans contain about 30% mu and nu carrageenans, respectively, randomly distributed within the molecular repeating structures. Subsequent alkaline modification converts mu and nu to kappa and iota structures to the point where modified kappa and iota carrageenans contain less than 5% mu and nu carrageenans, respectively. Lambda carrageenan occurs naturally and the degree

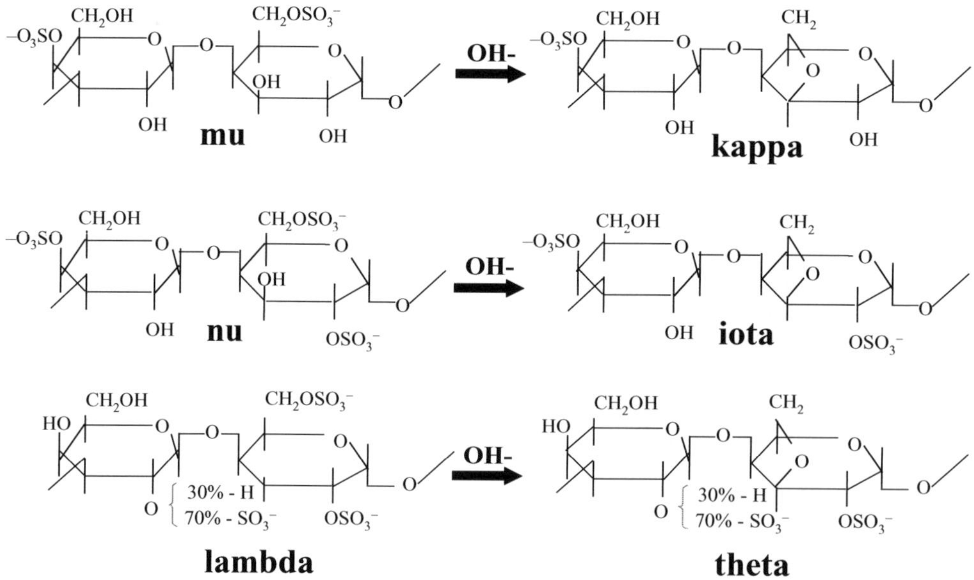

Fig. 5.3 Structures of primary carrageenans. (Reproduced with kind permission from FMC Corporation.)

of alkali modification to theta carrageenan varies depending on alkali concentration and extraction conditions.

The structures of carrageenan differ in 3,6-anhydrogalactose and ester sulphate content, and ester sulphate distribution, as shown in Fig. 5.3. Variations in these components in carrageenan influence the gel strength, texture, solubility, setting and melting temperatures, syneresis, synergies and interactions with other hydrocolloids and ingredients. These differences are controlled and created by seaweed selection and processing conditions and by blending different extracts.

The ester sulphate and 3,6-anhydrogalactose content of carrageenans are approximately 22% and 33%, respectively, for kappa carrageenan and 32% and 26%, respectively, for iota carrageenan. Lambda carrageenan contains approximately 37% ester sulphate with little or no 3,6-anhydrogalactose content. Furcellaran, which in the past has been rather misleadingly called Danish agar, contains 16–20% ester sulphate. These high ester sulphate levels contrast with the low ester sulphate of agar, typically from 1.5% to 2.5%. For food applications, carrageenan is best described as 'polygalactan extracts from *Rhodophyceae* with ester sulphate content of 18–40% and alternate α-(1,3) and β-(1,4) glycosidic linkages'.

The *Gigartina* family of seaweeds contains several commercial species that result in a continuum of molecular structures for kappa carrageenan, and more specifically to the amount of 2-sulphate associated with the 3,6-anhydrogalactose moiety of the idealised structure (Falshaw *et al.*, 2001). The 3,6-anhydrogalactose-2-sulphate content ranges from close to zero for *K. alvarezii* to about 5–10% for *C. crispus*, 10–15% for *Gigartina stellata*, 20–25% for *Gigartina chamisoi*, 30–40% for *Gigartina radula* and *Mastocarpus crispata* and 45–55% for *Gigartina skottsbergii*. This increase in ester sulphate results in a significant enhancement of protein reactivity and reduction of water gel strength. Kappa carrageenans with 3,6-anhydrogalactose-2-sulphate contents between about 30% and about 55% are known as 'kappa-2' carrageenans (Bixler *et al.*, 2001).

PES differs from traditional clarified carrageenan extracts in that it contains 8–15% acid insoluble matter (AIM) compared to 2% maximum for an extract. The AIM mainly comprises the structural network of plant cellulosic and proteinaceous materials, which maintain their integrity during the PES process. This means that in applications using PES, enough energy must be applied in the process to break down the AIM structure and release the carrageenan. Hence, the hydration and solubility profiles are different for carrageenan and PES. Only carrageenan extract can be used for applications requiring clear solutions and gels.

Carrageenan and PES are high-molecular-weight and polydisperse polysaccharides. Commercial carrageenans and PES materials normally have Mw between 200 000 and 800 000 daltons, but it can be as high as 1 500 000 daltons. All carrageenans contain minor fractions below 100 000 daltons, these lower Mw components being inherent to the native algal seaweeds. As pointed out earlier, these lower Mw fractions should not be confused with poligeenan made by deliberate acid hydrolysis.

5.6 FUNCTIONAL PROPERTIES

5.6.1 General properties

The combination of complex chemistry, interactions with other ingredients and natural variability may make carrageenan appear difficult to work with in food systems. However, carrageenan manufacturers reduce product variability and standardise interactions and

Table 5.1 Summary of carrageenan properties.

Solubility	**Lambda**	**Iota**	**Kappa**
Hot (80°C) water	Soluble	Soluble	Soluble
Cold (20°C) water	All salts soluble	Na^+ salt soluble Ca^{++} salt gives thixotrophic swollen particles	Na^+ salt soluble Limited swelling of K^+, Ca^{++} salts
Hot (80°C) milk	Soluble	Soluble	Soluble
Cold (20°C) milk	Thickens	Insoluble	Insoluble
Cold milk (TSPP added)	Increased thickening or gelling	Thickens or gels	Thickens or gels
50% sugar solutions	Soluble	Insoluble	Soluble hot
10% salt solutions	Soluble hot	Soluble hot	Insoluble
Gelation			
Effect of cations	Non-gelling	Strongest gels with Ca^{++}	Strongest gels with K^+
Gel texture	–	Elastic	Brittle
Syneresis	–	No	Yes
Hysteresis	–	5–10°C	10–20°
Freeze–thaw stable	Yes	Yes	No
Synergy with locust bean gum	No	No	Yes
Synergy with konjac flour	No	No	Yes
Synergy with starch	No	Yes	No
Shear-reversibile	Yes	Yes	No
Stability in acid	Hydrolysis	Hydrolysis, accelerated by heat, low pH and time	
		Gels are stable	
Protein reactivity	Strong protein interaction in acid		Specific reaction with kappa casein

properties to allow for easy application by users. The basic information in this chapter can be easily applied by the food formulator for effective utilisation of carrageenan and PES in food products and systems. Carrageenan is a proven and successful additive for stabilising, thickening and gelling food systems.

The primary physical properties of the three major carrageenan types, including solubility and gelation characteristics, are summarised in Table 5.1.

The thickening and gelling properties of the various types of carrageenan are quite different. For example, kappa carrageenan forms a firm gel with potassium ions while iota carrageenan interacts with calcium ions to give soft elastic gels. Cations have no effect on the non-gelling properties of lambda carrageenan, but lambda carrageenan will gel at very high salt concentrations. Application of these combinations requires experience and understanding of carrageenans, but this expertise is readily available from major suppliers.

Lambda carrageenan is the least utilised of the carrageenans, in most cases simply providing secondary functionalities to kappa carrageenan. Separation of diploid plants for the production of pure lambda carrageenan is expensive and not common practice. Diploid plants are normally co-processed with the haploid kappa plants. These normally comprise less than 20% of the mix and, hence, provide only secondary attributes, for example giving a kappa carrageenan milk gel a more creamy texture. Most commercial products labelled 'lambda carrageenan' are in fact 'non-gelling carrageenans' and comprise a co-processed mixture of unmodified kappa and lambda carrageenans.

5.6.2 Solubility

All carrageenans are soluble in hot water but, with the exception of lambda, only the sodium salts of kappa and iota are soluble in cold water. All carrageenans are soluble in hot milk, but in cold milk only lambda carrageenan has solubility, producing a thickening effect via protein interactions, this being enhanced by the presence of phosphates. Carrageenan solutions are viscous and show pseudoplasticity or shear thinning when pumped or stirred.

The influence of temperature is an important factor in deciding which carrageenan should be used in a food system. All carrageenans hydrate at high temperatures and kappa and iota carrageenans in particular exhibit a low fluid viscosity in both water and milk. On cooling, these carrageenans set to form a range of gel textures depending on the cations and other ingredients present.

5.6.3 Hydration

The presence of salts and sugars has a dramatic effect on the hydration temperature of the carrageenan and on its subsequent setting and remelting temperatures. For example, iota carrageenan will hydrate at ambient temperature in water but the addition of salt raises the gel point so that the solution is converted into a gel with a distinct yield point which is applied in cold-prepared salad dressings. Sodium salts of kappa carrageenan will hydrate at 40°C but the same carrageenan in a meat brine will only show full hydration at 55°C or above.

As a carrageenan dispersion is heated, there is no significant particle swelling or hydration until the temperature exceeds about 40–60°C. As the particles hydrate, the viscosity rises as the swollen particles offer more resistance to flow. Further heating to 75–80°C produces a drop in viscosity. On cooling, the solution shows a marked increase in viscosity followed by gelation below temperatures of 40–50°C. The hydration and gelation temperatures are strongly dependent on the salts associated with the carrageenan or added separately to the solution. For example, above about 4% sodium chloride can prevent full hydration of carrageenan in meat brines. This hydration profile is shown for kappa carrageenan in Fig. 5.4.

In contrast, very dilute levels of around 200 ppm of carrageenan, used to stabilise chocolate milks and other dairy beverages, may not form a stabilising gel network until the temperature drops below 20°C. The presence of high solids, in confectionery for example, effectively concentrates the carrageenan and cations on the aqueous phase so that gelation may occur at 80–85°C or higher, placing limitations on the levels and types of carrageenan suitable for such food applications.

Various methods may be used to ensure that the carrageenan particles are fully dispersed before the onset of hydration. These include mixing the carrageenan with 5–10 times its weight of inert filler, such as sugar or dextrose, slurrying the carrageenan in oil to provide a hydrophobic barrier around each particle or dispersing into a salt solution, sugar syrup or alcohol. Some carrageenans have been processed to include potassium salts for gelation, but

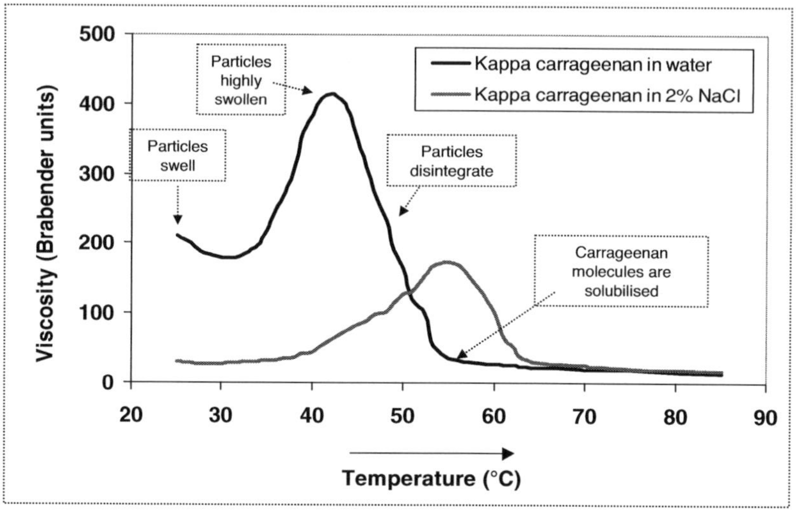

Fig. 5.4 Hydration profile of kappa carrageenan in water and in salt solution. (Reproduced with kind permission from FMC Corporation.)

these salts may assist with dispersion by avoiding rapid hydration. The use of high shear mixing can also assist by breaking up any lumps.

5.6.4 Water binding

Carrageenan particles not only have a high affinity for water, but also have structural 'memory'. This means that when water is added to a carrageenan particle, it is absorbed into the structure which swells back to the shape and dimensions of the pre-dried particle, not completely, but to a significant degree. This phenomenon results in strong water-binding properties being associated with these swollen carrageenan particles and, as hinted above, specific salts in this absorbed water can influence properties and applications.

A specific application of this water-binding property is the use of carrageenan in delicatessen meats, such as turkey breast and ham. The carrageenan is dispersed in brine before pumping into or tumbling with meat. The brine solubilises protein from the meat but the carrageenan only hydrates. When the meat is cooked, the carrageenan remains hydrated and continues to bind water, but the protein forms a gel, trapping the carrageenan particles in the gel matrix. Purge losses are minimised for improved cooked yield and moisture is retained for improved eating qualities.

5.6.5 Water gels

Hot solutions of kappa and iota carrageenans set when cooled below the gel point, which is between 30°C and 70°C, depending on the cations and other ingredients present, to form a range of gel textures. The two-step gel mechanism is shown in Figs 5.5a–5.5c, with Gel-1 phase being elastic (iota) and Gel-II phase being brittle (kappa).

The ionic composition of a food system is important for effective utilisation of the carrageenan. For example, kappa carrageenan selects potassium ions to stabilise the junction

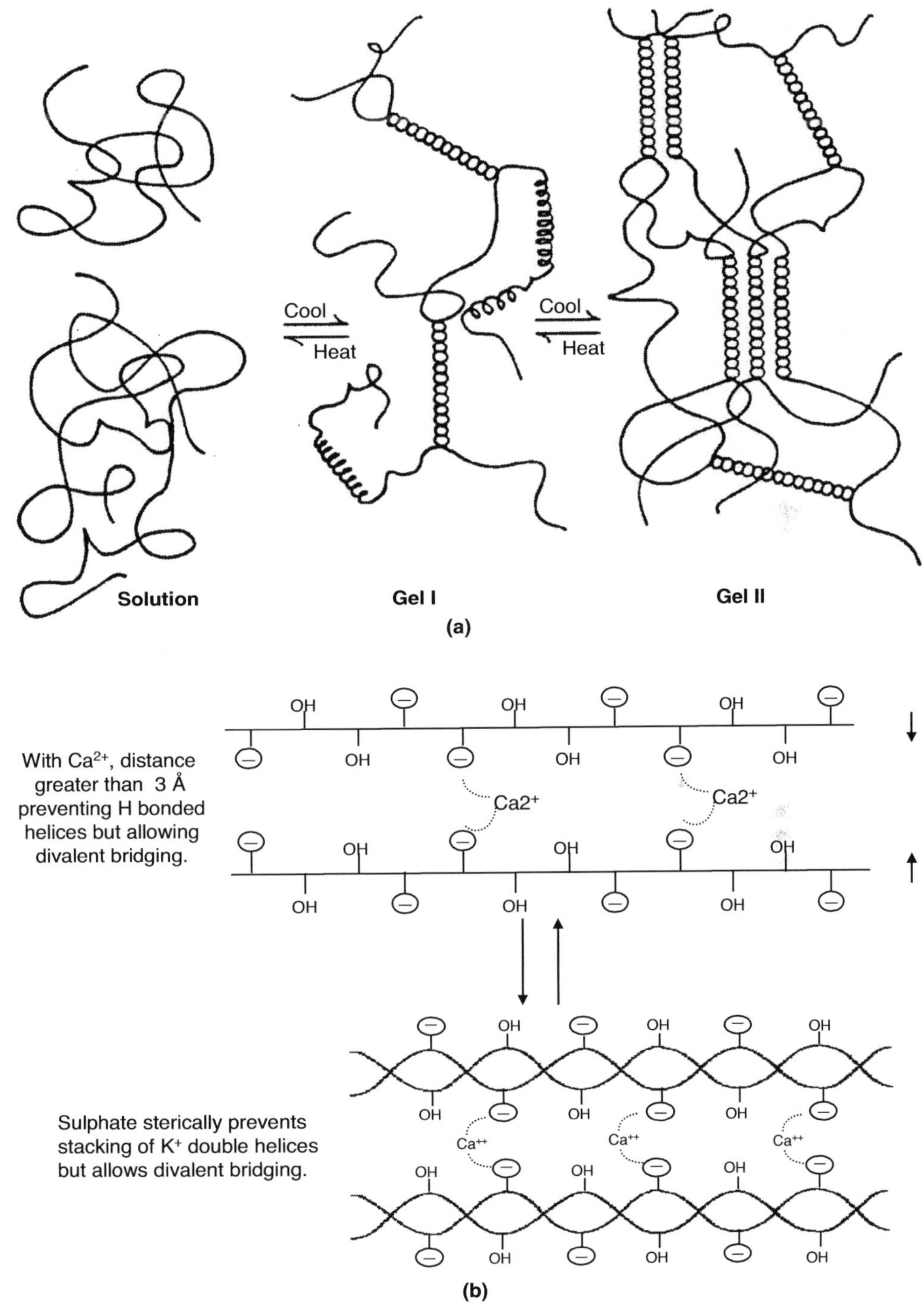

Fig. 5.5 (a) Carrageenan gelation mechanism. (b) Gel-I mechanism (iota carrageenan). (*Continued*)

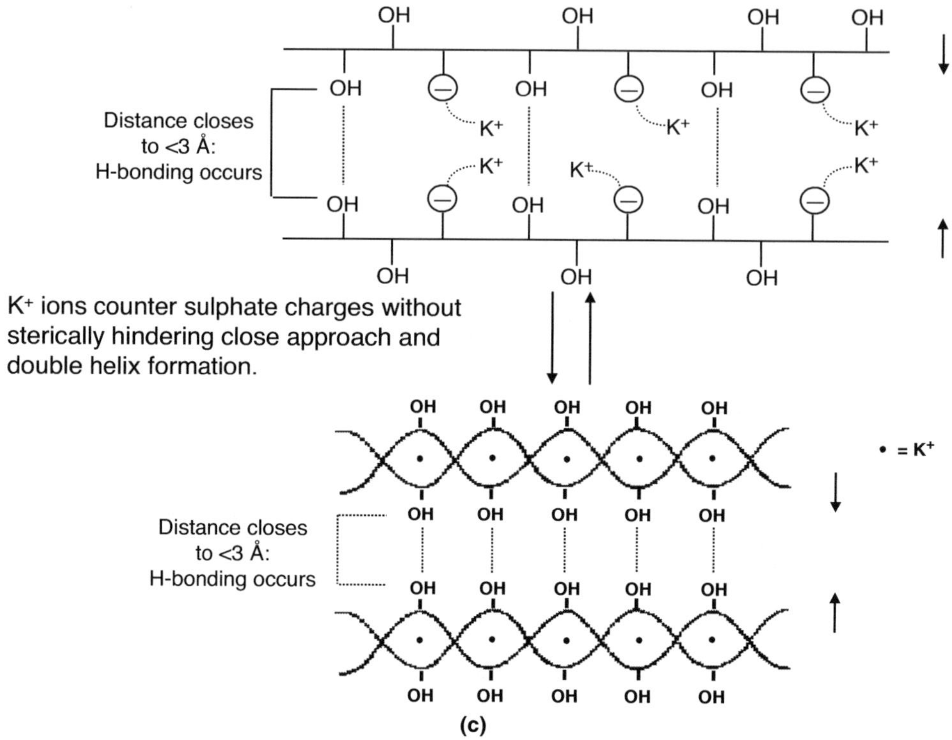

Fig. 5.5 (*Continued*) (c) Gel-II mechanism (kappa carrageenan). (Reproduced with kind permission from FMC Corporation.)

zones within the characteristically firm, brittle gel. Iota carrageenan selects calcium ions to bridge between adjacent chains to give typically soft, elastic gels.

Combinations of kappa and iota carrageenans give gel strengths and textures intermediate to the two extremes and in line with the ratio used, as indicated in Fig. 5.6.

Carrageenan gels exhibit hysteresis, which is the incremental energy between setting and melting temperatures. Gels are stable at room temperature but can be remelted by heating to 5–30°C above the gelling temperature. On cooling, the system will re-gel without loss of gel strength or change of texture in neutral conditions.

Iota gels break when subjected to shear but recover or re-gel after shear is removed, indicating thixotropic behaviour, but with longer time to recover fully, compared to xanthan gum for example. Kappa carrageenan gels break when sheared and the effect is permanent.

Syneresis is the elimination of water from a gel as the gel structure tightens and contracts. Kappa gels have high syneresis levels, iota gels show no syneresis. This syneresis property is directly linked to freeze–thaw stability, where freezing further irreversibly tightens the kappa gel structure, but has no influence on the iota gel, which fully recovers when thawed. Control of syneresis is essential to some applications.

Synergy occurs when two components combine to give advantages compared to either individual component. Kappa carrageenan forms highly synergistic gels with galactomannans, such as locust bean gum (LBG), and the glucomannan, konjac gum. In addition to increasing

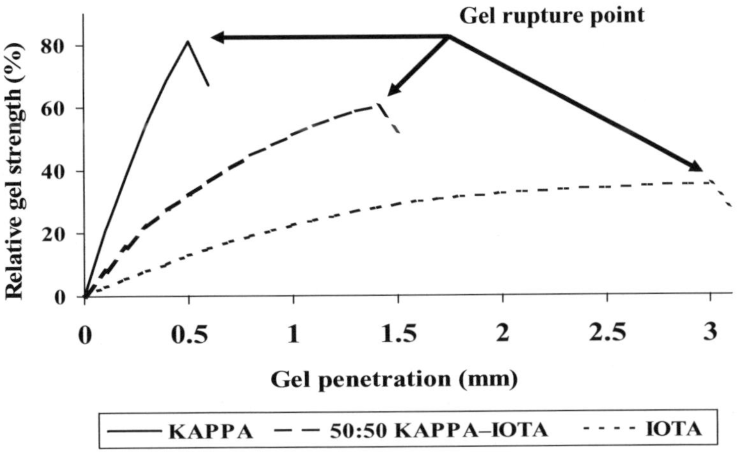

Fig. 5.6 Gel properties of pure and blended kappa and iota carrageenans. (Reproduced with kind permission from FMC Corporation.)

the gel strength, these polymannan gums also make the gel texture more elastic with reduced syneresis.

LBG is a galactomannan where the distribution of the galactose side chains on the mannan backbone is non-random and produces unsubstituted regions. These align with the length of the kappa carrageenan helical aggregates, the resulting strong hydrogen bonding giving the synergy. A similar mechanism with even stronger synergy occurs between kappa carrageenan and konjac glucomannan. In this case, the distance between ester groups along the konjac mannan also fits well with the length of the kappa carrageenan helical aggregates. The optimum ratio of kappa carrageenan to mannan gum is about 60:40 or 70:30, as shown in Fig. 5.7. These polymer combinations are used extensively in cooked meats and gelled petfoods.

The solution, hydration and gel properties of PES and kappa carrageenan obtained by traditional extraction processes are similar but there are some specific differences. The

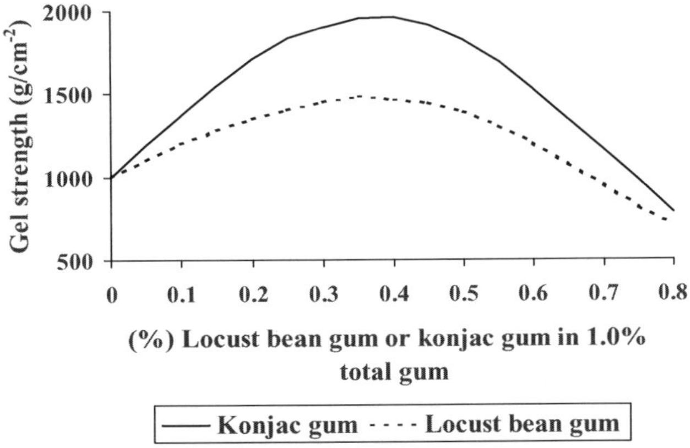

Fig. 5.7 Kappa carrageenan synergy with locust bean gum and konjac gum. (Reproduced with kind permission from FMC Corporation.)

cellulose network in PES reduces the rate of hydration so, in general, solutions develop viscosity after longer heating periods or after heating to higher temperatures. The presence of cellulose in the finished gel gives a lower rupture strength with a more brittle and fragile gel. The cellulose particles make the product cloudy and therefore unsuited to clear gel applications such as water dessert gels and cake glazes. The chemistry of the carrageenan in PES has not changed, but the presence of AIM can influence functionality or interfere with performance and aesthetics.

5.6.6 Acid stability

Carrageenan solutions lose viscosity and gel strength capability when subjected to pH values below about 5.5, but this effect is not significant until about pH 4.5, yet still manageable down to pH 3.5. This autohydrolysis occurs at low pH values as carrageenan in the acid form cleaves the molecule at the 3,6-anhydrogalactose linkage (Hoffman *et al.*, 1996). The rate of autohydrolysis increases significantly at elevated temperatures. However, once below the gelling temperature, carrageenan retains the sulphate-bound potassium ions and this prevents further autohydrolysis. Consequently, in acidic products, the carrageenan should be added at the last moment to avoid excessive acid degradation, and if appropriate, acid should be added to the food immediately before depositing and filling to minimise polymer breakdown.

Table 5.2 shows the approximate processing times at various pH values and temperatures, for a gel produced with 0.5% kappa carrageenan and 0.2% potassium chloride, such that no more than 20–25% of the original gel strength is lost when the solution is cooled. In general, each 0.5 pH unit reduction will decrease the available processing time by a factor of 3. Times will vary somewhat depending on carrageenan concentrations or other ingredients in the system, such as salts and sugars. In a continuous process, the processing time should be kept to a minimum. In systems above about pH 4.5, the process conditions become irrelevant as the carrageenan solution is stable to most food-processing times.

Carrageenan can be used effectively for low pH food applications when using the above guidelines to prevent excessive autohydrolysis during processing. Once the carrageenan has been gelled and is in 'gel-mode conformation' with a stable helical structure, the system is extremely robust. For example, in ready-to-eat water dessert gels, the finished product shelf life is well in excess of limits applied for other reasons, such as flavour or microbiology.

Table 5.2 Gel processing times.

	Final pH						
Temperature (°C)	**3.0**	**3.5**	**4.0**	**4.5**	**5.0**	**5.5**	**6.0**
120	2 s	6 s	20 s	1 min	3 min	10 min	30 min
110	6 s	20 s	1 min	3 min	10 min	30 min	1.5 h
100	20 s	1 min	3 min	10 min	30 min	1.5 h	5.0 h
90	1 min	3 min	10 min	30 min	1.5 h	5.0 h	15.0 h
80	3 min	10 min	30 min	1.5 h	5.0 h	15.0 h	2.0 days
70	10 min	30 min	1.5 h	5.0 h	15.0 h	2.0 days	6.0 days
60	30 min	1.5 h	5.0 h	15.0 h	2.0 days	6.0 days	20.0 days
50	1.5 h	5.0 h	15.0 h	2.0 days	6.0 days	20.0 days	60.0 days
40	5.0 h	15.0 h	2.0 days	6.0 days	20.0 days	60.0 days	200.0 days

Note: Gel process times are the approximate times at the various pH and temperatures to reduce the gel strength by 25%.

5.6.7 Protein interactions

Probably, the best-known carrageenan interaction is that involving milk proteins. Some of the first uses of carrageenan were in milk gels and flans, evaporated milk and ice cream mixes, where kappa-carrageenan–kappa-casein synergy allows for use levels as low as 0.01%. In these applications, the kappa carrageenan not only forms a weak gel in the aqueous phase, but also builds additional structure by interacting directly with positively charged amino acids and indirectly, via divalent cations, with negatively charged amino acids in the proteins at the surface of the casein micelles.

The most widespread use of carrageenan is in stabilising dairy products through the specific kappa-carrageenan–kappa-casein interaction as illustrated in Fig. 5.8, which shows

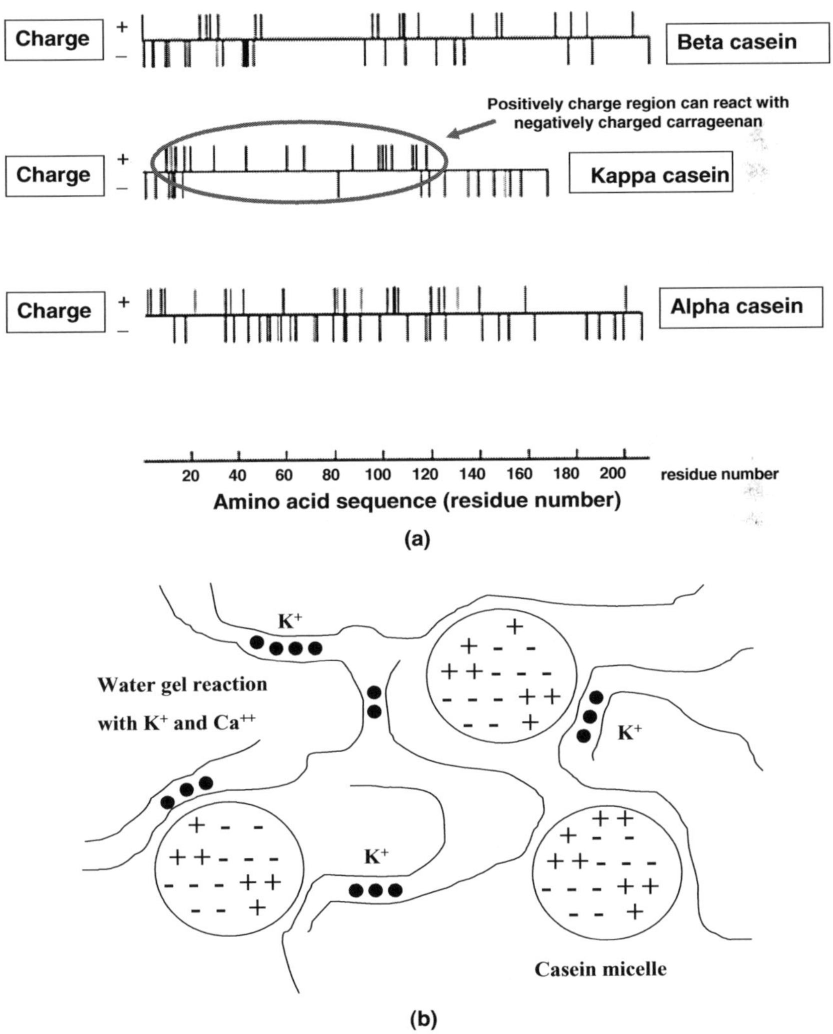

Fig. 5.8 (a) Charge differences between casein fractions. (b) Kappa-carrageenan–kappa-casein milk protein reaction. (Reproduced with kind permission from FMC Corporation.)

the differences between casein fractions, and Fig. 5.8, which illustrates the interaction mechanism.

Very low levels of 150–250 ppm of carrageenan are sufficient to prevent whey separation from a range of dairy products during manufacture and storage. These include ice cream and milk shake mixes, cream cheese and dairy desserts. In chocolate milks, this low level of carrageenan is able to prevent separation and also generate a stabilising network that maintains the cocoa particles in suspension.

5.7 FOOD APPLICATIONS

Carrageenan plays an important and valued role in modern-day formulations providing texture, structure and physical stability in food products. It is also used for cost reduction and added value. In meat products, carrageenan enhances the quality and/or increases the cooked yield of poultry, ham and sausage products. Water gels and cake glazes have used fast-gelling carrageenan for many years. Sauces, salad dressings and dips utilise carrageenan to impart body, provide thickness and stabilise emulsions. The utilisation of carrageenan has been established in fluid dairy and dairy dessert products as well. The stabilization of cocoa, as well as additional mouthfeel, can be attained using very small amounts of carrageenan due to protein reactivity, as discussed previously. Whipped creams and toppings retain their stable form due to carrageenan. Carrageenan can assist with the stability of frozen dairy products by preventing whey separation and ice crystal formation in ice cream and it is also used in puddings and pie fillings to create stable gels.

5.7.1 Water gelling applications

Water dessert gels and cake glazes are some of the more traditional uses of carrageenan. These products are based on the firm, brittle gel properties of kappa carrageenan with the texture modified as necessary for elasticity, cohesiveness and syneresis control using iota carrageenan or other gums, such as locust bean gum or konjac.

Recent improvements in the combinations used for these applications have resulted in vegetarian products which have a similar appearance and texture to traditional gelatin products while giving additional benefits of fast setting and stability at ambient temperatures. An example recipe is shown in Formulation 5.1.

Formulation 5.1 Fruit-flavoured water dessert jelly

Ingredients	Quantity (%)
Sugar	15.00–20.00
Carrageenan (kappa–iota blend)	0.60–0.90
Potassium citrate	0.20–0.35
Citric acid	0.30–0.45
Colour	as required
Flavour	as required
Water	to 100.00
Total	100.00

These gels can be used for aspics and gels in canned meats and petfoods and in cooked, sliced meats. In these latter products, the carrageenan is incorporated to improve moisture retention, cooking yields, slicing properties, mouthfeel and succulence. A typical formulation for a 30% added-water sandwich ham is given in Formulation 5.2.

Formulation 5.2 Cooked ham with 30% added brine

Ingredients	Quantity (%)
Meat, lean ham muscles	62.50
Carrageenan (firm gelling kappa)	0.60
Sodium tripolyphosphate	0.50
Nitrate salt[a]	1.67
Sodium chloride	0.53
Dextrose	1.20
Water	32.95
Total	100.00

[a]Sodium chloride containing 0.6% sodium nitrite, giving a 100 ppm sodium nitrite in the finished product. The total brine concentration of sodium chloride is 2.2%.

In this product, the carrageenan disperses readily in the meat brine when it is added after phosphate and salt. The brine has a very low viscosity and is easily injected and distributed throughout the meat tissue. During cooking, the carrageenan hydrates above 50–55°C to bind moisture as the ham is cooked to 72–74°C. When the product is cooled, a cohesive gel forms that maintains product integrity during high-speed slicing operations and binds moisture in the product throughout shelf life. Other water gel applications for carrageenan include gelled petfoods and meat and fish aspics.

PES has advantages in cooked, sliced meats, being very cost effective and dispersing readily without lumping in meat brines. During processing, some differences are apparent between this and traditional carrageenan extracts. The small PES particles do not swell in the brine and may cause less damage when injected into the meat (Philp *et al.*, 1998). The cellulose network in PES reduces the rate of hydration during heating so that solutions develop viscosity after longer heating periods or after heating to higher temperatures. The presence of cellulose in the finished gel gives a lower gel strength with a more brittle and fragile gel. The dispersed cellulose particles make the product cloudy and any gel spots in the injected meat will be masked against the background of the meat. As a consequence, meat brines may include both carrageenan and PES to optimise aesthetics, functional properties and ingredient costs.

The synergy of kappa carrageenan and locust bean gum is also used in water dessert gels and glazes, cooked, sliced hams and poultry products, canned meats and petfoods and similar firmly gelled products. The gel produced by this gum combination exhibits benefits of high gel strength, a cohesive, elastic texture, excellent syneresis control and cost-effectiveness. The gums also have an increased hot viscosity to retain meat juices better in cooked meats and to reduce emulsion separation and splashing during filling. Petfood is the largest single use for AMF in combination with locust bean gum for gelled products or with guar gum for gravies.

The strong interaction and synergy between kappa carrageenan and konjac gum is used to a lesser extent at present, mainly due to the relatively short time that purified konjac has been available. However, carrageenan/konjac combinations are already available for petfoods, surimi and water dessert gels, and available for product development.

The reversible-gel properties of dilute iota carrageenan can be used to stabilise suspended herbs and vegetables in vinaigrette-style salad dressings. The gel is formed by first dispersing carrageenan into water at ambient temperature and shearing to give a viscous solution, followed by the addition of salt (sodium chloride) which raises the setting point of the carrageenan. The solution is converted into a reversible gel with a distinct yield point which is very effective for suspending particulates over the long shelf life of such products. A recipe is given in Formulation 5.3.

Formulation 5.3 Vinaigrette-style salad dressing

Ingredients	Quantity (%)
7% spirit vinegar	12.50
Sugar	9.50
Salt	3.20
Carrageenan (iota)	0.30
Xanthan gum	0.15
Chopped spice pieces	1.00
Colour and preservative	as required
Water	to 100.00
Total	100.00

Other applications that utilise the stabilising properties of this reversible gel network include soy milks and sterilised milk drinks. Higher concentrations of this carrageenan give soft, elastic gels suited to gravies for canned meats and petfoods and for various toothpastes.

Beer and wine fining are applications which rely on the protein reactivity of carrageenan. Coarse particles of kappa carrageenan or PES are used to interact with proteinaceous materials and small protein fragments produced during pasteurisation to form aggregates which can be readily filtered to clarify the beer or wine and reduce chill haze. Applications for carrageenan in water-based products are shown in Table 5.3.

5.7.2 Dairy and protein applications

Milk-based puddings were one of the original uses for carrageenan from *C. crispus* harvested in Ireland. The seaweed was boiled in milk which formed a gel on cooling. These properties are now used worldwide for a large number of dry-mix and ready-to-eat flan, crème dessert and mousse applications. These products utilise the complete range of carrageenan types for thickening and gelling products. Textures may range from firm gels in crème caramel to soft gels in ready-to eat spoonable desserts and to thickened custards, vla and cream desserts.

Extremely low levels of carrageenan, around 100–200 ppm, are used to stabilise and prevent whey separation in a number of dairy products, including milk shake and ice cream mixes, chocolate milks and pasteurised and sterilised creams. In these applications, the

Table 5.3 Water-based applications.

Application	Use level (%)	
Hot-processed gelling applications		
Cake glaze	0.60–0.70	
Cheese, imitation block	2.20–2.70	
Cheese, imitation spread	0.30–0.35	
Desserts, water gels (dry mix)	0.50–0.80	
Desserts, water gels (RTE)	0.60–1.00	
Desserts, water gels (sugar free)	0.60–0.80	
Entrapment or encapsulation	1.00–2.00	
Fabricated or formed foods	2.00–3.00	
Fish gels	0.50–1.00	
Frozen dough	0.10–0.25	
Fruit-in-gel	0.80–1.20	
Ham, further processed	0.30–0.50	
Jelly, imitation (low sugar) dry mix	1.50–2.00	
Jelly, imitation (low sugar) RTE	1.00–1.50	
Mayonnaise, imitation	0.50–1.00	
Pasta	0.10–0.50	
Petfood	0.20–1.00	+ LBG
Petfood, gravy	0.10–0.20	+ Guar
Pie filling	0.30–0.50	
Poultry nuggets	0.40–0.70	
Poultry, further processed	0.25–0.50	
Red meats, further processed	0.25–0.50	
Seafood, further processed	1.25–1.75	+ starch
Sorbet	0.15–0.30	+ pectin
Sour cream	0.10–0.20	+ LBG
Surimi or kamaboko	0.20–0.30	+ starch
Tomato sauces	0.10–0.20	+ starch
Whipped cream	0.01–0.05	
Hot-processed thickening applications		
Batter mixes	0.10–0.30	
Coffee creamer	0.10–0.20	
Desserts, creamy whipped	0.15–0.30	
Fruit topping	0.30–0.50	
Mayonnaise, imitation	0.40–0.60	+ starch
Moisture barriers or meat glaze	0.80–1.20	
Salad dressing, hot process	0.20–0.50	
Syrups	0.10–0.30	
Variegates	0.30–0.80	+ starch
Cold-processed thickening applications		
Cheesecake (no bake)	0.60–1.00	
Fruit beverages	0.10–0.20	
Mayonnaise, imitation cold process	0.40–0.70	+ xanthan
Salad dressing (dry mix)	0.60–1.00	
Salad dressing (cold process)	0.20–0.50	

carrageenan interacts with the dairy proteins to form a network that is able to suspend particulates, such as cocoa in chocolate milk or insoluble calcium salts in calcium-fortified beverages. The network prevents protein–protein interaction and aggregation during storage. This avoids whey separation in fluid products and reduces shrinkage in ice cream. Formulation 5.4 shows a typical ice cream recipe.

Formulation 5.4 Typical ice cream mix

Ingredients	Quantity (%)
Butterfat	8.00–10.00
Milk solids non-fat	11.10–10.80
Sugar	10.00
Corn syrup solids	3.50
Carrageenan	0.015–0.025
Other hydrocolloids (guar gum, LBG, CMC, MCC, xanthan gum, sodium alginate)	0.10–0.20
Emulsifier (glyceryl monostearate)	0.20–0.50
Vanilla flavour	as required
Water	to 100.00
Total	100.00
(Total solids)	33.3–35.0

Processed cheese is another application where the protein reactivity and gelling properties of carrageenan are used. Processed cheese is made by incorporating 'melting' or 'emulsifying' salts into the cheese mix to control the melting temperature while maintaining firmness and mouthfeel, ensuring slice integrity and producing a strong block suitable for grating. It is possible to reduce the cheese content in such products and substitute a gel containing 0.5–3% carrageenan to give a product with excellent mouthfeel and good melting, grating and slicing properties.

Acidic dairy products, such as soft cheese and yoghurt, are generally unsuitable for carrageenan to be an effective stabiliser. The low pH increases the electrostatic interactions between protein and carrageenan producing unstable aggregates that flocculate and separate. However, careful selection of appropriate carrageenan–galactomannan blends can be used to control this aggregation to give effective stabilisation and prevent moisture separation while providing a smooth, creamy mouthfeel to the finished product.

When used in soy beverage formulations, carrageenan will provide a stable, uniform suspension of insoluble solids to ensure a high quality finished product. The stability provided by carrageenan ensures a consistent mouthfeel and uniform body in both extended shelf life (ESL) refrigerated and aseptic ambient-stored products. Carrageenan can also provide improved emulsion stability to reduce creaming. The result is better overall stability and higher quality taste and mouthfeel in soymilk beverages. The type and concentration of carrageenan used depends on the quality of soy protein being used, for example fresh soy milk versus soy concentrate.

The many dairy products which utilise the properties of carrageenan are shown in Table 5.4.

5.7.3 Applications summary

Currently, about 70–80% of all carrageenan products are utilised by the food industry. Consumers demand high-quality food products that maintain stability, suspension and visual appearance throughout the shelf life of the product, whether refrigerated or at ambient storage temperature. Consumers view defects in the stability of food products as spoilage or

Table 5.4 Protein-based applications.

Application	Use Level (%)	
Hot-processed milk thickening applications		
Calcium fortified milk	0.02–0.04	
Chocolate drink	0.02–0.04	
Chocolate milk (HTST)	0.02–0.04	
Chocolate milk (UHT)	0.02–0.05	
Cottage cheese dressing	0.01–0.05	
Cream cheese	0.05–0.08	
Egg nog	0.05–0.12	
Evaporated milk	0.005–0.020	
Ice cream (hard pack)	0.010–0.015	+ guar/CMC/LBG
Ice cream (soft serve)	0.02–0.03	+ guar/CMC/LBG
Infant formula	0.02–0.03	
Shakes (RTE)	0.02–0.03	
Sterilised milk	0.01–0.03	
Soy beverages	0.02–0.05	
Hot-processed milk gelling applications		
Custards (dry mix)	0.20–0.30	
Dutch vla	0.015–0.045	+ starch
Flans (dry mix)	0.20–0.30	
Flans (RTE)	0.20–0.30	
Flans (soy)	0.20–0.30	
Puddings (cold fill)	0.20–0.60	
Pumpkin pie	0.45–0.55	
Cold-processed milk thickening applications		
Beverages, nutritional	0.10–0.15	
Breads	2.00–3.00	
Cheese spreads, sauces	0.50–1.00	
Chocolate beverages, dry mix	0.08–0.12	
Chocolate syrups	0.20–0.40	
Desserts, dry mix	0.15–0.20	
Desserts, aerated (mousse)	0.50–1.00	
Ice cream (dry mix)	0.50–0.80	
Meringue topping	0.15–0.25	

improper processing. Over many years, the unique functionalities of carrageenans have been successfully applied to both food applications development and product stability.

References

Anderson, N.S., Dolan, T.C.S. and Rees, D.A. (1965) Evidence for a common structural pattern in the polysaccharide sulphates of the Rhodophyceae. *Nature*, **205**, 1060–1062.

Bixler, H.J., Johndro, K. and Falshaw, R. (2001) Kappa-2 carrageenan: structure and performance of commercial extracts II. Performance in two simulated dairy applications. *Food Hydrocolloids*, **15**, 619–630.

Cohen, S. M. and Ito, N. (2002) A critical review of the toxicological effects of carrageenan and processed *Eucheuma* seaweed on the gastrointestinal tract. *Critical Review in Toxicology*, **32**(5), 413–444.

Falshaw, R., Bixler, H.J. and Johndro, K. (2001) Structure and performance of commercial kappa-2 carrageenan extracts I. Structure analysis. *Food Hydrocolloids*, **15**, 441–452.

Hoffman, R.A., Russell, A.R. and Gidley, M.J. (1996) Molecular weight distribution of carrageenans. In: G.O. Phillips, P.J. Williams and D.J. Wedlock (eds), *Gums & Stabilisers for the Food Industry 8*. IRL Press at the Oxford University Press, Oxford, pp. 137–148.

IARC (International Agency for Research on Cancer) (1983) Monographs on the evaluation of the carcinogenic risk of chemicals to humans. *World Health Organization*, **31**, 79–94.

JECFA (2002) Safety evaluation of certain food additives and contaminants, prepared by the Fifty-seventh meeting of the Joint FAO/WHO Expert Committee on Food Additives (JECFA). Carrageenan and Processed *Eucheuma* Seaweed (addendum). *WHO Food Additives Series*, **48**, 91–101, available at: http://www.inchem.org/documents/jecfa/jecmono/v48je08.htm (accessed 5 August 2009).

Marrs, W.M. (1998) The stability of carrageenans to processing. In: P.J. Williams and G.O. Phillips (eds), *Gums & Stabilisers for the Food Industry 9*. The Royal Society of Chemistry, Cambridge, pp. 345–357.

McCandless, E.L., Craigie, J.S. and Walter, J.A. (1973) *Carrageenans in the Gametophytic and Sporophytic Stages of Chondrus Crispus*. Planta, Berlin.

Philp, K., DeFreitus, Z., Nicholson, D. and Hoffmann, R. (1998) Tiger striping in injected poultry. In: P.J. Williams and G.O. Phillips (eds), *Gums & Stabilisers for the Food Industry 9*. The Royal Society of Chemistry, Cambridge, pp. 276–285.

SCF (2003) European Commission, Health & Consumer Protection Directorate – General, Opinion of the Scientific Committee on Food on Carrageenan, SCF/CS/ADD/EMU/199 Final, 5 March 2003, available at: http://ec.europa.eu/food/fs/sc/scf/out164_en.pdf (accessed 5 August 2009).

Weiner, M.L., Nuber, D., Blakemore, W.R., Harriman, J.F. and Cohen, S.M. (2007) A 90-day dietary study of kappa carrageenan with emphasis on the gastrointestinal tract. *Food and Chemical Toxicology*, **45**(1), 98–106.

6 Cellulose Derivatives

Mary Jean Cash and Sandra J. Caputo

ABSTRACT

Cellulose derivatives are commonly used in food applications where they are effective as viscosifiers, stabilizers and rheology modifiers. The manufacture of cellulose gum, also known as carboxymethyl cellulose, methyl cellulose, hydroxypropylmethyl (methylhydroxypropyl) cellulose, hydroxypropyl cellulose and ethyl cellulose, are discussed. The chemistry of these hydrocolloids is reviewed and includes the impact of solutes, salts, acid, temperature, concentration and protein interactions. In food applications, cellulose gum is an effective thickener and moisture binder used for clear beverages, dairy products, such as ice cream, and in bakery and other prepared foods. Methyl cellulose and hydroxypropylmethyl cellulose exhibit thermogelation, utilized for bake-stable sauces, fillings and formed foods. Hydroxypropylmethyl cellulose and hydroxypropyl cellulose are surface-active agents used to stabilize foams and emulsions.

6.1 INTRODUCTION

As the name implies, cellulose derivatives are produced from cellulose. Cellulose is a naturally occurring substance which constitutes about one-third of all vegetable matter. It is the main constituent of cell walls of higher plants. Wood and cotton are the main cellulose sources used in the production of cellulose derivatives. Wood contains about 40–50% cellulose, while cotton contains about 85–97% cellulose. In this chapter, the derivatives discussed are as follows:

- Sodium carboxymethyl cellulose, also known as cellulose gum or CMC.
- Methyl cellulose and hydroxypropylmethyl cellulose, also known as modified cellulose gum.
- Hydroxypropyl cellulose.
- Ethyl cellulose.

The production of cellulose gum, the most common of the cellulosic derivatives, was patented in Germany in 1918. Since its commercial introduction into the USA in 1946 by Hercules Incorporated, cellulose gum can be found in an ever-increasing number of food applications. It was first introduced to food as a viscosifier in cake batters. Now, the annual global use of cellulose gum in food and related applications exceeds 20 000 metric tons. Manufacturing facilities can be found on each continent. It should be noted that, due to legislative differences, the level of purity may not be the same in each country, such as China.

Methyl cellulose and its derivative, hydroxypropylmethyl cellulose, were first produced commercially in Germany in the 1920s and then in the USA in 1938 (Grover, 1993). These polymers differ from cellulose gum and other common hydrocolloids because of their thermal gelling properties, that is, their ability to gel once heated. A multitude of industrial uses were developed capitalizing on this property; however, these gums were not used in the food industry until much later.

Today, the food industry is still learning about the usefulness of these unique cellulose derivatives in applications such as bakery creams, fried products, sauces and other prepared foods. Global consumption for food-related products is estimated to be about 6000 metric tons each year.

Despite the unique properties and potential benefits of hydroxypropyl cellulose, this hydrocolloid is the least known of the cellulose derivatives used in the food industry. Developed in the 1960s by Hercules, hydroxypropyl cellulose has thermoplastic properties. In the food industry, this property confers the ability to form films, extrude and reduce surface tension. The global annual use of hydroxypropyl cellulose in the food industry is approximately 300 metric tons.

Uses for these three cellulose derivatives are discussed in greater detail later in this chapter.

6.2 RAW MATERIALS AND PROCESSING

Cellulose derivatives for commercial use are usually based on cellulose from wood or cotton sources, as discussed above. Cotton has a higher molecular weight and may be necessary for some very high viscosity grades of cellulose ethers (Zecher and Gerrish, 1997). Otherwise, wood sources are preferred as they are less likely to be genetically modified and, therefore, may be certified with 'non-GMO' status.

Cellulose is made up of repeating cellobiose units; each unit is composed of two anhydroglucose units (AGUs). The number of AGUs, joined through β1–4 linkages, is known as the degree of polymerization (DP) of the cellulose. Each AGU contains three hydroxyl groups. These are the sites where substitution takes place to form the cellulose derivative. The number of hydroxyl groups that are substituted after reaction is known as the degree of substitution (DS).

In the reaction process, cellulose is first treated with alkali to swell the polymer. The alkali will also disrupt crystalline regions and the subsequent reactions will be more uniform. Reactions take place at elevated temperature. If molecular weight is to be preserved, a nitrogen atmosphere is used to prevent oxidative degradation. Cellulose ethers are formed through either Williamson etherification or alkoxylation. The alkali cellulose is reacted with sodium chloroacetate to form carboxymethyl cellulose, methyl chloride to form methyl cellulose, ethyl chloride to form ethyl cellulose or propylene oxide to form hydroxypropyl cellulose. Mixed derivatives, such as methylhydroxypropyl cellulose or hydroxypropylmethyl cellulose, may be formed with combinations of reactants.

6.3 COMPOSITION AND CHEMISTRY

6.3.1 Carboxymethyl cellulose

Carboxymethyl cellulose (CMC) is a water-soluble, anionic cellulose derivative. It is light tan to white, odorless, tasteless, free-flowing powder that is fairly hygroscopic. Highly purified

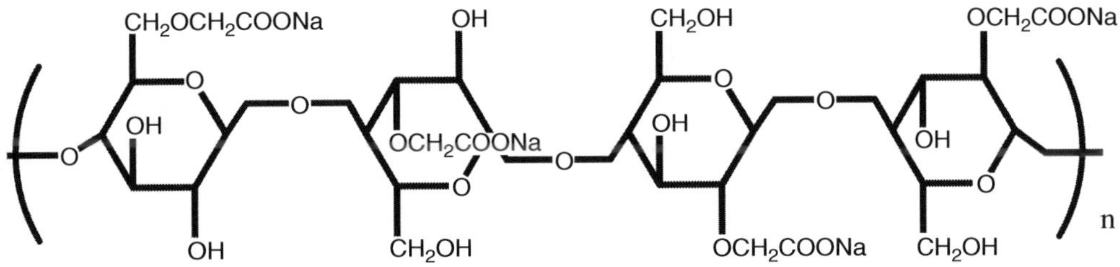

Fig. 6.1 Idealized structure of carboxymethyl cellulose. (Reproduced with kind permission of Hercules incorporated, a subsidiary of Ashland Inc.)

grades of CMC, at least 99.5% pure, are commonly known as cellulose gum. Cellulose gum is listed 'GRAS' (generally recognized as safe) under US Code of Federal Regulations (CFR) Title 21 Section 182.1745. Definitions for the identity and purity of cellulose gum are set out in the Food Chemical Codex as well as by the United Nations World Health Organization (FAO/WHO). The European Community Directives assign E466 to this food additive and the synonym cellulose gum may be used for labeling purposes.

Solution characteristics vary depending on the DP and degree of carboxymethyl substitution as well as uniformity of substitution. DS refers to the average number of carboxymethyl groups per AGU, as discussed above. Each AGU has three sites on which substitution may take place (see Fig. 6.1). Theoretically, CMC may have a maximum DS of 3. However, in reality, substitution levels this high are neither practical nor useful. Commercially available grades are available up to DS of 1.5. A degree of substitution of less than 0.4 will result in an insoluble CMC (Zecher and Gerrish, 1997).

Cellulose derivatives are long-chain polymers. Chain length or DP will also impact solubility and other solution characteristics. Molecular weight is determined by both the DP and amount of substitution. Small increases in molecular weight lead to large increases in viscosity. The average DP and molecular weight for some commercial CMC types are given in Table 6.1.

6.3.1.1 *Solubility*

Polymer chain length impacts solubility. Decreasing the DP will increase water solubility. Cellulose gum grades with the same substitution but of differing DP also differ in solubility. In addition, increasing carboxymethyl substitution results in increasing water solubility. Decreasing the DS will allow associative behavior along unsubstituted regions of the cellulose leading to reduced solubility in water. When substitution occurs uniformly along the backbone, associative behavior is prevented and solubility increases. Thus, cellulose gum with lower DS or less uniform substitution will be less soluble than a high DS CMC of the same DP.

Table 6.1 Typical molecular weights corresponding to the degree of polymerization (DP) for carboxymethyl cellulose (CMC), degree of substitution (DS) = 0.7.

DP	Molecular weight
3200	700 000
1100	250 000
400	90 000

6.3.1.2 *Viscosity*

The viscosity of cellulose gum is determined largely through controlling cellulose chain length or DP. Cellulose gum with a longer chain length has a higher viscosity. In addition, DS also impacts viscosity. Lower or less uniform substitution of the carboxymethyl groups leads to associative behavior as described above. Less soluble cellulose gum has a higher viscosity and a higher water-holding capacity. In this way, both DS and DP impact CMC solution viscosity.

The viscosity of cellulose gum is affected by the presence of nonsolvents. Water-miscible ingredients, such as glycerol and alcohol, promote cellulose gum viscosity by limiting the volume of water available as a solvent, leading to a more concentrated cellulose gum solution. In such solvent systems, viscosity can be greatly increased simply by increasing the nonsolvent portion of the mixture and any associative behavior that might be exhibited by the particular cellulose gum grade will also be increased.

Cellulose gum is highly tolerant of the presence of ethanol. In ethanol/water mixtures with 50% or more alcohol, solutions remain clear. A higher DS, more soluble cellulose gum, is even more tolerant of alcohol. Low-viscosity cellulose gum grades are also considerably more tolerant of the presence of organic solvents than medium or high viscosity grades.

6.3.1.3 *Dispersion and dissolution techniques*

Cellulose gum is soluble in either hot or cold water. Because of the rapid hydration of cellulose gum, particles tend to agglomerate and form lumps when the powder is introduced into water. Effective techniques used to create cellulose gum dispersions with minimal or no lumps include the use of agitating equipment or the addition of other ingredients to the formulation. High agitation, such as the vortex created with a high-speed mixer, will help keep particles apart when the powder is introduced to water and allow them to hydrate separately, thus reducing lumping. Pre-blending the powder with dry ingredients, such as sugar, will also separate particles and reduce lumping. Cellulose gum is insoluble in organic solvents, such as methanol, ethanol or glycerol, making any of these ingredients, if present in the formulation, effective dispersing aids. Other useful dispersing aids common in food formulations are sugar syrups such as corn, fructose or invert syrup. Oils may also be used, though hydration may be slowed due to the oil coating on the particle.

Commercial grades of cellulose gum are offered with varying particle size distributions. Coarse particle cellulose gum will take up water more slowly and reduce the risk of lumps. Fine particle material is useful in dry mix powder applications where the bulk phase ingredients separate the gum particles to eliminate lumping when added to water. Particle size is important in systems where there is competition for water from other ingredients; fine particle cellulose gum will completely hydrate when there is low water availability.

6.3.1.4 *Rheology of cellulose gum solutions*

Newtonian flow is characterized by viscosity that is independent of the shear rate. If viscosity increases in the presence of an increasing shear rate, the solution is dilatant; such behavior may be seen in mineral suspensions or nongelatinized starch pastes. Pseudoplastic flow is observed when the apparent viscosity decreases in response to an increase in the shear rate. As in the case of most aqueous polysaccharide solutions, cellulose gum solutions display this behavior, though low concentrations and low-molecular-weight cellulose gum

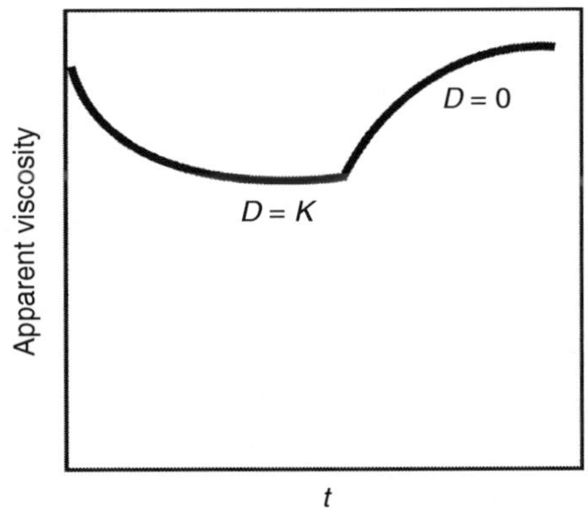

Fig. 6.2 Thixotropic flow is a time-dependent change in viscosity at constant shear ($D = K$). (Reproduced with kind permission of Hercules incorporated, a subsidiary of Ashland Inc.)

solutions show Newtonian-like flow behavior. If a critical shear stress must be exceeded before a solution flows, this is known as the yield value. Thixotropy is characterized by the decrease of viscosity at a constant shear rate over time. Once the shear is removed, viscosity significantly increases. This is demonstrated in Fig. 6.2. At a constant shear rate, $D = K$, viscosity decreases with time; when the shear is removed, $D = 0$, viscosity increases over time returning to the original value.

Thixotropic behavior is common in hydrocolloid solutions where the polymer chains interact and three-dimensional structures are formed. Unsubstituted regions of cellulose of low DS or nonuniformly substituted cellulose gum grades tend to associate and cause thixotropic behavior. These polymer solutions display a yield value as well. As discussed above, this associative behavior boosts viscosity. High DS cellulose gum grades and those with uniform substitution, where there is little associative behavior between polymer chains, tend to exhibit pseudoplastic flow with no thixotropy.

6.3.1.5 *Water holding*

The ability of cellulose gum to hold water is greater when the solubility is lower, such as for low DS or coarse particle-size materials. Water-holding capacity is also much higher in thixotropic grades of cellulose gum. In experiments to evaluate ingredients for water-holding capacity, dispersions were made under optimal conditions and allowed to sit for 30 min before centrifugation. It was found that the water-holding capacity for all-purpose flour was 0.8 g/g, cornstarch 1.0 g/g, soy fiber 8.5 g/g and a low DS high DP cellulose gum had a water-holding capacity of 42.3 g/g.

6.3.1.6 *Impact of acid*

Solutions of cellulose gum maintain viscosity over a wide pH range. However, maximum viscosity and stability will be experienced at near neutral pH. At pH values below 3.0, the acid form of CMC predominates and is less soluble. High DS or more uniformly substituted cellulose gum grades are more resistant to any hydrolytic degradation that may take place. The order of addition of ingredients is important: allowing the cellulose gum to fully hydrate prior to adding other ingredients will greatly enhance utility in low pH systems. After the initial acidification, solution viscosity is stable over time.

6.3.1.7 *Impact of salt*

The effects of salt on cellulose gum solutions will vary with concentration and salt type. In general, more highly substituted grades are more tolerant of the presence of salt. Also, full viscosity will be obtained if the cellulose gum is fully hydrated before the salt is added. Monovalent salts are compatible with cellulose gum and full viscosity may be obtained if the proper order of addition of ingredients is followed. Divalent salts may impact the solution characteristics. At high concentrations, calcium ions will inhibit the hydration of cellulose gum to such an extent that full viscosity may not be obtained and the solution may be hazy. Trivalent ions will interact with cellulose gum and cause hazy dispersions. Aluminum ions, for example, may be used to create gels.

6.3.1.8 *Synergies*

Cellulose gum is often used in conjunction with other hydrocolloids. Blends of ingredients are commonly offered by compound manufacturers or blending houses to simplify the formulation and manufacture of the final product. Blenders often provide expertise in ingredient interactions and optimization. Blends may contain one or more hydrocolloids as well as emulsifiers, surfactants, conditioners and flavors. Such blends are often used in the dairy and baking industries and are of increasing importance in all food manufacturing. Blends are often created to utilize particular attributes of several hydrocolloids and may also take advantage of particular synergies between ingredients. Cellulose gum combined with locust bean gum is known for giving a synergistic viscosity increase. It is found that this synergy is further emphasized at higher overall concentrations of polymer (Kaletunic-Gencer and Peleg, 1986). Cellulose gum/starch synergies have been reported; Sudhakar *et al.* (1992) found higher synergies in CMC/cornstarch viscosity than CMC/amaranth and a higher resistance to mechanical breakdown. These workers also reported improved freeze–thaw stability due to possible interactions of amylose with CMC resulting in reduced starch retrogradation. Cellulose gum blends with hydroxypropyl cellulose (HPC) also show synergy. Enhanced viscosity increases may also be seen in cellulose gum mixtures with hydrocolloids such as guar, tragacanth, karaya and xanthan (Hoefler, 2004).

6.3.1.9 *Protein interactions*

Cellulose gum is useful in stabilizing low pH protein beverages. Fermentation leading to acid formation or direct acidification will cause the precipitation of protein in milk or soymilk. The anionic nature of cellulose gum results in an interaction with the positive charges on protein at or near its isoelectric point (Ganz, 1974). CMC–protein complex formation promotes the stabilization of protein in acidified dairy or soy beverages giving fruit-flavored protein beverages and stabilized yogurt drinks.

6.3.1.10 *Impact on sugar and ice crystal growth control*

Cellulose gum is used to maintain the quality of high sugar applications such as syrups, frostings and fondants. Such applications may be compromised by the formation of sugar crystals causing a grainy texture. The rate of crystallization or adjustment depends on both the diffusion of small crystals and the growth of larger crystals as they serve as nuclei for further crystal growth. In these applications, cellulose gum controls water in the continuous phase surrounding the crystals. In this way, cellulose gum is effective at maintaining a predominance of small crystals, thus reducing the rate of adjustment.

Cellulose gum is effective at controlling ice crystal growth in ice cream and frozen desserts. Because of its anionic nature, cellulose gum will destabilize milk proteins at or near neutral pH. Regand and Goff (2002) theorized that some thermodynamic instability leads to a concentration of proteins at the ice crystal surface, thus leading to increased water-holding performance. These features of cellulose gum make it quite effective in controlling the growth of ice crystals during the initial freezing process as well as during subsequent freeze–thaw cycling or heat shock.

6.3.2 Methyl cellulose and methylhydroxypropyl cellulose

Methyl cellulose (MC) and hydroxypropylmethyl cellulose (HPMC) or methylhydroxypropyl cellulose (MHPC) are nonionic water-soluble cellulose ethers. The terms HPMC and MHPC are both used in industry. These cellulose derivatives are light-colored free-flowing powders with neutral taste and odor but less hygroscopic than cellulose gum. MC and HPMC are listed as 'GRAS' (generally recognized as safe) and requirements may be found in US CFR Title 21 Section 182.1480 (MC) and section 172.874 (HPMC). Definitions for the identity and purity are set out in the Food Chemical Codex as well as by the United Nations World Health Organization (FAO/WHO). Alternative names are sometimes used and include 'vegetable gum'. E461 (MC) and E464 (HPMC) have been assigned to these additives in European directives.

Solution characteristics are impacted by DP and substitution level. Cellulose is derivitized with methyl chloride in the case of MC or both methyl chloride and propylene oxide in the case of HPMC. DS or the relative number of substituted sites on the AGU is normally 1.4–2.2 in MC and 1.0–2.3 in HPMC (Fig. 6.3). DS is, however, less important when discussing the impact of substitution than the molar ratio of substitution to the number of AGU, the molar substitution (MS). Often the weight percent of substituted groups is quoted for these cellulose ethers.

6.3.2.1 *Solubility and thermal gelling*

MC and HPMC are soluble in cold water. Polymer solubility is impacted by DP as well as substitution type and level. Higher DP grades will have lower solubility compared to grades of lower DP with the same MS. A wide range of solution viscosities is available commercially.

Of great interest in food applications is the thermal insolubility of MC and HPMC. This is characterized by thermal gelation and possibly eventual flocculation of the polymer at elevated temperatures. Increases in temperature of MC or HPMC solutions will first lead to a minor decrease in viscosity. When the gelling temperature is reached, there will be a sharp increase in viscosity. The viscosity or gel strength will then remain at a constant value (see Fig. 6.4). Gel temperature and texture are dependent on the substitution level and type. For example, MC types will gel at lower temperatures and form hard, brittle gels, whereas HPMC types gel at higher temperatures and the resultant gels are softer in nature.

Thermal gelling is caused by dehydration of the polymer with heating. This phenomenon can be explained as the weakening of water–polymer interactions and the strengthening

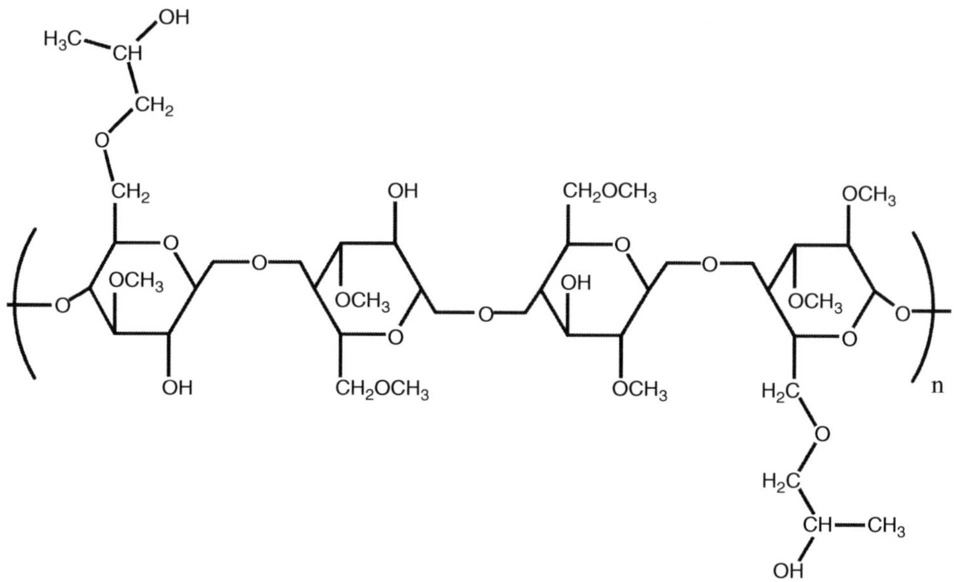

Fig. 6.3 Idealized structure of hydroxypropylmethyl cellulose. (Reproduced with kind permission of Hercules incorporated, a subsidiary of Ashland Inc.)

of polymer–polymer interactions. Water molecules form hydrogen bonds around the hydrophobic polysaccharide. Heating effectively disrupts this water structure allowing polymer–polymer interaction. Hague and Morris (1993) found that there was a two-phase gelling of MC, first an initial melting of structures found in solution and then a reordering

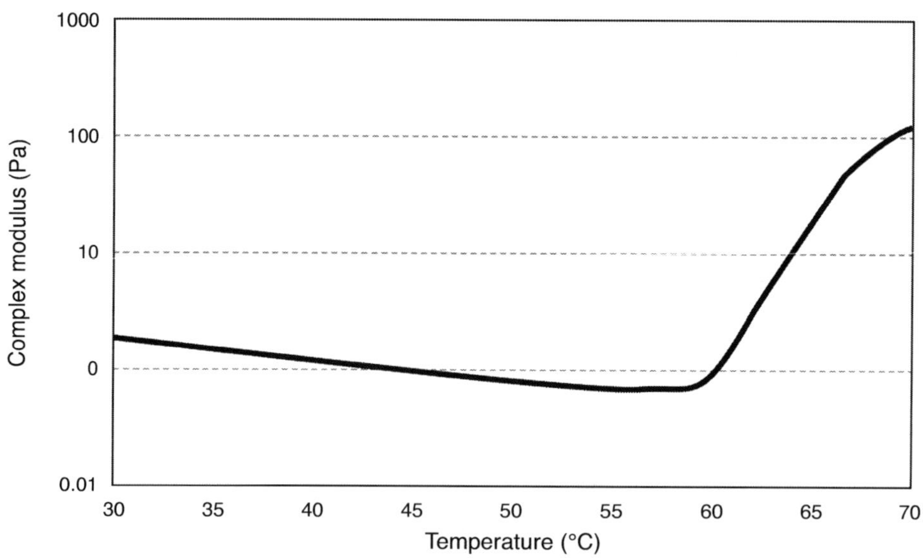

Fig. 6.4 Complex modulus versus temperature for a typical methyl cellulose solution. (Reproduced with kind permission of Hercules incorporated, a subsidiary of Ashland Inc.)

to allow a different structure at high temperature. Thermal gelation is completely reversible. The gel will return to the initial solution viscosity when the gel is cooled and this thermal gelling and cooling to a solution can be repeated.

Insolubility at elevated temperatures allows hot water to act as a useful medium for dispersing MC and HPMC. A portion of water required in the formulation is heated above the gelling temperature of the grade being used and the polymer is added and mixed to create a dispersion. A portion of cold water is then added and the mixture cooled under agitation to create the solution.

6.3.2.2 *Rheology*

Aqueous solutions of MC and HPMC are pseudoplastic and show little thixotropy; solutions are smooth. Pseudoplasticity increases with concentration and molecular weight. Low concentrations of the polysaccharide in water exhibit almost Newtonian behavior.

6.3.2.3 *Surface activity*

MC and HPMC are surface active due to their slight hydrophobic character. Solutions of MC and HPMC show increases in surface activity depending on the level and type of substitution (see Table 6.2) and these properties may be utilized in emulsion applications. The surface-active nature of MC and HPMC also allows these solutions to entrap air easily and deaerate quite slowly. This is a useful function in the stabilization of foams.

6.3.2.4 *Impact of solutes*

MC and HPMC solutions are generally tolerant to the types and amounts of salts in food. The presence of salts lowers the thermal gelling temperature of MC and HPMC. The addition of 2% sodium chloride lowers the gel temperature of a 2% solution by 10–15°C. Higher levels of salts may produce a 'salting out' effect. MC as well as high-molecular-weight HPMC has lower tolerance to salt than medium viscosity HPMC. For example, greater than 5% sodium chloride may cause precipitation of 2% MC but not the lower viscosity HPMC grades. HPMC with a 2% viscosity of 70 000 cP will also be precipitated with 5% sodium chloride, whereas HPMC of the same substitution and a 2% viscosity of 20 000 cP is stable.

Other solutes also impact gel temperatures. The addition of 10% sucrose will depress gelling temperatures up to 10°C. High amounts of sugar have a more severe impact. Solutions containing 40% sucrose will reduce gel temperatures as much as 30°C below the gelling point obtained with no sugar.

Table 6.2 Surface and interfacial tension reduction in methyl cellulose (MC) and hydroxypropylmethyl cellulose (HPC) solutions.

	Surface tension (dynes/cm)	**Interfacial tension vs. paraffin oil (dynes/cm)**
Water	72	45
MC	50–55	18–21
HPMC	46–51	17–18.5

6.3.3 Hydroxypropyl cellulose

Hydroxypropyl cellulose (HPC) is a nonionic water-soluble cellulose ether available commercially in powder or granular form. Its color is off-white and taste is neutral. Regulations and definitions are set out in CFR title 21 Section 121.1160. The European Community Directives have assigned E463 to HPC and it is also described in the Food Chemical Codex and specifications are listed by the WHO/FAO.

6.3.3.1 *Rheology and solution characteristics*

HPC solutions are exceptionally clear and smooth flowing. They show pseudoplastic behavior with decreased viscosity under shear, exhibiting almost no thixotropy, and returning to the original viscosity once the shear is removed.

Viscosity is impacted by DP. Substitution occurs through propylene oxide forming an ether linkage with the AGU. In addition to the available sites on the AGU, secondary hydroxyls are present on the side chain allowing additional substitution. In this way, branched side chains are formed. The molar substitution (MS) of typical commercial HPC is 3.0–4.5 (see Fig. 6.5) (Zecher and Gerrish, 1997).

HPC has the additional unique property of being thermoplastic. It may be used in injection and compression molding as well as injection molding, blow molding and injection foam molding operations. It has excellent film properties and may be heat sealed.

At typical commercial MS levels, HPC is insoluble in water above 45°C. Precipitation will appear in water solutions raised to temperatures between 40°C and 45°C. There is no formation of a gel, unlike MC and HPMC. Lowering the temperature while stirring will restore the polysaccharide solution to its original viscosity. To make HPC solutions, it is useful to initially disperse the powder in hot water and then cool under agitation either by adding the balance of cold water or by cooling by other means. HPC may also be dispersed in glycerin. The polymer is soluble in many organic solvents including ethanol and propylene glycol.

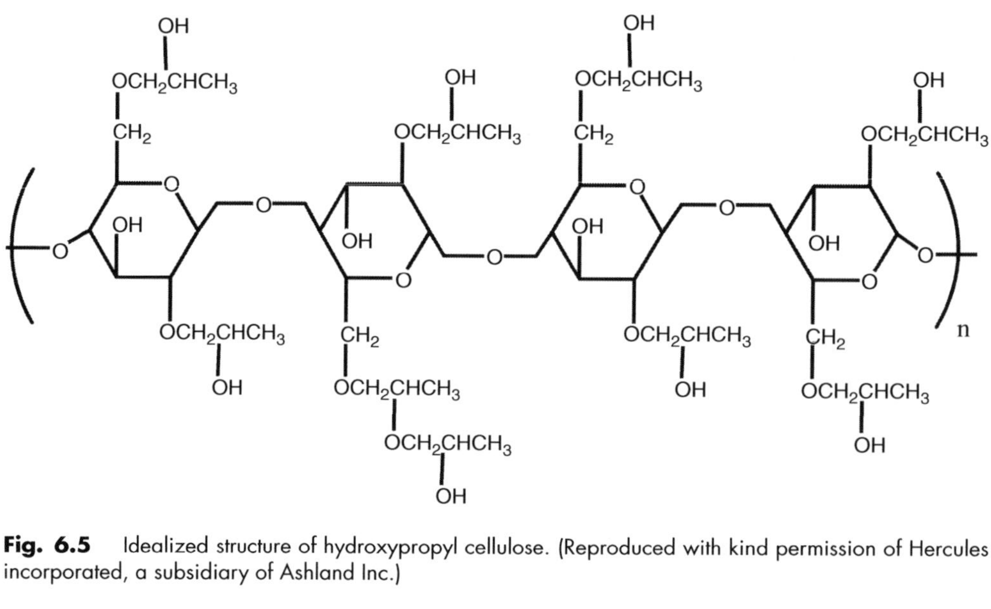

Fig. 6.5 Idealized structure of hydroxypropyl cellulose. (Reproduced with kind permission of Hercules incorporated, a subsidiary of Ashland Inc.)

Table 6.3 Surface and interfacial tension reduction in hydroxypropyl cellulose (HPC) solutions.

	Surface tension (dynes/cm)	Interfacial tension vs. paraffin oil (dynes/cm)
Water	72	31.6
0.01% HPC	45.0	–
0.1% HPC	43.6	12.5
0.2% HPC	43.0	–

6.3.3.2 *Impact of solutes*

The precipitation temperature will be depressed by the presence of solutes. For example, 10% sucrose will lower the precipitation temperature by 5–8°C and 40% sucrose can lower the temperature by 20°C. Salts have similar effects. Sodium chloride added at 5% will lower the precipitation temperature by 10°C. High levels of salt will tend to cause 'salting out'.

6.3.3.3 *Surface and interfacial activity*

HPC solutions exhibit greatly lowered surface tension making them good foam promoters. HPC demonstrates emulsification properties and will also lower the interfacial tension of oil and water (see Table 6.3). These properties, as well as the protective colloid action of HPC, give it great utility as a whipping aid and foam stabilizer for foamed food applications.

6.3.4 Ethyl cellulose

Ethyl cellulose (EC) is a nonionic ethyl ether of cellulose, soluble in a wide range of organic solvents. It is not water soluble. US regulations for its use are set out in 21CFR 172.868 where EC is described for use as a binder and coating in vitamin preparations as well as a flavor fixative. EC is also described in the Food Chemicals Codex. The European Community Directives have assigned E462 to this additive and it may be used in food applications in general.

6.3.4.1 *Rheology and film-forming characteristics*

As EC is completely hydrophobic, in food applications it is solvated in ethanol. EC viscosity is measured and reported in 80:20 toluene:ethanol and products are available in a wide range of viscosities from 5 cP to exceeding 5000 cP. EC is resistant to changes in pH conditions with particular resistance to alkalinity. EC has particular utility as a moisture barrier and it is an excellent film former. The grades of EC available differ in molecular weight and substitution; changes in molecular weight will impact the film tensile strength, elongation and flexibility, while DS impacts softening point and solubility in ethanol.

EC may be used in food to stabilize flavors and to protect nutritional ingredients against active interactions, hydrolysis and oxidation. It may also be used to retard the release of active ingredients by encapsulation. EC has been used as barrier to reduce moisture migration in foods prepared in two stages, such as pizzas and snacks.

6.4 FOOD APPLICATIONS

6.4.1 Applications in beverages

6.4.1.1 *Beverages: ready to drink, dry mixes and concentrates*

Hydrocolloids are commonly used in beverages to improve mouthfeel. They may be used to increase the perception of richness or to restore mouthfeel when another ingredient is removed. Mouthfeel may be described as a combination of sensations in the mouth during and after ingestion of food or beverage. The mouthfeel of beverages is not entirely due to viscosity; it is related to density, viscosity, surface tension and other mechanical properties (Bourne, 2002).

Hydrocolloid selection for beverages is based on a number of factors including solubility, clarity and the desired sensation in the mouth. Cellulose gum, being of high solubility and excellent clarity in solution, is widely used in beverages. Szczesniak and Farkas (1962) found that gum solutions exhibiting a high degree of thixotropy were characterized by panelists as having some degree of 'sliminess' whereas gum solutions with low thixotropy were not considered slimy. Hydrocolloids found to exhibit thixotropic behavior include, among others, pectin, CMC and locust bean gum, all used commercially in beverages. The dose levels in the study were quite high, 1–2 and even 5% gum by weight. Typical use levels in beverages are less than 0.1%. Although a slimy mouthfeel could be perceived in concentrated gum solutions, in practice a pleasant mouthfeel is produced at low gum concentrations. Cellulose gum is used at low use levels, to target apparent viscosities in the range of 5–30 cP.

Cellulose gum is used to add richness or to impart a juice-like feeling in the mouth in fruit-flavored beverages. This cellulose derivative also is utilized to replace sugar in light or diet drinks. Cellulose gum is used to impart a more smooth or rounded flavor in the liquid beverage. HPC has been found to effectively decrease the tastes of sourness and saltiness in tomato juice while having less impact on aroma and flavor intensity (Pangborn *et al.*, 1978). In this study, it was found that a medium viscosity cellulose gum in an orange-flavored drink had little impact on sweetness while effectively reducing sourness.

In dry mixes, cellulose gum grades with medium molecular weight are often selected due to their quicker hydration compared to high-molecular-weight grades; however, high viscosity grades are still quite popular in many beverage applications. Higher DS or more smoothly substituted types are recommended for fast hydration. In addition, a fine particle size gum will hydrate more quickly than regular or coarse particle sizes.

Many beverages are acidic for both flavor and microbial stability reasons. For ready-to-drink fruit-flavored beverages and beverage concentrates, high-DS and medium-molecular-weight grades may be chosen for the maximum tolerance of low-pH conditions. There are cellulose gum grades on the market that are designed for low-pH conditions. Figure 6.6 demonstrates the viscosity of one such cellulose gum in the presence of common food acids. Because of the anionic nature of cellulose gum, it may also exhibit some buffering capacity in beverages resulting in increases in pH compared to formulations with no CMC. Therefore, formulation adjustments may be needed to give the desired flavor profile and mouthfeel.

6.4.1.2 *Protein beverages*

Because of its anionic nature, cellulose gum will interact with protein. This interaction is used to stabilize protein, such as casein or soy protein, at or near the isoelectric point.

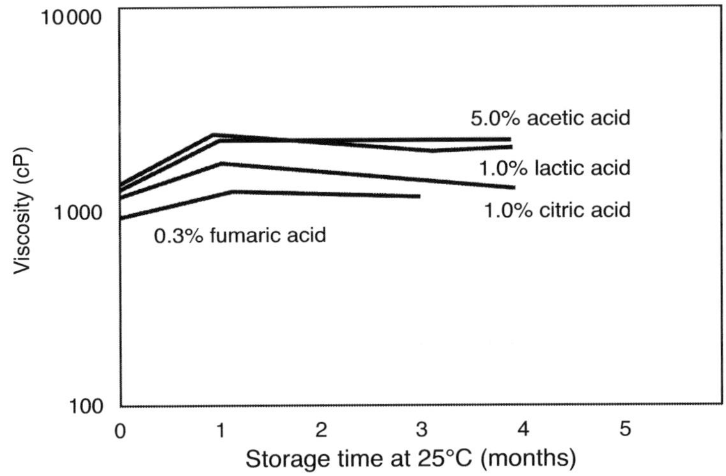

Fig. 6.6 Stability of cellulose gum in organic acids. (Reproduced with kind permission of Hercules incorporated, a subsidiary of Ashland Inc.)

Fruit-flavored dairy or soy drinks, yogurt-based drinks and other acidified protein beverages require the use of a stabilizer to prevent the precipitation of protein. In the production of such beverages, cellulose gum should be present at the time of acidification. Homogenization is a useful technique to facilitate the hydrocolloid–protein interaction. This step is followed by heat treatment to ensure microbial stability. In applications containing live cultures, care must be taken due to the possible presence of cellulase, as this enzyme will degrade the cellulose backbone leading to viscosity loss. High DS grades will be more suitable as they are less likely to be degraded.

Most cellulose gum grades may be used to stabilize acidified protein beverages. Uniformly substituted or high DS grades are generally chosen for acid tolerance. The viscosity grade is selected for the desired performance; high viscosity grades may be used to create additional texture and the higher viscosity will also aid in stabilizing the beverage. When beverages with cellulose gum are processed, viscosity loss will be experienced during the acidification and heat-process steps but the viscosity remains stable during shelf storage.

Protein will precipitate from low-pH beverages unless properly stabilized; this precipitation leads to a much lower viscosity. The cellulose gum–protein complex stabilizes viscosity as the protein remains suspended in the beverage. Therefore, the stability of acidified protein beverages may be evaluated using viscosity measurement. In Fig. 6.7, various cellulose gum types are compared in a dairy beverage (6% MSNF) over several months of storage. These beverages were processed with homogenization and heat pasteurization treatment. All beverages remained stable for up to 1 year (Hercules Incorporated, 2005a).

6.4.2 Applications in bakery

6.4.2.1 *Tortillas*

Cellulose gum is the most widely used hydrocolloid in the manufacture of corn tortillas through its inclusion in dry masa flour. Masa is made through the alkali cooking and steeping of corn (nixtamalization) and is the basis for many Mexican-style cereal foods and snacks. Dry

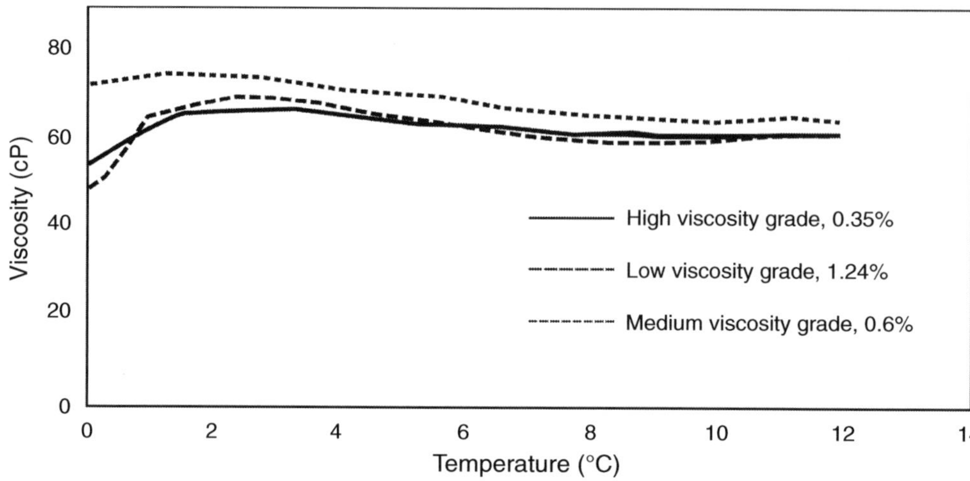

Fig. 6.7 Viscosity versus time in acidified dairy beverages stabilized with cellulose gum grades of various viscosities. (Reproduced with kind permission of Hercules incorporated, a subsidiary of Ashland Inc.)

masa flours have replaced the use of fresh masa in much of commercial tortilla production. Dry masa flour offers the advantages of long shelf life and convenience but products made from dry flour may have inferior textures compared to those made with fresh masa flour (Gomez *et al.*, 1987). For this reason, ingredients are added to the dry masa flour to improve tortilla quality. Cellulose gum increases water absorption and amylograph peak viscosity; it is also suitable as an anti-sticking agent in tortilla production (Serna-Saldivar *et al.*, 1990). Cellulose gum maintains the rolling and folding properties of tortillas during shelf life. High viscosity cellulose gum grades with low to mid-range DS are most often used. During production, additives, including hydrocolloids, preservatives and acidulants, are mixed into the dry masa flour at the mill. The appropriate particle size cellulose gum is important for determining the performance in the tortilla dough and final product.

Wheat flour tortillas are made from flour, fat, water and salt and include other added ingredients such as leavenings, flavors, preservatives and hydrocolloids. In the marketplace, much more variation in these other additives may be seen in wheat tortillas compared to corn tortillas. Cellulose gum is commonly used in wheat flour tortillas to improve dough machinability and help counter natural variations in flour (Serna-Saldivar *et al.*, 1988). Cellulose gum will also reduce staling and moisture loss, improving the flexibility of wheat flour tortillas (Qarooni, 1993). Friend *et al.* (1993) found that both HPMC and cellulose gum improved the quality of wheat flour tortillas. Cellulose gum most improved the 'rollability' of the tortillas and the tolerance to freeze/thaw cycling. Cellulose gum also improves the quality of wheat flour tortillas made with added sorghum flour (Torres *et al.*, 1993).

6.4.2.2 *Bread*

Consumer expectations for bread variety and quality continue to increase. In addition, recent dietary recommendations include increasing the use of whole grains as a part of a healthy diet. These demands from the marketplace present challenges to bakers and formulators. Hydrocolloids may be used effectively to improve the quality and consistency of breads.

Cellulose gum has been found to improve bread quality. In simple dough formulations, Sidhu and Bawa (2000) found that the water adsorption increased significantly with the addition of cellulose gum. Dough extensibility and elasticity also increased as did gas retention, producing greater dough volume. The improved dough quality resulted in increased loaf volume as well as increased yield in batches containing cellulose gum. The addition of cellulose gum was also found to reduce the firming of the crumb during storage. Collar *et al.* (2001) found lower firming rates in bread with added cellulose gum which may be attributed to the increase in starchy lipids through the preferential binding of cellulose gum to gluten leading to the displacement of nonstarchy lipids from the starch. Cellulose gum has also been found to counter some of the deleterious nature of freezing on frozen dough (Sharadanant and Khan, 2003).

Armero and Collar (1998) found HPMC and cellulose gum effective in decreasing crumb firming over time in whole wheat, white and sourdough breads. Softening effects from HPMC were more evident in whole grain breads, while cellulose gum was most effective in reducing firming in sourdough breads. The same workers (Armero and Collar, 1996) found that cellulose gum had positive effects on the crumb, eating quality and elasticity in white bread.

Gluten-free bread formulations have been made possible using cellulose derivatives. Cato *et al.* (2004) found that highly acceptable bread could be formulated from rice flour and potato starch using HPMC alone or in combination with cellulose gum. The thermal-gelling properties of HPMC aid in the retention of gas during proofing and baking, allowing a porous crumb structure that is similar to the structure formed when gluten is present. Cellulose gum also increased dough viscosity contributing to the improved structure and volume of the bread.

6.4.2.3 *Cakes and sweet goods*

Cake mixes are widely used in home and institutional baking. These mixes are expected to tolerate over- or under-mixing. Cellulose derivatives add viscosity to batter and help maintain leavening. In addition, cellulose gum will help retain moisture during baking and storage. Cellulose gum increases cake volume and improves crowning and crumb structure. Low-DS, high-viscosity types, as well as specialty CMC types made to absorb water, also allow additional water to be included in the formulation, improving both yield and moistness in the baked goods. Cellulose gum is also used in dry mix applications where quick hydration is useful.

MC and HPMC improve crumb structure and volume in cakes during baking due to the thermal-gelling properties. MC and HPMC have proven especially useful in microwaved bakery goods leading to improved cake height and eating quality (Coffey *et al.*, 1995). During baking, the thermal-gelling nature of MC and HPMC aids the suspension of inclusions, such as fruit or nuts, to maintain well-dispersed pieces throughout the cake or muffin.

6.4.2.4 *Instant noodles*

Instant noodles, also known as Ramen-style noodles, are an important cereal food in Asian markets. This type of noodle is also growing in popularity in western diets. The texture of the noodles is an important component of consumer acceptance. Texture preference may vary from region to region. Cellulose gum may be formulated into noodles to improve dough strength, which aids in machinability and extrusion of the dough into different noodle forms.

During the frying stage, the cellulose gum will reduce oil uptake, leading to lower frying oil costs and improvements in shelf life due to reduced rancidity. It has been shown that the cellulose gum grade influences the eating quality of the finished product (Boukerchi, 1995). In this application, high-viscosity grades are most often used. For a more springy texture, desired in some markets, lower DS grades may be used.

6.4.2.5 *Icings, frostings and glazes*

The terms icing, frosting and glaze are interchangeable to some bakery professionals, while others see them as very distinct products. For the purposes of this discussion, a glaze is a thin coating containing no fat, while frosting and icing are both thick coatings that may contain fats. Cellulose derivatives are used at low levels in icings and frostings to control quality and modify texture. Uniformly substituted grades are used for increased solubility. Cellulose gum is used to reduce sugar crystallization that will cause sandiness in fondant and rolled frostings and icings; low-molecular-weight grades are most effective in reducing the rate of adjustment (see Section 6.3.1.10). Because of its film-forming properties, cellulose gum will also help prevent the frostings on snack cakes sticking to packaging. Creaminess and spreading quality can also be improved with cellulose gum or HPMC. Cellulose gum may also be used in glazes for water control and to reduce the effects of 'sweating' during storage.

6.4.3 Applications in dairy desserts

6.4.3.1 *Ice cream and frozen desserts*

Ice cream is most often stabilized with a combination of hydrocolloids and other ingredients, typically supplied as a stabilizer blend. The hydrocolloids and emulsifiers contained in these blends are used to contribute body and give a creamy texture, increase overrun and reduce ice crystal growth during freeze–thaw cycling.

Cellulose gum is widely used and known for its superior performance in heat-shock protection by controlling ice crystal growth in ice cream. The slight whey-off that is sometimes seen in ice cream mixes prior to freezing is easily stirred back into the mixture. Cellulose gum appears to be more effective in reducing ice crystal growth in solutions containing proteins (Wang *et al.*, 1998). In ice cream, high-viscosity cellulose gum grades typically are used at 0.3% by weight. In addition to ice crystal growth control, cellulose derivatives are used to entrap air and increase overrun. MC and HPMC are used in stabilizer blends for ice cream and frozen desserts. The surface-active behavior of these cellulose derivatives promotes foam formation during freezing and mixing leading to high overrun products.

6.4.3.2 *Whipping cream and whipped toppings*

HPC, MC and HPMC are highly surface active. As previously discussed, these cellulose derivatives are very effective in foamed food applications, such as nondairy whipped toppings and aerated desserts. Nondairy whipped toppings are formulated with partially hydrogenated fats, emulsifiers, water, flavors and colors. They may or may not contain proteins. HPC, MC or HPMC may be used to increase the viscosity of the continuous phase as well as lowering the interfacial tension at oil/water and air interfaces, thereby promoting foam formation and stability.

HPC may be used in dairy cream for whipping, allowing the formulation of reduced fat and low-fat whipping creams. In a recent study, it was found that foam stiffness and stability were greatly enhanced by using a medium-molecular-weight HPC in reduced (31%) fat whipping cream. Cake decorations made from whipping cream containing HPC were stable for up to one week. HPC also allowed the formulation of very low-fat creams with reasonable whipping properties. Cream containing HPC and a fat level of 22% could be whipped, whereas the control formed no foam (Hercules Incorporated, 2005b).

6.4.4 Applications in processed foods

6.4.4.1 *Meat analogues*

Meat analogues and vegetarian meat replacements are increasingly popular with consumers due to both personal beliefs and health concerns. Vegetarians and many other consumers include meat-free meals in their diet. Some of the most popular items are vegetarian patties and hot dogs that are similar to their meat counterparts in appearance, flavor and texture. Grains and soy are most often used in the formulation of these products and binders are necessary to help the product keep its form and have an appropriate texture. Egg may be used as a binder but this can be expensive and it is rejected by vegans. Methyl cellulose is used to bind ingredients, especially during the cooking process. Thermal gelation holds the product together and contributes a firm meat-like texture. Because of the stronger gel texture and lower gel temperature, MC grades are preferred over HPMC grades. MC also contributes to the emulsification of fat and prevents fat separation. Because of the cold solubility of MC, keeping the mixture cold during processing will aid dissolution and the effective use of the polymer. Methods for preparing meat analogues include the preparation of emulsified bases with MC and are described in the patent literature (Hargarten *et al.*, 2005; Howse *et al.*, 2005). Cellulose gum may also be included in meat analogue formulations. Grades with high water absorption retain moisture leading to better eating qualities and moisture perception.

6.4.4.2 *Meat, fish and poultry products*

Cellulose derivatives allow the formulation of novel meat and fish products, such as extruded shrimp pieces and patties. Methyl cellulose aids in the retention of water during extrusion and cooking and helps maintain product shape. Meat and vegetable snacks are also formulated with MC to maintain shape integrity and improve mouthfeel at high temperatures as the product is consumed.

MC and HPMC are used in formulations for dysphagia sufferers who have difficulty swallowing. Entrée and side dishes are created with meat, fish, poultry and vegetables with familiar flavors and good nutrition while also being easy for these patients to consume and swallow.

6.4.4.3 *Syrups, sauces and soups*

Table syrups are often designed to imitate maple syrup in appearance and flavor. Syrups with fruit juices and fruit flavors are also popular. Table syrups are sweetened and thickened with corn syrup or invert syrup and they may be considered full or reduced sugar formulations depending on the sugar and calorie level. Most formulations contain cellulose gum whether they are reduced-calorie or not. These products are generally thicker than the natural counterpart in accordance with consumer preference. Cellulose gum provides a clear solution with a consistency of sugar syrup for the correct 'pour' from the bottle and minimal dripping. The high water solubility of cellulose gum makes it especially suitable for use in the high-solids contents of these products (40–70° Brix). Higher DS and medium viscosity grades provide the correct texture, best clarity and shelf life.

Fluid formulations such as sauces and soups benefit from the rheology modification and stabilizing features of cellulose derivatives. The emulsification properties of MC, HPMC and HPC stabilize dispersed oil droplets and help prevent fat separation. Starches are often used to give the desired texture; however, they are sensitive to increases in temperature. MC and HPMC provide high-temperature viscosity, thus ameliorating temperature effects. Cellulose derivatives also increase freeze–thaw stability, aiding the formulation of increasingly popular frozen or refrigerated soups and sauces.

Adding cellulose gum to tomato ketchup was found to reduce or eliminate serum loss during storage (Gujral *et al.*, 2002). Cellulose gum may be added either as a solution or mixed with dry ingredients such as sugar and spices. MC and HPMC will aid in the emulsification of oils in dressings and may be especially useful in dry mix dressings due to rapid hydration and migration to the oil/water interface (Coffey *et al.*, 1995).

6.4.4.4 *Fried foods*

Fried food formulations present unique challenges in balancing food texture and eating quality with nutritional and shelf life concerns. Cellulose derivatives reduce oil uptake during frying, improve eating quality and maintain product integrity. Cellulose gum in doughnut batters increases the viscosity of the batter for a more uniform doughnut shape with lower surface area, thus lowering oil penetration. Cellulose gum will also bind water, contributing to lower fat uptake during frying.

MC and HPMC are used in formed foods, such as reformed onion rings, to maintain the food product shape during frying as well as to add texture for a pleasant eating quality while the product is warm. MC and HPMC are also used in filled pastries to prevent boil-out during baking or frying. MC types are found to be most suitable for batters for fried foods (Sanz *et al.*, 2005) due to the lower gelling temperature and strong gelling nature of methyl cellulose. Batter viscosity will also influence fat uptake during frying. A higher viscosity results in greater batter coatings and lower oil penetration, thus high viscosity MC types may be useful in this area (Naruenartwongsakul *et al.*, 2004). In dry batter mixes, MC hydrates well as it is more soluble in cold water than in hot water.

In fried applications, a balance must be maintained between moisture retention and moisture lost at the surface to give a crisp coating. Retaining moisture leads to lower oil uptake; however, if moisture retention is too high, the cooking process will take longer or the product will be too soft and the quality will be inferior. Methyl cellulose solution is used to coat potato fries to reduce oil uptake and allow the escape of sufficient moisture for a well-cooked, crisp product (Kuntz, 1995). Potato croquettes and other formed products are produced with MC or HPMC to maintain product integrity during frying as well as to create consistently shaped croquettes and a predictable process.

6.4.4.5 *Encapsulation and films*

Cellulose derivatives have been used in encapsulation applications for flavors and other ingredients. Work has been conducted using HPC to encapsulate probiotic dairy cultures using a direct compression method resulting in an effective vehicle for viable cells (Chan and Zhang, 2002).

Cellulose derivatives are used to form films as coatings or in unsupported films. Unsupported films may be made by casting cellulosic solutions. HPC, as a thermoplastic, may also be molded. Cellulosic-based films are good oxygen barriers compared to plastic films and could be used effectively to protect food against oxidative degradation while also being edible (Park *et al.*, 1993). This research suggests that HPC and MC films might be used to protect fatty foods against moisture transmission and oxidative rancidity. Cellulose gum films are stiffer and more brittle than HPC, MC and HPMC films but have superior oxygen-barrier properties. Plasticizers, such as polyethylene glycol, glycerin or propylene glycol, may be used to increase film flexibility. HPC is used as a coating for candies to prevent color leaching; these candies are used as inclusions in a frosting. In coating and cast film applications, low-molecular-weight cellulose derivatives are used to maximize the polymer solids in solution and reduce the amount of water to be removed.

Ethyl cellulose is an effective film former. Changes in the molecular weight affect the film tensile strength, elongation and flexibility, while the DS changes the softening point and solubility in ethanol. EC is particularly useful as a moisture barrier and is an excellent film former; it has been used to reduce moisture migration in foods prepared in two stages, such as pizzas and snacks. Coatings of EC also may be used to stabilize flavors and protect nutritional ingredients against active interactions, hydrolysis and oxidation and to retard the release of active ingredients by encapsulation.

6.5 FUTURE DEVELOPMENTS

As eating patterns and expectations change over time, new market trends emerge. In addition, continually increasing competition in the marketplace results in increased cost pressures on food manufacturers. The formulator must create products to meet both demands. Currently, healthy eating as well as easy-to-prepare meals interest consumers, while cost control challenges manufacturers.

The benefits of whole grains are of increasing interest for a healthy diet. Bakery formulators will need ways to maintain softness and texture that consumers appreciate while incorporating whole grain ingredients in their products. Meatless alternatives have become mainstream and formulations should satisfy vegetarians and nonvegetarians alike; texture will play a key role in these products. Although there is increasing awareness of portion control, remnants of the 'low and no' alternatives remain. High-quality, no- and low-fat alternatives in salad dressings and low- or no-calorie beverages, for example, have become expected on grocery store shelves. Cellulose derivatives play important roles in such formulations.

Convenience is here to stay. Easy-to-prepare meals that are tasty and healthy will reap rewards in the marketplace. Consumers are willing to pay a premium for such products. The areas where cellulose derivatives play a role are in water control in multiphase systems, freeze–thaw stability and maintaining texture over product shelf life.

And finally, cost control will always be a driver for manufacturers. Cellulose derivatives have long offered economic benefits in the reduction or replacement of more expensive ingredients, such as fats and sugar solids. It is clear that cellulose derivatives, and hydrocolloids in general, play essential roles in current food applications and will continue to enable formulators to create new products in the future.

References

Armero, E. and Collar, C. (1996) Antistaling additive effects on fresh wheat bread quality. *Food Science and Technology International*, **2**(5), 323–333.

Armero, E. and Collar, C. (1998) Crumb firming kinetics of wheat breads with anti-staling additives. *Journal of Cereal Science*, **28**(2), 165–174.

Boukerchi, S.A. (1995) *Evaluation of Different Types of Gums on the Texture and the Oil's Absorption in Instant Noodle Application*. Master's Thesis, Chemical Process and Biotechnology, Singapore Polytechnic.

Bourne, M. C. (2002) *Food Texture and Viscosity*, 2nd ed. Academic Press, New York.

Cato, L., Gan, J.J., Rafeal, L.G.B. and Small, D.M. (2004) Gluten free breads using rice flour and hydrocolloid gums. *Food Australia*, **56**(3), 75–78.

Chan, E.S. and Zhang, Z. (2002) Encapsulation of probiotic bacteria *Lactobacillus acidophilus* by direct compression. *Food Bioproducts and Processing*, **80**(Part C), 78–82.

Coffey, D.G., Bell, D.A. and Henderson, A. (1995) Cellulose and cellulose derivatives. In: A.M. Stephen (ed), *Food Polysaccharides and Their Applications*. Marcel Dekker, New York, pp. 123–153.

Collar, C., Martinez, J.C. and Rosell, C.M. (2001) Lipid binding of fresh and stored formulated wheat breads. Relationships with dough and bread technological performance. *Food Science and Technology International*, **7**(6), 501–510.

Friend, C.P., Waniska, R.D. and Rooney, L.W. (1993) Effects of hydrocolloids on processing and qualities of wheat tortillas. *Cereal Chemistry*, **70**(3), 252–256.

Ganz, J. (1974) How cellulose gum reacts with proteins. *Food Engineering*, **46**, 67–69.

Gomez, M.H., Rooney, L.W. and Waniska, R.D. (1987) Dry corn masa flours for tortilla and snack food production. *Cereal Foods World*, **32**(5), 372–377.

Grover, J.A. (1993) Methyl cellulose and its derivatives. In: R.L. Whistler and J.N. BeMiller (eds), *Industrial Gums Polysaccharides and Their Derivatives*, 3rd ed. Academic Press, Inc., New York, pp. 475–504.

Gujral, H.S., Sharma, A. and Singh, N. (2002) Effect of hydrocolloids, storage temperature and duration on consistency in tomato ketchup. *International Journal of Food Properties*, **5**(1), 179–191.

Hague, A. and Morris, E.R. (1993) Thermogelation of methylcellulose. Part 1: Molecular structures and processes. *Carbohydrate Polymers*, **22**, 161–173.

Hargarten, P.G., Joenke, J. and Cavalline, V. (2005) Vegetable protein and meat analogue. European Patent Application EP 1,493,337 A2, 4 pp.

Hercules Incorporated (2005a) Acidified dairy beverages stabilized with Blanose® cellulose gum, Bulletin VC635, 3 pp.

Hercules Incorporated (2005b) AeroWhip™ whip-optimized solutions for use in whipping cream, Bulletin VC622B, 8 pp.

Hoefler, A.C. (2004) *Hydrocolloids*. Eagan Press, St. Paul.

Howse, G.T., Sihu, K. and Grex, D.D. (2005) Methods of preparing meat analogues and food stuffs comprised of meat analogues. US patent application US 2005/0008758 A1.

Kaletunic-Gencer, G. and Peleg, M. (1986) Rheological characteristics of selected food gum mixtures in solution. *Journal of Texture Studies*, **17**, 61–70.

Kuntz, L.A. (1995) Building better fried foods. *Food Product Design*, **5**, 129–145.

Naruenartwongsakul, S., Chinnen, M.S., Bhumiratana, S. and Yoovidhya, T. (2004) Influence of methylcellulose on oil absorption of wheat batter coated cut potatoes. *Journal of Food Processing Preservation*, **28**, 223–239.

Pangborn, R.M., Gibbs, Z.M. and Tssan, C. (1978) Effect of hydrocolloids on apparent viscosity and sensory properties of selected beverages. *Journal of Food Texture Studies*, **9**, 415–436.

Park, H.J., Weller, C.L., Vergano, P.J. and Testin, R.F. (1993) Permeability and mechanical properties of cellulose-based edible films. *Journal of Food Science*, **58**(6), 1361–1364.

Qarooni, J. (1993) Wheat flour tortillas. *AIB Research Department Technical Bulletin*, **XV**(5), 7.

Regand, A. and Goff, H.D. (2002) Effects of biopolymers on structure and ice recrystalization in dynamically frozen ice cream model systems. *Journal of Dairy Science*, **85**, 2722–2732.

Sanz, T., Fernandez, M.S., Salvadoe, A., Munoz, J. and Fiszman, S.M. (2005) Thermalgelation properties of methylcellulose (MC) and their effect on batter formulation. *Food Hydrocolloids*, **19**, 141–147.

Serna-Saldivar, S.O., Gomez, M.H. and Rooney, L.W. (1990) The chemistry, technology and nutritional value of alkaline-cooked corn products. *Advances in Cereal Science and Technology* **10**, 243–307.

Serna-Saldivar, S.O., Rooney, L.W. and Waniska, R.D. (1988) Wheat flour tortilla production. *Cereal Foods World*, **33**, 856–863.
Sharadanant, R. and Khan, K. (2003) Effect of hydrophilic gums on quality of frozen dough: II. Bread characteristics. *Cereal Chemistry*, **80**(6), 773–780.
Sidhu, J.P.S. and Bawa, A.S. (2000) Incorporation of carboxymethylcellulose in wheat flour: rheological, alveograph, dough development, gas formation/retention, baking and bread firmness studies. *International Journal of Food Properties*, **3**(3), 407–419.
Sudhakar, V., Singhal, R.S. and Kulkarni, P.R. (1992) Starch–gum interaction: formulations and functionality using Amaranth/Corn Starch and CMC. *Starch/Stärke*, **44**(10), 369–374.
Szczesniak, A.S. and Farkas, E. (1962) Objective characterization of the mouthfeel of gum solutions. *Journal of Food Science*, **27**, 381–385.
Torres, P.I., Ramirez-Wong, B., Serna-Saldivar, S.O. and Rooney, L.W. (1993) Effects of sorghum flour addition on the characteristics of wheat flour tortillas. *Cereal Chemistry*, **70**(1), 8–13.
Wang, S.T., Barringer, S.A. and Hansen, P.M.T. (1998) Effects of carboxymethylcellulose and guar gum on ice crystal propagation in a sucrose-lactose solution. *Food Hydrocolloids*, **12**, 211–215.
Zecher, D. and Gerrish, T. (1997) Cellulose derivatives. In: A. Imeson (ed), *Thickening and Gelling Agents for Food*. Blackie Academic & Professional, London, pp. 60–85.

7 Gelatine

Paul Stevens

ABSTRACT

Gelatine is a proteinaceous material obtained from animal connective tissue using hydrolysis in acidic (type A) or basic (type B) solution followed by hot water extraction. Gelatine is characterised according to the 'Bloom' gel strength and extraction process. It hydrates readily in warm or hot water to give low-viscosity solutions that have good whipping and foaming properties. Concentrated solutions containing up to 40% gelatine be made for use in confectionery. After cooling, the network of polypeptide chains associates slowly to form clear, elastic gels that are syneresis free. The slow setting rate enables processes such as aeration and depositing to be completed before gelation commences. The thermoreversibility of gelatine gels gives them unique properties: the melting point is below 37°C so they melt in the mouth to give smooth textures with excellent flavour release. Gelatine is used in a wide range of food, pharmaceutical and photographic applications. Food products include jelly candies and aerated confectionery, yogurt and other cultured desserts, dairy desserts and creams, low-fat spreads, canned meat products, water dessert gels and decorative aspics.

7.1 INTRODUCTION

The history of gelatine goes back 4000 years, to the Egyptians, when wooden objects were bound with glues made from gelatinous collagen (Delarosa, 2005). The collagen compound from animal connective tissue such as skin, bone and sinews was converted into a soluble, gel-like material.

Through the years, mankind discovered that gelatine is a multifunctional protein molecule, used as an ingredient for the food, pharmaceutical and photographic industries in a wide range of applications.

The photographic industry initiated research on gelatine. Sheppard *et al.* (1921, 1932) studied the rheological properties of gelatine and, in particular, the influence of pH, temperature and the presence of certain ions on the degradation of gelatine and collagen molecules.

The manufacturing processes were also found to have a large impact on gelatine characteristics. Research by Ames (1952) clearly made the distinction between two major manufacturing processes: the acid or A-type process, where the raw materials are pre-treated in acid, and the alkaline or B-type (basic) process, where the raw materials are pre-treated in an alkaline solution.

The basic characteristics of gelatine, such as gel strength, viscosity and isoelectric point (IEP), are mainly determined by which process is used. Gel strength is primarily determined by the proportion of proline and hydroxyproline present in the total amino acids. A high content of these two amino acids indicates a high gelling power (Eastoe and Leach, 1958). Viscosity is primarily determined by the molecular weight distribution of the gelatine. A high molecular weight is linked to a high viscosity (Pouradier and Venet, 1952). The alkaline process results in IEP values between 4.5 and 5.5, and the IEP from an acid process can vary from 6 to 9.5 (Ames, 1952). These three characteristics are important in determining the appropriate type of gelatine required for a particular application.

Religion can also determine the type of gelatine used for certain applications. Food products consumed by strict adherents to the Muslim and Jewish faiths can only contain Halal and Kosher gelatines, respectively. This means that, as a general rule, porcine gelatines are excluded and most of the gelatine used in these circumstances will be of bovine origin.

In the early days, most of the studies on gelatine and collagen were made by the gelatine industry in order to understand its behaviour in different applications. Nowadays, the importance of scientific research on collagen and gelatine has been brought to the fore again as medical research studies have shown that collagen and gelatine play an important role in our metabolism.

7.2 MANUFACTURING PROCESS

7.2.1 Raw materials

Gelatine is obtained by hydrolysis of the collagen that is contained in animal connective tissues. Commercially, skins or bones of different animal species, such as beef, pork, fish and poultry, form the main raw material for gelatine production. These raw materials are collected from animals approved for human consumption by ante- and post-mortem veterinary inspection. Skins can be frozen if there is a long delay between sourcing and eventual use in the gelatine manufacturing process. Some raw materials, mainly bones, are pre-treated by grinding, degreasing and drying before they are used in gelatine production.

7.2.2 Production processes

Two main processes are used to extract gelatine from the raw material: the acid process for gelatine type A and the alkaline process for gelatine type B. Some other processes can also be used, but these are on a smaller scale.

The first step of the process is to reduce the size of raw material, if needed, by cutting or grinding. For bones, the grinding operation is done during the degreasing operation to reach a bone chip size from 2 to 25 mm. Skins are reduced to 30–150 mm. The ground or cut pieces of raw material are washed with clear water for some hours before any further specific treatment begins. For bones, this washing is done after the degreasing operation.

Degreased bone chips are pre-treated by hydrochloric acid to remove the mineral matter, calcium phosphate, inside the bone. At the end of this operation, only organic matter containing gelatine remains. This is called ossein.

7.2.2.1 *Acid process*

An acid pre-treatment on skins or bones is affected by adding the raw material to an acid bath at ambient temperature. The acidulation time and the concentration of acid vary with the type of raw material used. Different acids, including hydrochloric acid, phosphoric acid, sulphuric acid or even organic acids, may be used.

7.2.2.2 *Alkaline process*

The second type of process consists of a pre-treating the raw material with an alkaline chemical agent. Lime or sodium hydroxide solutions are usually used for this operation. Depending on the alkali and the concentration used, the treatment time takes from 30 days to more than 70 days at ambient temperature, which is normally less than 18°C.

After the chemical pre-treatment, the raw materials are washed to remove all excess acid or alkali and adjust the pH for extraction. Following pH adjustment, the raw material is put into a cooking tank with hot water, for continuous or batch extractions, to render the gelatine soluble. In batch processes, different extracts are removed as temperatures are progressively increased from 50°C to 90°C over 3–6 h.

The solution from the cooking tanks contains around 4–7% gelatine with some fines of bones or skins. These physical impurities are removed by filtration, including earth, cardboard or membrane filters, or by centrifugation. After removing solids, the salt content is reduced by passing the gelatine solution through ion-exchange resins. This operation removes the cations and anions of mineral salts contained in the gelatine. This step can also be done by specific ultra-filtration treatment.

After purification, the gelatine solution is concentrated to 30–50% dry matter using vacuum multi-effect evaporators and/or membrane technologies, depending on the viscosity of the solution. Before being gelled, the solution is treated in an ultra-high temperature (UHT) system at a minimum temperature of 138°C for 4 s to guarantee the bacteriological properties of the final product.

The concentrated solution is cooled in a votator and extruded in the form of gelatine 'noodles'. The noodles are spread on the conveyor of a continuous belt dryer and the water is removed through hot air flow to a moisture content of around 10%. After drying, the gelatine is ground and then homogenised, which results in semi-finished lots of dry gelatine. At this point, all the physical, chemical and bacteriological properties are checked in the laboratory. Different lots of semi-finished gelatine are blended and packaged for shipment and delivery.

7.2.3 Commercial products

Gelatine is available in sizes ranging from a fine powder to coarse granules. It is also available in the form of sheets or flakes. In addition to the description of its physical appearance, gelatine is sold commercially according to its major characteristic of gel strength, or 'Bloom'. It is the Bloom strength, the raw material and manufacturing process that determine the choice of gelatine used in applications. Indeed, as gelatine can be obtained from a wide range of raw materials (skin or bones of pork, cattle, fish and poultry) and from two distinct processes (acid or alkaline), several combinations are possible, resulting in different types of gelatine characterised by the ratio of Bloom to viscosity and the IEP. Table 7.1 provides an overview.

Table 7.1 Overview of gelatine types, production processes and product properties.

	Raw material pig skin	**Fish skin**	**Bovine bone**	**Pig bone**	**Pig bone**	**Bovine hide**	**Bovine hide**	**Poultry**
Process	Acid	Acid	Alkaline	Alkaline	Acid	Alkaline	Acid	Acid
Duration	~1 week	~1 week	3 months	3 months	~1 week	3 months	~1 week	~1 week
Type of gelatine	A	A	B	B	A	B	A	A
Bloom (g)	100–260	200–270	100–250	100–250	100–250	100–250	100–240	100–240
Viscosity (mPa s)	1.8–4.8 + hydrolysed	2.8–4.5 + hydrolysed	2.8–5.2	2.8–5.2	1.8–3.2	2.0–5.0 + hydrol-ysed	1.8–3.2	0.75–3.5
Isoelectric point	>7.5	7.0–9.5	4.7–5.2	4.7–5.2	6.5–8.5	4.7–5.2	5.0–7.0	>8.0
Main usages	Food, pharma	Food, pharma	Photo, pharma	Photo, food, pharma	Food, pharma	Food, pharma	Food, pharma	Food, pharma

It is to be noted that for each gelatine type, the higher the Bloom, the higher the viscosity. However, at the same Bloom value, type B gelatine shows a higher viscosity than type A gelatine.

Depending on the final use of gelatine, other characteristics and quality criteria can be required, such as colour, pH, moisture content and microbiological limits. Moreover, to comply with current standards and regulations, food and pharmaceutical gelatines are screened for specific items such as ash content, conductivity, heavy metals and the presence of particular microorganisms.

7.3 REGULATIONS: EUROPEAN UNION AND THE USA

Effective from 1 January 2006, edible gelatine is regulated in Europe by Section XIV of Annex III of Regulation (EC) No. 853/2004, recently updated in Regulation (EC) No. 1243/2007, which details specific hygiene rules for food of animal origin. However, the provisions of this new regulation do not vary greatly from its predecessor, which has been in place since 1999.

This European regulation identifies the raw materials which can be used for gelatine manufacture as bones, hides and skins of farmed ruminant animals, pig skins, poultry skins, tendons and sinews, wild game hides and skins, fish skins and bones. All these raw materials must come from animals deemed fit for human consumption after veterinary inspection. Traceability with the relevant documentation must be kept during transport and storage.

The gelatine process is described as acid or alkali treatment followed by one or more rinses, pH adjustment, extraction by heating and purification by filtration and sterilisation. For bovine bones, in most cases, the gelatine must be alkali treated for more than 20 days above pH 12.5 or treated with acid (pH < 3.5) for 10 h minimum. In both cases, the gelatine is then sterilised at 138°C for 4 s minimum. Regulation (EC) No. 853/2004 amended in (EC) No. 1243/2007 provides some additional requirements in terms of maximum levels for heavy metals, sulphur dioxide and hydrogen peroxide. *Salmonella* must also be absent in 25 g. Gelatine packaging must bear the words 'gelatine fit for human consumption' and must indicate the 'best before date'.

In the USA, the Food Chemical Codex (FCC) specification is similar to the European regulations: raw materials from connective tissues of animals, including fish and poultry, extracted by acid, alkali or enzymatic hydrolysis processes. Further requirements for bovine gelatine are detailed by the Interim Final Rule, 21 CFR Parts 189 and 700 on the use of materials derived from cattle in human food and cosmetics of 14 July 2004. These requirements prohibit the use of some raw materials such as skull, spinal cord and vertebrae of cattle over 30 months. The FCC also specifies maximum limits for heavy metals and sulphur dioxide. *Escherichia coli* and *Salmonella* must be absent in 25 g. There is also a specific limit for pentachlorophenol of 'not more than 0.3 mg/kg'.

For gelatine to be used as a food ingredient, it must comply with several additional international standards. Most of the international regulations set limits on chemical and microbiological characteristics. It is accepted worldwide that gelatine must comply with the limits shown in Tables 7.2 and 7.3.

Testing methods for these parameters are not gelatine specific and therefore it is recommended to use internationally accepted standard methods (ISO, AOAC, USP).

Table 7.2 Chemical characteristics of food-grade gelatine.

Chemical characteristics	Upper limits
Arsenic (As)	1 ppm
Lead (Pb)	5 ppm
Cadmium (Cd)	0.5 ppm
Mercury (Hg)	0.15 ppm
Chromium (Cr)	10 ppm
Cupper (Cu)	30 ppm
Zinc (Zn)	50 ppm
Moisture (105°C)	15%
Ash (550°C)	2%
SO_2	50 ppm
H_2O_2	10 ppm

7.4 CHEMICAL STRUCTURE AND REACTIVITY

Gelatine is obtained by thermal denaturation of collagen, the main fibrous component of skin, bones, sinews and cartilage (Ramachandran, 1967; Ward, 1977). Collagen is an extra-cellular protein containing three helical chains of polypeptides, each containing 1000 amino acids (see Fig. 7.1). The sequence of amino acids is very regular with glycine present in every three residues.

Several kinds of collagen exist and are well described as a precise sequence of amino acids (Ward, 1977). An example of collagen I from calf skin is given in Fig. 7.2.

The extracted gelatine is a group of molecules of different molecular weight (Fig. 7.3). The molecular weight profile depends on the process.

The crucial key in the chemistry of gelatine is the amino acid profile, given in Table 7.4, in particular the content of arginine, aspartic acid/asparagine, glutamic acid/glutamine, histidine, lysine, hydroxylysine and hydroxyproline. The amino acid profile determines hydrogen bond formation and reactivity via side groups such as amine, imidazole, alcohol, amide and carboxylic acid.

7.4.1 Hydrogen bonds

Hydrogen bonds (Cooper, 1971; Finer, 1975; Engel *et al.*, 1977; Ledward, 1986) play a predominant role in gelatine structure; they stabilise the collagen and are involved in the gel-forming process, for example in the shrinkage of tissues.

Table 7.3 Microbiological limits for food-grade gelatine.

Microbiological characteristics	Upper limits
Total aerobic count	10^3 per g
Escherichia coli	0 per 10 g
Anaerobic sulphite-reducing bacteria (no gas production)	10 per g
Salmonella	0 per 25 g

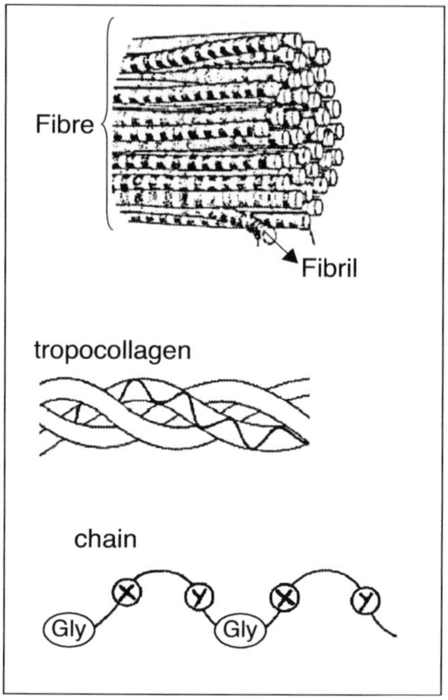

Fig. 7.1 Units of collagen/gelatine.

pGlu-Met-Ser-Tyr-Gly-Tyr-Asp-Glu-Lys-Ser-Ala-Gly-Val-Ser-Val-Pro-

1 Gly-Pro-Met-Gly-Pro-Ser-Gly-Pro-Arg-Gly-Leu-Hyp-Gly-Pro-Hyp-Gly-Ala-Hyp-Gly-Pro-Gln-Gly-Phe-Gln-Gly-Pro-Hyp-
28 Gly-Glu-Hyp-Gly-Glu-Hyp-Gly-Ala-Ser-Gly-Pro-Met-Gly-Pro-Arg-Gly-Pro-Hyp-Gly-Pro-Hyp-Gly-Lys-Asn-Gly-Asp-Asp-
55 Gly-Glu-Ala-Gly-Lys-Pro-Gly-Arg-Hyp-Gly-Gln-Arg-Gly-Pro-Hyp-Gly-Pro-Gln-Gly-Ala-Arg-Gly-Leu-Hyp-Gly-Thr-Ala-
82 Gly-Leu-Hyp-Gly-Met-Hyl-Gly-His-Arg-Gly-Phe-Ser-Gly-Leu-Asp-Gly-Ala-Lys-Gly-Asn-Thr-Gly-Pro-Ala-Gly-Pro-Lys-
109 Gly-Glu-Hyp-Gly-Ser-Hyp-Gly-Glx-Asx-Gly-Ala-Hyp-Gly-Gln-Met-Gly-Pro-Arg-Gly-Leu-Hyp-Gly-Glu-Arg-Gly-Arg-Hyp-
136 Gly-Pro-Hyp-Gly-Ser-Ala-Gly-Ala-Arg-Gly-Asp-Asp-Gly-Ala-Val-Gly-Ala-Ala-Gly-Pro-Hyp-Gly-Pro-Thr-Gly-Pro-Thr-
163 Gly-Pro-Hyp-Gly-Phe-Hyp-Gly-Ala-Ala-Gly-Ala-Lys-Gly-Glu-Ala-Gly-Pro-Gln-Gly-Ala-Arg-Gly-Ser-Glu-Gly-Pro-Gln-
190 Gly-Val-Arg-Gly-Glu-Hyp-Gly-Pro-Hyp-Gly-Pro-Ala-Gly-Ala-Ala-Gly-Pro-Ala-Gly-Asn-Hyp-Gly-Ala-Asp-Gly-Gln-Hyp-
217 Gly-Ala-Lys-Gly-Ala-Asn-Gly-Ala-Hyp-Gly-Ile-Ala-Gly-Ala-Hyp-Gly-Phe-Hyp-Gly-Ala-Arg-Gly-Pro-Ser-Gly-Pro-Gln-
244 Gly-Pro-Ser-Gly-Ala-Hyp-Gly-Pro-Lys-Gly-Asn-Ser-Gly-Glu-Hyp-Gly-Ala-Hyp-Gly-Asn-Lys-Gly-Asp-Thr-Gly-Ala-Lys-
271 Gly-Glu-Hyp-Gly-Pro-Ala-Gly-Val-Gln-Gly-Pro-Hyp-Gly-Pro-Ala-Gly-Glu-Glu-Gly-Lys-Arg-Gly-Ala-Arg-Gly-Glu-Hyp-
298 Gly-Pro-Ser-Gly-Leu-Hyp-Gly-Pro-Hyp-Gly-Glu-Arg-Gly-Gly-Hyp-Gly-Ser-Arg-Gly-Phe-Hyp-Gly-Ala-Asp-Gly-Val-Ala-
325 Gly-Pro-Lys-Gly-Pro-Ala-Gly-Glu-Arg-Gly-Ser-Hyp-Gly-Pro-Ala-Gly-Pro-Lys-Gly-Ser-Hyp-Gly-Glu-Ala-Gly-Arg-Hyp-
352 Gly-Glu-Ala-Gly-Leu-Hyp-Gly-Ala-Lys-Gly-Leu-Thr-Gly-Ser-Hyp-Gly-Ser-Hyp-Gly-Pro-Asp-Gly-Lys-Thr-Gly-Pro-Hyr-
379 Gly-Pro-Ala-Gly-Gln-Asp-Gly-Arg-Hyp-Gly-Pro-Ala-Gly-Pro-Hyp-Gly-Ala-Arg-Gly-Gln-Ala-Gly-Val-Met-Gly-Phe-Hyp-
406 Gly-Pro-Lys-Gly-Ala-Ala-Gly-Glu-Hyp-Gly-Lys-Ala-Gly-Glu-Arg-Gly-Val-Hyp-Gly-Pro-Hyp-Gly-Ala-Val-Gly-Pro-Ala-
433 Gly-Lys-Asp-Gly-Glu-Ala-Gly-Ala-Gln-Gly-Pro-Hyp-Gly-Pro-Ala-Gly-Pro-Ala-Gly-Glu-Arg-Gly-Glu-Gln-Gly-Pro-Ala-
460 Gly-Ser-Hyp-Gly-Phe-Gln-Gly-Leu-Hyp-Gly-Pro-Ala-Gly-Pro-Hyp-Gly-Glu-Ala-Gly-Lys-Hyp-Gly-Glu-Gln-Gly-Val-Hyp-
487 Gly-Asp-Leu-Gly-Ala-Hyp-Gly-Pro-Ser-Gly-Ala-Arg-Gly-Glu-Arg-Gly-Phe-Hyp-Gly-Glu-Arg-Gly-Val-Glu-Gly-Pro-Hyp-
514 Gly-Pro-Ala-Gly-Pro-Arg-Gly-Ala-Asn-Gly-Ala-Hyp-Gly-Asn-Asp-Gly-Ala-Lys-Gly-Asp-Ala-Gly-Ala-Hyp-Gly-Ala-Hyp-
541 Gly-Ser-Gln-Gly-Als-Hyp-Gly-Leu-Gln-Gly-Met-Hyp-Gly-Glu-Arg-Gly-Ala-Ala-Gly-Leu-Hyp-Gly-Pro-Lys-Gly-Asp-Arg-
568 Gly-Asp-Ala-Gly-Pro-Lys-Gly-Aln-Asp-Gly-Ala-Pro-Gly-Lys-Asp-Gly-Val-Arg-Gly-Leu-Thr-Gly-Pro-Lle-Gly-Pro-Hyp-
595 Gly-Pro-Ala-Gly-Ala-Hyp-Gly-Asp-Lys-Gly-Glu-Ala-Gly-Pro-Ser-Gly-Pro-Ala-Gly-Thr-Arg-Gly-Ala-Hyp-Gly-Asp-Arg-
622 Gly-Glu-Hyp-Gly-Pro-Hyp-Gly-Pro-Ala-Gly-Phe-Ala-Gly-Pro-Hyp-Gly-Ala-Asp-Gly-Gln-Hyp-Gly-Ala-Lys-Gly-Glu-Hyp-
649 Gly-Asp-Ala-Gly-Ala-Lys-Gly-Asp-Ala-Gly-Pro-Hyp-Gly-Pro-Ala-Gly-Pro-Ala-Gly-Pro-Hyp-Gly-Pro-Ile-Gly-Asn-Val-
676 Gly-Ala-Hyp-Gly-Pro-Hyl-Gly-Ala-Arg-Gly-Ser-Ala-Gly-Pro-Hyp-Gly-Ala-Thr-Gly-Phe-Hyp-Gly-Ala-Ala-Gly-Arg-Val-
703 Gly-Pro-Hyp-Gly-Pro-Ser-Gly-Asn-Ala-Gly-Pro-Hyp-Gly-Pro-Hyp-Gly-Pro-Ala-Gly-Lys-Glu-Gly-Ser-Lys-Gly-Pro-Arg-
730 Gly-Glu-Thr-Gly-Pro-Ala-Gly-Arg-Hyp-Gly-Glu-Val-Gly-Pro-Hyp-Gly-Pro-Hyp-Gly-Pro-Ala-Gly-Glu-Lys-Gly-Ala-Hyp-
757 Gly-Als-Asp-Gly-Pro-Ala-Gly-Ala-Hyp-Gly-Thr-Pro-Gly-Pro-Gln-Gly-Ile-Ala-Gly-Glr-Arg-Gly-Val-Val-Gly-Leu-Hyp-
784 Gly-Gln-Arg-Gly-Glu-Arg-Gly-Phe-Hyp-Gly-Leu-Hyp-Gly-Pro-Ser-Gly-Glu-Hyp-Gly-Lys-Gln-Gly-Pro-Ser-Gly-Ala-Ser-
811 Gly-Glu-Arg-Gly-Pro-Hyp-Gly-Pro-Met-Gly-Pro-Hyp-Gly-Leu-Ala-Gly-Pro-Hyp-Gly-Glu-Ser-Gly-Arg-Glu-Gly-Ala-Hyp-
838 Gly-Ala-Glu-Gly-Ser-Hyp-Gly-Arg-Asp-Gly-Ser-Hyp-Gly-Ala-Lys-Gly-Asp-Arg-Gly-Glu-Thr-Gly-Pro-Ala-Gly-Ala-Hyp-
865 Gly-Pro-Hyp-Gly-Ala-Hyp-Gly-Ala-Hy-Gly-Pro-Val-Gly-Pro-Ala-Gly-Lys-Ser-Gly-Asp-Arg-Gly-Glu-Thr-Gly-Pro-Ala-
892 Gly-Pro-Ile-Gly-Pro-Val-Gly-Pro-Ala-Gly-Ala-Arg-Gly-Pro-Ala-Gly-Pro-Gln-Gly-Pro-Arg-Gly-Asx-Hyp-Gly-Glx-Thr-
919 Gly-Glx-Glx-Gly-Asx-Arg-Gly-Ile-Hyl-Gly-His-Arg-Gly-Phe-Ser-Gly-Leu-Gln-Gly-Pro-Hyp-Gly-Pro-Hyp-Gly-Ser-Hyp-
946 Gly-Glu-Gln-Gly-Pro-Ser-Gly-Ala-Ser-Gly-Pro-Ala-Gly-Pro-Arg-Gly-Pro-Hyp-Gly-Ser-Ala-Gly-Ser-Hyp-Gly-Lys-Asp-
973 Gly-Leu-Asn-Gly-Leu-Hyp-Gly-Pro-Ile-Gly-Hyp-Hyp-Gly-Pro-Arg-Gly-Arg-Thr-Gly-Asp-Ala-Gly-Pro-Ala-Gly-Pro-Hyp-
1000 Gly-Pro-Hyp-Gly-Pro-Hyp-Gly-Pro-Hyp-Gly-Pro-Pro-

Fig. 7.2 Collagen I extracted from calf skin (Ward, 1977).

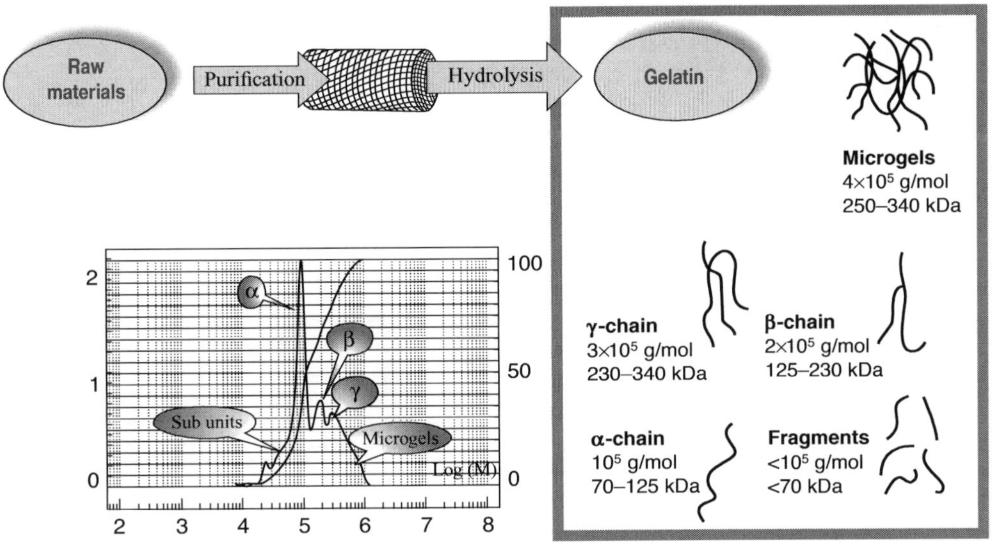

Fig. 7.3 Gelatine is a group of different molecular weight molecules.

Collagen is insoluble in water and its hydrolysis results in gelatine. The thermal denaturation of collagen takes place by heating the collagen after an acid or alkali treatment. Then the fibres and fibrils dissociate into tropocollagen units, mainly by the loss of hydrogen bonds. Now it is commonly acknowledged that the thermal stability of the different collagens is closely dependent on the total amount of pyrrolidine (proline and hydroxyproline). However, the role of hydrogen bonds in the stability of collagen is arguable (Cooper, 1971).

In tropocollagen, the presence of proline residues gives the peptide chain a helical structure that differs from the α-helix of globular proteins by the lack of hydrogen bonds inside the α-chain. However, glycine residues are located inside the α-chain and are able to create hydrogen bonds.

X-ray studies by Bella (1998) showed that the formation of hydrogen bonds involves the participation of water as well as the hydroxyproline groups (Fig. 7.4). Water plays a critical role in maintaining the conformation of collagen molecules and the mechanical properties of collagen fibrils.

The thermo-reversibility of gelatine gels makes them unique. They have a melting point below 37°C, which means that they melt in the mouth and are easily dissolved. The gelatine gel is a network of polypeptide chains with junction zones (Fig. 7.5). In a heated solution of gelatine, the triple helices are widely disorganised. Cooling this solution results in the nucleation of helical regions. Gelation is the consequence of a partial return to the collagen-like triple helix structures. It is generally accepted that the junction zones (triple helices) are stabilised by inter-chain hydrogen bonds similar to those found in native collagen.

7.4.2 NH_2 reactivity

The reactivity of gelatine increases with the level of lysine and hydroxylysine because the ε-amino group can react as a traditional amine (Fig. 7.6). The amino groups from histidine

Table 7.4 HPLC amino acid profiles (S. Guedj, Rousselot, personal communication).

Amino acid	Reactivity	Code	Molecular weight (free AA)	Side group	Amino acid (AA) residues (g/ 100 g protein)		
					Alkaline pig bones	Alkaline bovine hide	Alkaline bovine bones
Alanine	nH	Ala	89.09		8.0	8.3	8.5
Arginine	B	Arg	174.20	–NH–C(=NH)–NH_2	8.2	8.1	7.7
Aspartic acid/asparagine	A/P	Asp/Asn	133.10	–COOH/–$CONH_2$	5.5	5.4	5.2
Cysteine	P	Cys	121.16	–SH	n.d.	n.d.	n.d.
Glutamic acid/glutamine	A/P	Glu/Gln	147.13	–COOH/–$CONH_2$	11.7	11.2	9.9
Glycine	P/nH	Gly	75.07		20.2	19.3	21.2
Histidine	B	His	155.16	Imidazole	1.0	1.4	1.0
Hydroxylysine	B	OH-L	162.19	–NH_2 et –OH	1.1	0.5	1.0
Hydroxyproline	P	OH-P	131.13	–OH	12.5	12.1	12.3
Isoleucine	nH	Ile	131.18		1.3	1.6	1.3
Leucine	nH	Leu	131.18		2.9	2.8	2.9
Lysine	B	Lys	146.19	–NH_2	3.6	3.6	3.6
Methionine	nH	Met	149.21	–S–CH_3	0.5	0.9	0.7
NH_3	B	NH_3	17.03		0.3	0.5	1.0
Ornithine[a]	B	Orn	132.16	–NH_2	0.5	0.3	0.4
Phenylalanine	nH	Phe	165.19	–Φ	2.0	2.0	2.0
Proline	nH	Pro	115.13		12.8	13.4	13.5
Serine	P	Ser	105.09	–OH	3.1	3.1	2.8
Threonine	P	Thr	119.12	–OH	2.0	2.0	1.8
Tryptophan	nH	Trp	204.23	indole	n.d.	n.d.	n.d.
Tyrosine	P	Tyr	181.19	–Φ–OH	0.3	0.4	0.4
Valine	nH	Val	117.15		2.4	3.2	2.6

A, acidic; B, basic; P, polar uncharged; nH, non-polar hydrophobic.
[a] Ornithine obtained by guanidine loss from arginine in the case of alkali-treated gelatines.

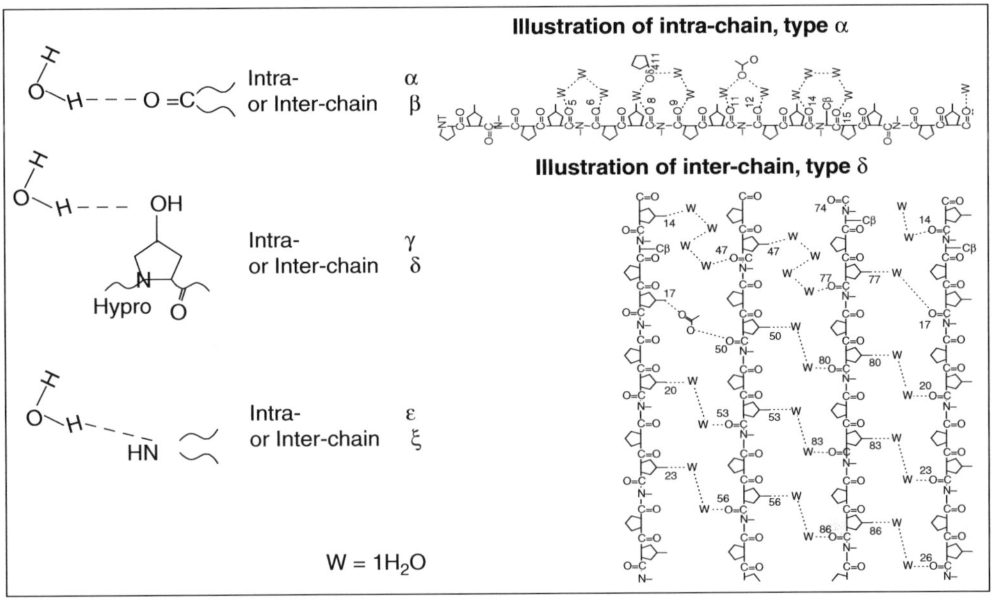

Fig. 7.4 Different models of hydrogen bonds involving water molecules and their role in cross-linking tropocollagen chains.

residues are generally thought to play a less important role than those on lysine. The highly basic guanidino group of arginine is protonated at neutral pH and is active only under some conditions. In general, the amines (Vollhardt, 1999) are slightly basic but may behave as weak acids. They are involved in hydrogen bonds and play the role of nucleophile in nucleophilic substitutions.

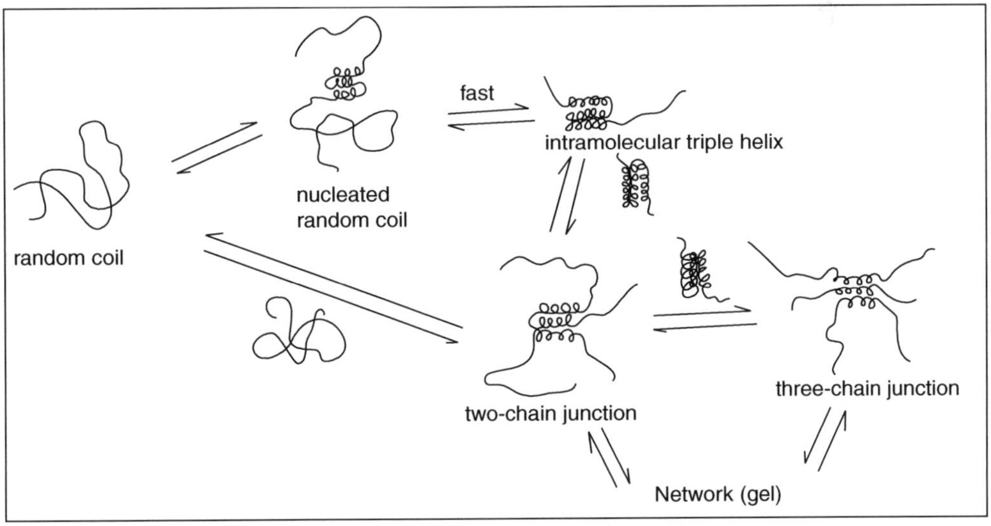

Fig. 7.5 Schematic representation of the possible conformational changes during the formation of a gel. (Reproduced from Finer, 1975, with permission from John Wiley & Sons, Inc., copyright 1975.)

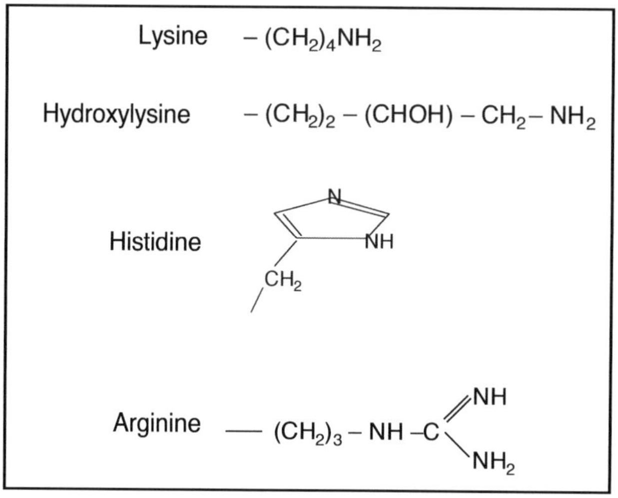

Fig. 7.6 Amino acids involved in NH_2 reactivity.

This reactivity is involved in the process of cross-linking and modifying gelatine. It is possible to inhibit the reactivity of amino groups by blocking them through a nucleophilic substitution and the most widely used is the modification of gelatine with anhydrides, as shown in Fig. 7.7.

Three categories of cross-links have been described by Guedj (2005). The first category of cross-links is involved in the mechanisms of ageing and the development of tissues whose manifestations are, for example, wrinkled skin, stiffened joints and shortened stature (Bailey, 1998).

The second category of cross-links concerns the reactions of amino groups with glucose or other sugars, known as glycation. Glucose or other sugars are included in formulations or are naturally present at trace levels in collagen. The aldehyde functional groups may react with lysine to lead to a series of products as shown in Fig. 7.8. In humans, these glycation reactions are enhanced in diabetic subjects due to hyperglycaemia and this process accelerates the ageing process.

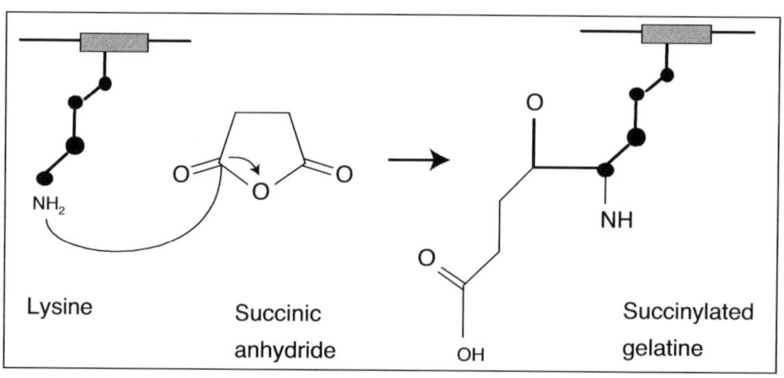

Fig. 7.7 Lysine modification with succinic anhydride.

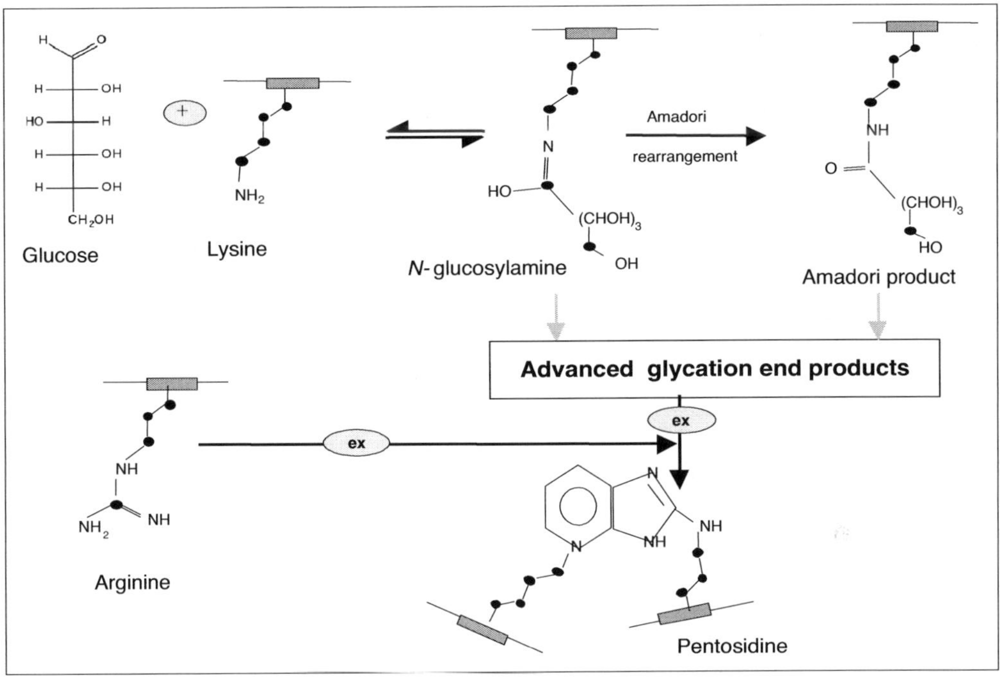

Fig. 7.8 Glycation cross-links, main steps (Bailey, 1998).

The reaction between sugars and amines is commonly named Maillard reactions and they are largely involved in colour development. Another problem linked to the presence of sugars in gelatine is the acceleration of insolubilisation (Marks, 1968).

Finally, the third kind of reaction concerns those that involve aldehydes (Taylor, 1978). The reactivity of amine has been used for a long time to improve some characteristics of films in photographic and pharmaceutical applications and to target some innovative applications, such as bio-adhesives and controlled-release capsules. Formaldehyde has the longest history because of its high reactivity and low cost. It is widely used to cross-link protein in the leather and photographic industries in order to increase the melting point and the mechanical strength of films. The reactions involved are not fully understood: it could take place either between two amino groups or between an amino group and some other group, such as guanidine, hydroxyl, amide and many components that can be formed during hardening. The presence of 10^{-4} mole of formaldehyde per gram of gelatine in a gelled and dried layer reduces its swelling in water from 600–800% to 200–300% at 20°C. The melting point of such a layer is increased to 100°C. This can be useful, or conversely, aldehydes could be considered a contaminant depending on the application.

Some other aldehydes, such as glutaraldehyde, are widely used as a cross-linking reagent for biomaterials (Tomihata, 2001). In fact, the efficiency of glutaraldehyde is due to its polymerisation which facilitates intermolecular reactions.

Other hardeners have been described, such as those listed by Pouradier (1977), and used in the photography industry, including *N*-methylol, diketones, sulfonate esters, triazines and aziridines. In addition to this list, other sophisticated cross-linking agents are being developed for medical applications. A Japanese team (Saito, 2004) developed a tissue adhesive by

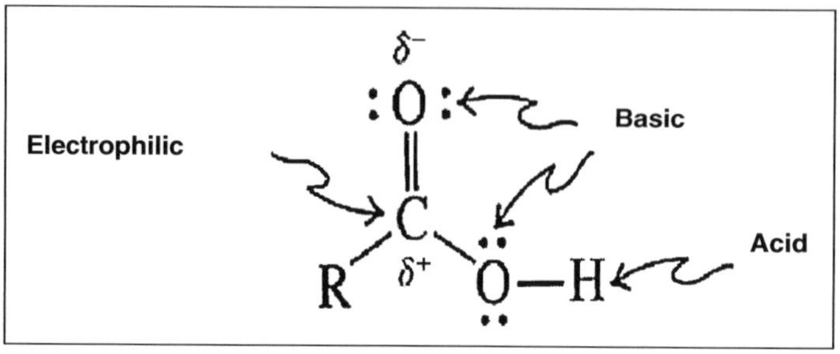

Fig. 7.9 Carboxylic acid site of reactivity.

cross-linking gelatine with a mixture of citric acid and *N*-hydroxy succinimide (NHS). Genipin (Liang, 2004) obtained via enzymatic hydrolysis from geniposide, isolated from the tropical fruit *Genipa america* and *Gardenia jasminoides Ellis*, which has been used in Chinese medicine for many years, reacts with primary amines to give a blue pigment. The genipin-cross-linked gelatine is being studied in microspheres as a drug carrier for intramuscular administration.

In conclusion, the reactivity of the amide groups allows a large range of cross-links and networks for simple and sophisticated applications.

7.4.3 COOH reactivity

The second reactive functional group in collagen is the carboxylic acid of aspartic and glutamic acids shown in Fig. 7.9 (Vollhardt, 1999). In collagen, the amount of carboxylic acid is very low, but an alkaline process (desamidation) transforms the aspargine and glutamine to the corresponding acids. Acid gelatines that have amide groups are less affected by this reactivity.

The process of chromium tanning involves the carboxylic acid function and is successfully employed throughout the leather industry; 90% of leather produced worldwide is chromium tanned (Covington, 2001). The addition of chromium salts increases the shrinkage temperature. In compliance with current regulations, the raw materials for gelatine are never hardened with chromium.

While chromium has the highest reactivity compared to other metals, other trivalent ions can react with gelatine. The carboxylic acid reactivity of collagen is largely used in the leather industry, but in the presence of gelatine, it represents a problem because of the interaction with dyes or their additives.

Another use of carboxylic acid reactivity is cross-linking with carbodiimide for biodegradable materials, anti-adhesive materials and artificial skins (Tomihata, 2001). Compared to modifications formed with aldehydes, this is more biosafe because it is possible to totally remove all of the reagent that does not react, and the secondary product is urea which is not toxic.

7.5 PHYSICOCHEMICAL PROPERTIES

Gelatine shows a wide range of physicochemical properties which confer great versatility and give a large number of functional properties.

7.5.1 Physical properties

7.5.1.1 *Solubility and viscosity*

Because of its structure and the presence of ionic charges along the chain, gelatine has a very good solubility and exhibits high water-binding capacity. Gelatine swells either in cold or warm water and can bind up to 10 times its own weight of water. Highly concentrated solutions, up to 50%, can be obtained using various methods. To reach total solubilisation, gelatine requires only low temperatures between 50°C and 60°C or a short time at 90–95°C.

Gelatine solutions have a low viscosity and are easy to handle, even at low temperatures of 50–70°C, compared to other hydrocolloids which have more limited conditions for use. The viscosity of gelatine solutions depends on temperature, concentration, ionic content, pH and molecular weight. Gelatine solutions exhibit the properties of a Newtonian fluid above 40°C.

A standard method is used to characterise the viscosity of gelatine. The viscosity of a 6.67% solution of gelatine is determined at 60°C by measuring the flow time of 100 mL of the solution through a standard pipette and ranges from 1 to 7 mPa s.

7.5.1.2 *The Bloom gel strength*

A gelatine solution sets below a given temperature dependent on the gelatine type, the concentration, its gel strength and viscosity. This gelation process is due to a rearrangement of individual molecular chains into an ordered network of helical arrangements (Cuvelier and Michon, 2003; de Wolf, 2003).

A standard AOAC method is used to characterise the gel strength of gelatine. A 6.67% solution of the gelatine sample is prepared in a standard test bottle at 60°C, cooled to 10°C and kept for 17 h for maturation at this temperature. The resulting gel is tested using a texture analyser. The gel strength, often expressed as Bloom, is the mass necessary to depress a standard plunger with a plane surface and diameter 12.7 mm (1/2 inch) to a distance of 4 mm into the gel. Gel strength measured in these conditions is called Bloom and expressed in grams. Bloom ranges generally from 50 to 300 g.

For commercial products, gelatine is commonly divided in three categories:

- Low Bloom gelatine (less than 125 Bloom)
- Medium Bloom gelatine (150–200 Bloom)
- High Bloom gelatine (above 220 Bloom)

The gel strength is dependent on concentration, pH, thermal history (temperature and time) and ionic content. The lower the ionic content, the higher the physicochemical properties such as viscosity and gel strength.

Gelatine gels are viscoelastic materials with the properties of a solid and a liquid. Resilience and hardness of gelatine are given, respectively, by Bloom and viscosity, which can be modified by the presence of other substances, such as sugar, proteins and plasticiser.

7.5.1.3 *Stability*

As gelatine is a protein, it can undergo hydrolysis, which can be caused by numerous factors such as acid, alkali, temperature, bacteria, enzymes and irradiation. Gelatine is not affected by UHT sterilisation at 120°C for a few seconds. A concentrated gelatine solution can be kept at 55–60°C for 4–6 h without significant degradation. Beyond 6 h, and for temperatures above 60–65°C, gelatine properties can be degraded, particularly if the solution is acid, below pH 5, or alkaline, above pH 8. At lower concentrations, degradation occurs more quickly. After cooling, an acidic gelatine food product remains stable when stored at low temperature.

7.5.1.4 *Setting point*

Above a minimum concentration of 0.8% gelatine solutions set when cooled. The setting point is commonly determined by evaluating the viscosity of a 10% solution, and observing the temperature at which gelation starts. However, the setting point can be estimated at different concentrations using more sophisticated methods based on the viscoelastic character of gelatine. The setting points of commercial gelatines vary from 20°C to 29°C.

7.5.1.5 *Melting point*

When heated to temperatures over 25–35°C, for example when placed in the mouth, gelatine gels melt. The melting point of gelatine is above the setting point by 2–5°C. The thermo-reversibility of gelatine is illustrated by the fact that the 'gel-to-sol' and 'sol-to-gel' changes can be repeated several times without loss of gel characteristics.

7.5.1.6 *Colour and clarity*

Colour and clarity depend on the raw materials, treatment and extraction level of gelatine from collagen. The colour is generally evaluated by visual observation, but standard measurements for colour and clarity are made by comparing the absorption of a 6.67% solution with water at 450 and 620 nm, respectively. Generally speaking, colour does not influence the properties of gelatine or reduce the commercial value.

7.5.2 Chemical properties

7.5.2.1 *Isoelectric point*

Because of the presence of functional groups on the amino acids, and the terminal amino and carboxyl groups, gelatine exhibits an amphoteric behaviour. In the presence of strong concentrations of hydrogen ions (acid media), gelatine carries a positive net charge, whereas in the presence of hydroxyl ions (basic media) gelatine shows a negative net charge (Jones, 2004).

The IEP corresponds to the pH at which gelatine has a neutral net charge and thus shows no movement in an electric field. At this point, the positive charges from amine ($NH3^+$) groups equal the negative charges on the carboxyl (COO^-) groups. The IEP is an intrinsic property of gelatine determined by raw materials (skin, bones, pork, beef, etc.), raw material pre-treatment and manufacture by the acid (A) or alkaline (B) process:

- Type A gelatine exhibits an IEP in the range 6–9.5. However, the exploitation of new raw materials for which pre-treatment causes an IEP fall can give type A gelatine showing an IEP between 5 and 6.
- Type B gelatine has an IEP in the range 4.5–5.6.

Low IEP gelatine is related to the deamidation of amino acids. Type A and B gelatines from raw materials treated with alkali, for instance in the dehairing process, are deamidated resulting in low IEP in the range 4.5–6.5. Additionally, gelatine pH may be adjusted during processing to give IEP from 4.5 to 6.5.

Functional properties of proteins are affected near their IEP because of electrostatic attraction of oppositely charged groups. Thus, at the IEP, gelatine properties coincide with either maximum or minimum values: swelling, viscosity and gelation show minimum values at the IEP, whereas turbidity, gel strength, foaming power and syneresis show a maximum value. This is more obvious in dilute solution or in the absence of sugars but can be neglected for the majority of gelatine uses. IEP is of interest to explain the interactions of gelatine with other ingredients, in particular anionic polymers.

7.5.2.2 *Compatibility*

Gelatine is compatible with most hydrocolloids, sugar, corn syrup, starch, glucose, dextrose, and common edible acids and flavours. Known examples of gelatine interactions with other polymers are coacervation reactions with acacia gum, a property used in microencapsulation, and the interaction of type A gelatine with carrageenan, controlled by substituting B-type gelatines for A-type gelatines in food applications such as mousse (Cuvelier and Michon, 2003).

7.5.3 Functional properties

The physical and chemical properties previously described are the basis of the use of gelatine in various applications. In addition, the sensory characteristics of gelatine have to be emphasised: its thermo-reversible behaviour, responsible for the mouthfeel, and its neutral taste make gelatine widely associated with food products. Another outstanding feature, probably the origin of the popular use of gelatine, is its 'ease of use'. Thermo-reversibility, the low viscosity and the moderate melting temperature make gelatine compatible with all the processes found within the food industry.

With regard to the functional properties of the gelatine, its water-binding behaviour and its ability to form a protein network are central. Thus, gelatine is used in confectionery for its gelling and stabilising properties, such as the regulation of crystal growth. The dairy industry utilises gelatine to avoid syneresis and adsorb the water released by other hydrocolloids. Gelatine also helps control the texture of yogurts and mousses by reinforcing the milk protein network (Jones, 1977).

The amphoteric properties of gelatine lead to emulsifying and foaming properties: gelatine is used in the confectionery and dairy industries to produce aerated products with low density, such as marshmallows, mousses and whipped creams. Emulsifying properties are used when fat is included in the formulation, for example in chewy candy and meat products. The binding effect of gelatine provides cohesiveness to a wide range of nutritional products, including bars and tablets. The interactions between gelatine and other compounds have an important application in the juice and wine industry where gelatine is used to aid clarification and fining.

This list of functional properties is not exhaustive since gelatine has properties valued in other areas. For example, in the pharmaceutical industry, gelatine is used for its film-forming properties in the manufacture of capsules and in the flavour and colour industries it acts as a protective colloid in encapsulation and coacervation.

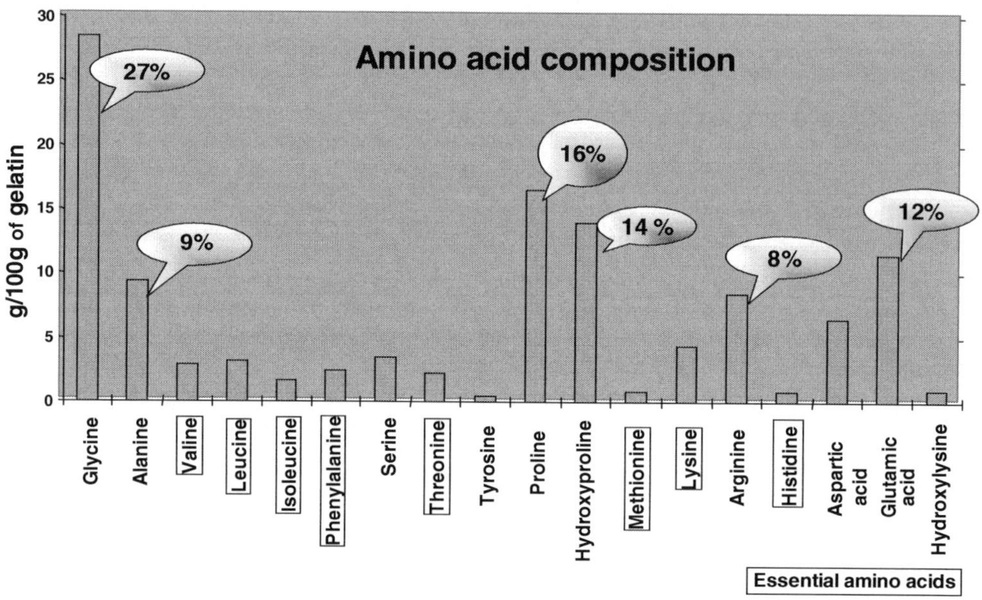

Fig. 7.10 Amino acid profile of gelatine.

In conclusion, food manufacturers often use gelatine because this ingredient provides several technological benefits at the same time. A good illustration of this is the use of gelatine in low-fat spreads, where it provides emulsifying and stabilising properties and gives a pleasant mouthfeel.

7.5.4 Nutritional features

Gelatine contains 86–90% protein and 1–2% mineral salts with the remainder being water. A total of 18 of the 20 amino acids are present, including 8 of the 9 essential amino acids. Gelatine contains high levels of glycine, proline and hydroxyproline, which are the main constituents of collagen (Fig. 7.10).

Recently, several investigations and clinical studies indicate that the favourable amino acid composition of gelatine has a positive health benefit on bones and joints (Moskowitz, 2000; de Wolf, 2003).

Most regulatory authorities consider gelatine as a fully digestible food material with an energy value ranging from 350 to 450 kcal per 100 g.

7.6 FOOD APPLICATIONS

7.6.1 Sugar confectionery

In the early days of sugar confectionery, products were made by skilled craftsman working empirically and the science of sugar confectionery came later. The link with more scientific technology for the production of cough sweets and similar products in the pharmaceutical industry was a big step in understanding the basic aspects of sugar confectionery. The sugar confectionery market offers a wide variety of products around the world. Cultural differences, habits and eating customs can be completely different, so apparently identical sugar confectionery products can have a completely different taste and texture. A good example is the famous gummy bear. A German gummy bear, with a hard texture, resembles the appearance of a British jelly baby, which has a very soft texture. These are different from the soft Asian gummy bear which has unusual exotic flavours. This means that there is no standard reference for sugar confectionery, but similar basic principles are applied throughout the industry.

Gelatine is one of the most versatile texturising ingredients for sugar confectionery and is perfectly adaptable to most industrial sugar confectionery processes as a result of a number of features:

- the thermo-reversible gel allows confectionery products to be re-melted for easy recycling and rework of finished products,
- solutions are low viscosity and easy to deposit without tailing,
- it is a protein with good whipping and emulsifying properties,
- it is fully compatible with many other texturising agents,
- strong water binding prevents the finished product from drying out,
- different mesh sizes from 8 mesh (2.35 mm) up to 60 mesh (0.25 mm) allow combination with other ingredients of the same particle size and processing is easy.

Gelatine sugar confectionery products can be divided into two classes: jelly confectionery and aerated confectionery. Jelly confectionery is characterised by the formation of a gel after depositing into starch and cooling down to ambient temperature. Aerated confectionery is made by the introduction of air into the liquid phase and the amount of air is characteristic of the aerated product.

7.6.1.1 *Jelly confectionery*

These products are usually called jellies. Occasionally, they are also referred to as gums, although the name 'gum' is generally reserved for products made with modified starch. Sometimes modified starch is combined with gelatine, for example in a so-called wine gum.

In general, jelly candies are made according to the recipe shown in Formulation 7.1.

Formulation 7.1 Basic jelly candy recipe

Recipe for 100 kg of finished product:

(1)	Gelatine (175–275 Bloom)	7 kg
	Water	14 L
(2)	Granulated sugar	37 kg
	Glucose syrup 40 D.E.	45 L
	Water	13 L
(3)	Citric acid (50% solution)	2 kg
	Flavour and colour	as desired

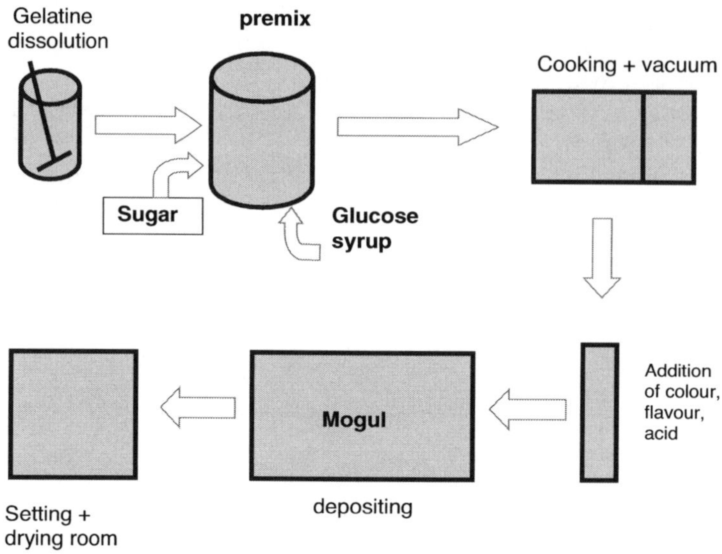

Fig. 7.11 Flow diagram of jelly candy production.

Variations in texture will be obtained when using gelatine of different gel strengths at the same concentration. Higher gel strength gelatine gives harder textures. Sugar and glucose syrup are the bulking agents in the recipe and are responsible for the sweet taste and the shelf life of the product. Citric acid is used to lower the pH to about 3.2, which will enhance the typical acid fruit flavour.

The jelly production process is shown in Fig. 7.11. The gelatine solution is made in a separate tank. Solutions of gelatine up to maximum 40% can be made; at higher concentrations, part of the gelatine will not be dissolved. Gelatine dissolves rapidly in hot water (80–90°C) and can be kept for 4–6 h maximum at 50–60°C without significant degradation.

Sugar and glucose syrup are dosed into a second tank and then mixed together with the dissolved gelatine. From the premix tank, the mass is transferred to a continuous cooker followed by a vacuum step. Only then, acid, flavour and colour are added. The cooking temperature will determine the total soluble solids (TSS) at depositing which should be in between 78% and 80% before depositing. The gel mass is deposited in dry, cool moulding starch in trays. The starch trays are kept for 24 h at room temperature (20–23°C) before demoulding. After removing from the moulding starch, the jelly candies are sugar sanded or oiled before packing.

From the basic recipe above, a number of modifications can be made to give a range of different finished products in terms of texture, calorie content and taste:

- Hard gum/jelly: increased concentration of gelatine and drying at 50°C to reach a TSS of 88–90%.
- Grained jelly: part of the glucose syrup is substituted by sugar in the recipe. The finished product is then dried at 40°C for 48–72 h and crystallisation of the sugar is obtained at the surface.
- Wine gum: modified starch is added to the recipe.
- Sugarless jelly: sugar and glucose syrup are substituted by sugar alcohols and/or fibre.

Table 7.5 Ingredients used in chewy candy.

Gelatine	0.5–2.0%
Reducing sugars	15–23%
Sugar/Glucose syrup ratio	50/50 or 40/60, depending on other ingredients
Fats	2.5–10%
Acid	0–1.5%
Moisture	6–10%
Flavour and colour	as desired

7.6.1.2 *Aerated confectionery*

In aerated confectionery, gelatine gives the air cell wall the required mechanical resistance to avoid deformation. The basic principles and the recipes are very similar to jelly confectionery. The big difference is that air is introduced in the gel mass, which results in a mass with densities that can vary from 1.0 to 0.25. The aeration can be done with a planetary beater or a continuous beater under pressure. It is the final density which determines the texture and type of finished product. Four main types can be defined:

- Chewy candy: density 1.0–0.9
- Aerated jelly candy: density 0.9–0.8
- Deposited marshmallow: 0.55–0.5
- Extruded marshmallow: 0.30–0.25

Chewy candy is not produced using a standard recipe. Table 7.5 gives an overview of the ingredients used and their variation. The recommended gel strength for chewy candy is from 100 to 150 Bloom gelatine.

Aerated jelly candy is based on the formulation shown in Formulation 7.1 with the additional step of aeration of the gel mass just before depositing.

Deposited and extruded marshmallows are made using the same ingredients as for the jelly candy but the concentration of gelatine is lower; concentrations of only 3–4% gelatine are used. Most marshmallows are vanilla flavoured and the use of acid is rare. The recommended gel strength for deposited and non-deposited marshmallow is 200–250 Bloom gelatine.

Formulation 7.2 Deposited and extruded marshmallows

Recipe for 100 kg of finished product:

(1)	Gelatine	3–4 kg
	Water	6–8 L
(2)	Sugar	44–49 kg
	Glucose syrup	44–38 kg
	Water	14 L
(3)	Colour and flavour	as desired

The production process for aerated candy is the same as for jelly candy from dissolution until the addition of colour and flavour. The cooking temperature for chewy candy and marshmallows will be regulated to obtain the final TSS because no additional drying steps follow. The additional step for the aerated product is the introduction of air in the gel mass. This can be done by different techniques.

For chewy candy, a specific technique called 'pulling' is used. The whole mass, which has a pastry structure, is pulled over and over until enough air is introduced in the mass. The mass is then cut and wrapped into its final packaging. New techniques, such as aeration under continuous pressure, are becoming more and more popular.

For deposited marshmallow and most industrially processed marshmallow, air is introduced into the mass by a continuous pressure beater. Once the desired pressure is obtained, the marshmallow is extruded. The rope is cut and the marshmallow pieces are rolled in powdered sugar or a mix of powdered sugar and starch. Deposited marshmallow is deposited into dry, cool moulding starch, stored for 24 h at ambient temperature (20–23°C), and removed from the starch the next day.

From the basic recipe, a wide range of modifications can be made:

- Chewy candy with shorter or longer texture can be obtained by varying TSS, sugar concentrations or gelatine gel strength.
- Grained marshmallow is made by substituting part of the glucose syrup by sugar and drying the finished product in the starch for 24–48 h at 40°C. Sugar crystals develop at the surface.
- Sugar-free aerated products are produced by replacing sugar and glucose syrup by sugar alcohols and/or fibres.

Gelatine is a multifunctional ingredient, compatible with most materials. Therefore, it is likely that gelatine will play a major role in the development of new sugar confectionery products and in the growing trend towards more healthy and functional food. Gelatine will remain an irreplaceable ingredient for this type of sugar confectionery because of its unique texture.

7.6.2 Dairy products

In order to respond to the demands of the market for dairy products, industry is applying more technology. Milk products are combined with ingredients and additives to thicken, stabilise or aerate. One of these ingredients is gelatine, which is able to confer a wide range of benefits to dairy products. These properties include:

- thermo-reversible gels that are elastic and smooth and melt agreeably in the mouth;
- complete compatibility with milk, casein, other components of milk and principal colloids used in milk products;
- thickening and stabilising without precipitating casein and without salt addition;
- binding 5–10 times its weight of water to avoid exudation or syneresis in milk products;
- acting as a protective colloid so that the coagulation of milk or casein is finer and more homogeneous in the presence of gelatine;
- providing good foaming capacity;
- providing some emulsification properties.

Gelatine is readily incorporated into dairy products. Firstly, the gelatine powder is mixed with the other powder ingredients such as sugar, flavour and other stabilisers. The mixture is added with agitation to the cold milk, where the gelatine swells and absorbs up to 10 times its weight of water. During pasteurisation the gelatine hydrates.

If the milk is filtered before pasteurisation, it is better to add the gelatine as a solution, otherwise the filters may retain the grains of swollen gelatine. The solution is prepared by adding gelatine directly to hot water at 80–90°C with vigorous agitation for a few minutes. Solutions of 30% concentration can be prepared easily and then maintained at a temperature of 50–60°C for several hours without degradation. In addition, gelatine is not affected by UHT sterilisation at 120°C for a few seconds.

The gelatine solution is added to the cold milk under constant agitation and then pasteurised. In order to compensate for the water added in the gelatine solution, milk protein, whole milk powder (for full fat yogurt) or skim milk powder (for low-fat yogurt) is added.

7.6.2.1 *Yogurt*

Yogurt is produced by inoculating milk with two typical bacilli – *Lactobacillus bulgaricus* and *Streptococcus thermophilus* – to develop the flavour of the yogurt. The lactic bacilli exert both direct and indirect action on the fermentation of lactose and citric acid and form the basis for the consistency and taste of the final product.

Skimmed milk with a minimum dry solids content of 10% is most often used as the starting material. Pasteurisation at 90–95°C for 5–10 min denatures serum proteins and increases their water-binding capability. In general, the milk used for yogurt production has standardised fat content and dry solids in order to obtain consistent finished products for low-, half- or full-fat yogurt. This standardisation plays an important role in obtaining the consistency, viscosity, stability and nutritional value. The dry solids content can be standardised by concentration or ultrafiltration, by adding powdered milk or milk protein, or by means of thickeners such as gelatine. It should be noted that the composition of yogurts can vary depending on local legislation.

It is preferable to homogenise milk to produce yogurt. Homogenisation at 180–200 bar and 55–70°C prevents the separation of fat in a full-fat yogurt. This step modifies the physical characteristics of the fat and casein to ensure consistent gel firmness. Homogenisation also shortens the incubation time: the acidification is quicker and the coagulation period shorter.

The pH at which the incubation is stopped and the cooling occurs is very important. The most favourable pH is generally between 4.3 and 4.7. For stirred yogurt, the process is monitored and at a pH of 4.5 the yogurt is cooled rapidity to 15°C over approximately 20 min by means of a plate heat exchanger. After cooling, no more acid is produced. Two points are of the utmost importance: the pH of the yogurt when cooling starts, and the temperature of the yogurt after cooling. A pH of 4.5 is considered to be ideal as this will produce a pH of 4.2–4.3 in a soft yogurt. If the cooling starts at 4.7, the texture is less smooth and there is a risk of syneresis.

Stabilisers have specific characteristics so the quantity and combination of stabilisers determine the final texture of the yogurt. This can vary from a short viscous texture to a more unctuous one. The characteristics of gelatine ensure that it is a prime choice for the stabilisation of different types of yogurt.

Introducing gelatine into milk before incubation does not affect the activity of the lactic fermentation so the temperature and time to develop acidity are unchanged. In full-fat yogurt, gelatine improves the texture of the final product obtained by the lactic fermentation without modifying the characteristics of taste. Gelatine binds water well and prevents syneresis. For low-fat yogurts, made with skimmed milk and standardised to 10% solids, concentrations of 2–4 g/L gelatine give products with a soft texture and with 6–8 g/L firm yogurts are obtained.

Gelatine is particularly useful in the production of fruit yogurts where some syneresis is almost inevitable without the addition of stabilisers. Gelatine can be incorporated into the yogurt or in the fruit preparation when this is not mixed with the yogurt. By binding the fruit juice, gelatine prevents its diffusion into the mass.

7.6.2.2 *Quark*

A multitude of products are based on curdled milk. These products may be aerated and contain flavours, glacé fruits, fruit jellies and other inclusions. Gelatine plays an indispensable role as a binder and it influences texture and prevents syneresis. Gelatine levels vary between 0.1% and 2.0% and, in the form of a solution at 40°C, it can be added to the curdled milk.

7.6.2.3 *Thermally treated fermented milks*

In certain countries, the shelf life of certain types of yogurt and fermented milk may be extended by pasteurisation after the incubation process. This operation prevents prolonged action by the fermenting agents but does destabilise the product structure leading to exudation. The addition of a mixture of gelatine and starch before pasteurisation produces a good texture and minimises any exudation. The texture of a product stabilised with gelatine alone is sensitive to changes in storage temperature. A combination with 0.4–0.6% modified starch gives a highly satisfactory formulation in which the starch stabilises the viscosity between 5°C and 20°C and the gelatine avoids the risk of exudation.

7.6.2.4 *Flavoured gelled milk desserts*

These dessert products have a semi-solid consistency and are prepared from flavoured, sweetened milk. The stabilising agent must hydrate during thermal treatment of the milk, produce no increase in viscosity at high temperature and gel when the product is cooled in the pot. Gelatine can be used alone or in combination with other gelling agents, such as carrageenan. For example, replacing one to two parts of carrageenan with two to three parts of gelatine gives a soft, more elastic gel texture. Higher gelatine levels allow the preparation of light, aerated products that are very pleasant to eat cold. In every case, a soft, smooth gel is obtained which has no syneresis.

7.6.2.5 *Dessert creams*

Dessert creams have a relatively thick consistency and are made from flavoured, sweetened milk. Gelatine is used to achieve a smooth gel texture and prevent exudation during freezing or as a result of major temperature variations during storage. The amount of gelatine in these preparations is quite variable and can be as much as 2% or even higher, depending on the characteristics of the desired product.

7.6.2.6 *Ice cream and water ices*

The presence of a stabilising agent is essential to adjust mix viscosity, stabilise the emulsion until the ice cream is consumed, facilitate aeration and improve expansion, avoid deterioration during storage and prevent the formation of large ice crystals during prolonged storage.

Used in association with other stabilising agents, gelatine provides the finished product with a remarkably slow melting rate and characteristic texture.

7.6.2.7 *Low-fat spreads*

Low-fat spreads have reduced fat content and are exclusively milk-fat based, vegetable-fat based or a combination of the two. The preparation of a stable emulsion requires the presence of stabilisers and emulsifiers. The stabiliser must ensure good water binding and improve the structure, consistency and spreadability of the finished product. Good stabilisation can be obtained with high Bloom gelatine, added at doses varying from 1% to 2%, because the melting point of that type of gelatine is 31–33°C, which is very near to the melting point of the different fats used. Other hydrocolloids, such as alginate, carrageenan and xanthan, may be used with gelatine in order to improve water binding.

7.6.2.8 *Mousse*

The foaming action of gelatine is used in the preparation of a wide range of aerated products. The presence of butyric or other fats tend to inhibit this foaming capability so it is necessary to inject air, CO_2 or nitrogen to obtain the required increase in volume of an aerated product. The low setting temperature of gelatine stabilises the foam, gives strength to the aerated structure and prevents collapse during storage. Used in combination with other hydrocolloids, gelatine provides excellent emulsion stability, facilitating aeration and good mix stability, avoiding separation of the ingredients and maintaining foam stability, before it gels. When used in combination with carrageenan, it is necessary to use low isoelectric point (type B) gelatine to avoid excessive interaction

7.6.2.9 *Cheese*

Gelatine can be used in cheese production to increase water binding and thus achieve better yield and lower fat content. Gelatine will also strengthen the texture of the product and enhance flavour release.

The properties of gelatine in dairy desserts are summarised in Table 7.6. In view of the low dosages used, normally medium- or high-grade gelatines, 150 Bloom or higher, are chosen, as shown in Table 7.7, but other types may be used according to the characteristics required by the manufacturer.

7.6.3 Meat products

Meat products have an important position in the food industry in many countries. Gelatine is obtained from animal raw materials and can be considered a natural complement of meat-based proteins. Gelatine is frequently incorporated in the following:

Table 7.6 Functional properties of gelatine in dairy products.

Main characteristics of gelatine	Consequence on the products	Types of dairy products
Gelling/thickening power	–Gives consistency – Improves the texture	All dairy products
Suitable melting point	–Enhances thickness (in mouth sensation); in low/non-fat products, mimics fat because melting point of gelatine is similar to that of milk fat (for high Bloom gelatine) – Improves flavour especially in low/non-fat products	Yogurts, soft white cheese, cheese
Binding water	–Improves syneresis especially in low/non-fat products and low MSNF content products. –Milk protein reduction in low cost cheese –Increases the yield of cheese (retained whey)	Yogurts, cheese
Foaming water	–Ability to aerate even in the presence of fat – Stabilises the foam	Mousse

- cooked pressed ham, cooked shoulder,
- canned meat products,
- meat emulsions (pork products, pâtés),
- jellies and aspics.

The typical quality and dose of gelatine for each product are given in Table 7.8.

7.6.3.1 *Cooked pressed ham and cooked shoulder*

Gelatine is added to the ham during preparation, when the bone has been removed. It is used to bond the meat where the rind and excess fat have been trimmed. The gelatine powder absorbs moisture from the meat during cooking and forms a film which seals the meat after cooling. Gelatine also gels the liquid that is lost during cooking and holds it in and around the ham. This added gelatine also stiffens the jelly obtained directly from the connective tissues

Table 7.7 Gelatine characteristics for dairy products.

Application	Gel strength Bloom	Rate of use
Yogurts	150	0.2–2%
Thermally treated fermented milk	150	0.2–2%
Gelled milk dessert	150	0.2–2%
Dessert creams	150	0.2–0.6%
Ice cream, water ices	150	0.2–1%
Low calorie spreads	250	0.5–3%
Mousses	150–250 hydrolysed gelatine	0.2–3%

Table 7.8 Gelatine characteristics for meat products and miscellaneous applications.

	Characteristics					
Application	**Gel strength (Bloom)**	**High viscosity**	**Average viscosity**	**Low viscosity**	**Others**	**Rate of use**
Meat industry						
jellies	150–250	X	X		Transparency	3–15%
binders for meat emulsions	150–250	X	X			0.5–3%
hams, canned meat products	150–250	X	X		Transparency	1–2%
coating	150–250		X		Transparency	5–20%
Fish products						
binders	150–250		X			0.5–3%
aspics	150–250		X		Transparency colour	3–15%
Sauces, soups	150–250	X	X			0.5–2%
Wine clarification	75–125 hydrolysed gelatine		X	X		

Note: X indicates the property conferred by the specific grade of gelatine in the application.

during the cooking process to give an attractive presentation and bind the ham together for improved slicing.

7.6.3.2 *Canned meat products*

Gelatine is used to gel the juices lost from meat products during cooking or pasteurisation. Gelatines with high gel strength (200–250 Bloom) are used for this application at levels of 0.5–2%. As sterilisation temperatures are relatively high, gel strength and viscosity losses during processing should be taken into account as these depend on the duration of thermal treatment.

7.6.3.3 *Meat emulsions*

As a result of their high fat and water content, meat emulsions present particular stabilisation problems, mainly from loss of water or fat and texture irregularities after cooking. The use of a cutter partially eliminates these problems, but gelatine is still used to achieve better water binding, to stabilise the emulsion and to obtain a homogeneous batch texture. Gelatine content varies considerably, depending on the presence of other binding agents, the amount of collagen present in the other ingredients and local regulations.

7.6.3.4 *Decorative jellies*

Jellies are widely used for coating and decorating hams and pâtés. Conventional jellies based on gelatine or jellies prepared with type B gelatine and carrageenan are used for this application. The latter combination gives a clear jelly with reduced setting time, increased gel strength and a higher melting point as the addition of 10% carrageenan, based on the weight of gelatine, increases the melting point from 30°C to 53°C.

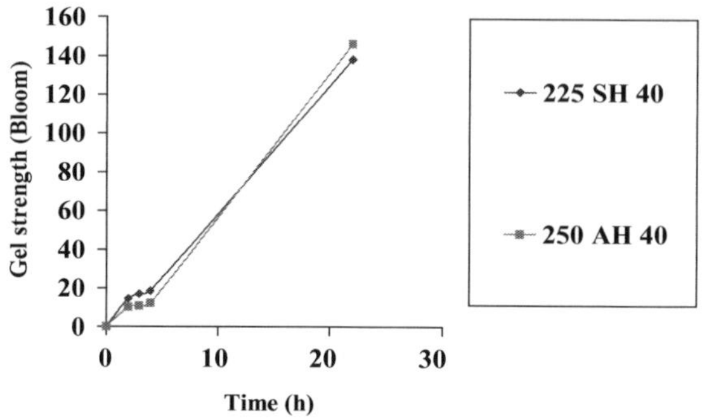

Fig. 7.12 Gel strength evolution in water gel desserts.

7.6.4 Water gel desserts

Water gels are popular desserts around the world. Prepared at home, by food services or purchased ready to eat, people consume this preparation after meals or as snacks.

The powdered mix is prepared with gelatine, sugar, acids, salts, flavour and colour.

Gelatine forms water jellies that are elastic, soft, bright and transparent; transparency is a particular characteristic that is important to the final customer. Some points should be considered for these dry mixes:

- Gelatine types are primarily pigskin or bovine hide.
- Sweeteners include sucrose, dextrose, polyols and intense sweeteners.
- Acids may be citric, fumaric or malic and are recommended to give a good flavour.
- Colour and flavour must be stable in hot solutions and at low pH.
- Buffers or acidity regulators include sodium citrate or sodium/potassium tartrate.
- Anti-caking agents are usually phosphates.

All components must be of a similar particle size to ensure a homogeneous mix. Sieve 30 or 40 (400–500 μm) is a good mesh size for gelatine particles. Hygroscopicity is a problem for this kind of product, so choosing raw materials with minimal hygroscopicity and using packaging coated with a moisture barrier is the key to solving this problem.

The evolution of gel strength over time shows that bovine hide gelatine from the alkaline process (SH) forms a gel faster than acid-processed gelatine (AH), in the first 3 h of the gelling process. However, the gel strength is similar after 24 h (see Fig. 7.12).

The final texture depends on the dosage and Bloom of gelatine, the type and quantity of sugar and the temperature. The dose for a standard water gel is 1.8–2.0% of 225 Bloom gelatine in the final preparation and for a sugar-free water gel the dose is 2.0–2.4% of 175 Bloom gelatine.

7.6.5 Wine fining and fruit juice clarification

Fining is the name given to the clarification step in the making of wine and juices. The object of fining is to improve the taste, colour, bouquet and clarity of juice or wine. Solids in the extracted juice, such as pieces of pulp or skins, should be removed because they occupy volume, affect flavour, cause foam and slow the rate of fermentation. After fermentation, young wines contain precipitated proteins, salts, dead yeast cells, etc. Precipitates develop and could affect the final flavour if they are not removed.

Different proteins are used in the fining step including albumen, casein or gelatine. Fining agents work on the principle that all particles responsible for the clouding or haze in a wine or juice have an electrical charge. Gelatine has a positive charge so it can attract negatively charged materials. Usually, gelatine is combined with a negatively charged agent such as bentonite. The gelatine is precipitated by the tannin or by the acidity of wine, forming aggregates which readily settle to form sediment.

Two types of gelatine are used for this application, low Bloom gelatine or hydrolysed gelatine, which is readily soluble in cold water, using the following doses:

• Juice:	20–120 g per hectolitre
• Wine:	white: 2–7 g
	rose: 3–8 g
	red: 7–15 g per hectolitre
• Beer:	1–3 g per hectolitre

Gelatine reduces turbidity and astringency in red wines by lowering tannin levels. It tends to remove more of the higher molecular weight tannins, which are perceived as astringent, than bitter-tasting lower molecular weight tannins. Gelatine also improves colour.

7.7 FUTURE DEVELOPMENTS

Gelatine is a multifunctional ingredient, compatible with most materials. Therefore, it is likely that gelatine will play a major role in the development of new sugar confectionery products and in a growing trend towards more healthy and functional food. Gelatine will remain an irreplaceable ingredient for sugar confectionery, dairy products, meat applications and other water gel products due to its unique texture.

References

Ames, W.M. (1952) The conversion of collagen to gelatine and their molecular structures. *Journal of the Science of Food and Agriculture*, **3**(10), 454–463.

Bailey, A. (1998) Mechanisms of maturation and ageing of collagen. *Mechanisms of Ageing and Development*, **106**, 1–56.

Bella, J. (1998) X-ray crystallographic determination of a collagen-like peptide with the repeating sequence (Pro-Pro-Gly). *Journal of Molecular Biology*, **280**, 623–638.

Cooper, A. (1971) Thermal stability of tropocollagens: are hydrogen bonds really important? *Journal of Molecular Biology*, **55**, 123–127.

Covington, A. (2001) Chrome tanning: exploding the perceived myths, preconception and received wisdom. *Journal of American Leather Chemist's Association*, **96**, 467–480.

Cuvelier, G. and Michon, C. (2003) *Gelatine Substitution: What Functionalities?* Polymerix, Rennes, pp. 1–9.

de Wolf, F.A. (2003) Collagen and gelatine. In: W.Y. Aalbersberg, R.J. Hamer, P. Jasperse, H.H.J. de Jongh, C.G. de Kruif, P. Walstra and F.A. de Wolf (eds), *Industrial Proteins in Perspective* (Progress in Biotechnology), Elsevier Sciences, **23**, Chapter V, pp. 133–218.

Delarosa A. (2005) Adhesives. Avaiable at www.mnsu.edu/emuseum/prehistory/ancienttech/adhesives.html (accessed 4 August 2009).

Eastoe, J.E. and Leach, A.A. (1958) A survey of recent work on the amino acid composition of vertebrate collagen and gelatine. In: G. Stainsby (ed), *Recent Advances in Gelatine and Glue Research*. Pergamon Press, London, p. 193.

Engel, J., Tan, H.-T., Prockup, D.J. and Frump, H. (1977) The triple helix to coil conversion of collagen like polytripeptides in aqueous and non aqueous solvents. Comparison of the thermodynamic parameters and the binding of water to (L-Pro-L-pro-Gly)*n* and (L-Pro-Hypro-Gly)*n*. *Bioploymers*, **16**(3), 601–622.

Finer, E.G. (1975) Gel formation from solutions of single-chain gelatine. *Biopolymers*, **14**, 1995–2005.

Guedj, S. (2005) *Crosslinks and Their Role in the Chemistry of Gelatine*. IAG Conference, Heidelberg, September 2005.

Jones, B.E. (1977) Uses of gelatine in edible products. In: A.G. Ward and A. Courts (eds), *The Science and Technology of Gelatine*. Academic Press, London, pp. 365–394.

Jones, B.E. (2004) Gelatine: manufacture and physico-chemical properties. In: F. Podczeck and B. Jones (eds), *Pharmaceutical Capsules*, 2nd edn. Pharmaceutical Press, London, Chapter 2, pp. 23–60.

Ledward, D.A. (1986) Gelation of gelatin. In: J.R. Mitchell and D.A. Ledward (eds), *Functional Properties of Food Macromolecules*. Elsevier Applied Science, London, pp. 171–201.

Liang, H.C. (2004) Crosslinking structures of gelatine hydrogels crosslinked with genipin or a water-soluble carbodiimide. *Journal of Applied Polymer Science* **91**, 4017–4026.

Marks, E.M. (1968) The Phenomenon of gelatine insolubility. *Food Technology*, **22**, 1433–1436.

Moskowitz, R., (2000) Role of collagen hydrolysate in bone and joint disease. *Seminars in Arthritis and Rheumatism*, **30**(2), 87–99.

Pouradier, J. (1977) Organic hardeners. In: T.H. James (ed.), *The Theory of the Photographic Process*, 4th edn. Macmillan publishing, Inc., New York and Collier Macmillian Publishers, London, pp. 79–87.

Pouradier, J. and Venet, A.M. (1952) Contribution a l'etude de la structure des gelatines. 4. Comparaison de gelatines d'origines diverses. *Journal de Chimie Physique et de Physico-Chemie Biologique*, **49**, 85–92.

Ramachandran, G. (1967) *Treatise on Collagen*. Academic Press, London and New York, 556 pp.

Saito, H. (2004) Physicochemical properties of gelatine gels prepared using citric acid derivative. *Materials and Science and Engineering*, **C24**, 781–785.

Sheppard, S.E. and Houck, R.C. (1932) The structure of gelatine sols and gels. *Journal of Physical Chemistry*, **36**, 2319–2324.

Sheppard, S.E. and Sweet, S.S. (1921) The elastic properties of gelatin jellies. *Journal of the American Chemical Society*, **43**(3), 539–543.

Taylor, S.K. (1978) 13C NMR studies on gelatine crosslinking by formaldehyde. *Photographic Science and Engineering*, **22**(3), 134–137.

Tomihata, K. (2001) Crosslinking and degradation of biopolymer. *Recent Research Developments in Biotechnology & Bioengineering*, **4**, 35–49.

Vollhardt, P. (1999) *Traité de chimie organique*, De Boeck Université, pp. 1297 (traduction de la 3e édition américaine par Paul Depovere).

Ward, A. (1977) *The Science and Technology of Gelatine*. Academic Press, London, pp. 564.

8 Gellan Gum

Raymond Valli and Ross Clark

ABSTRACT

Gellan gum is a fermentation polysaccharide produced by the microorganism *Sphingomonas elodea*. It has a straight chain structure based on repeating glucose, rhamnose and glucuronic acid units with side groups of acyl groups. Gellan gum hydrates in hot water and the low-acyl form also hydrates in cold water with sequestrants. On cooling, native high-acyl gellan gum gives gels that are soft and elastic. Low-acyl gellan gum gels at very low concentrations using both monovalent and divalent cations to give firm, brittle textures with excellent thermal stability. Combinations of gellan gum can be used to control syneresis and form a range of textures from soft and elastic to firm and brittle. A major food application is water dessert gels, particularly for Asian desserts. Other significant applications include confectionery, dairy desserts and bakery fillings. At levels too low to form a demoldable gel, gellan gum can form fluid gels that can suspend particulates in sauces and dressings and fruit pulp in beverages.

8.1 INTRODUCTION

Gellan gum is a fermentation polysaccharide produced by the microorganism *Sphingomonas elodea* (previously identified as *Pseudomonas elodea,* but later reclassified). CP Kelco found this organism when it undertook a worldwide screening program to discover interesting and useful new gums that could be made by fermentation. At the time of its discovery, gellan gum was thought to be a 'universal' gelling agent. The fact that gellan gum gel textures can range from soft and elastic to hard and brittle is unusual. Adding to this appeal, gellan gum can form gels using both monovalent and divalent cations. Gellan gum can be used at very low use levels, and its gels show excellent thermal stability. With its unique combination of properties, it is easy to understand why gellan gum has high appeal to the food industry.

8.2 MANUFACTURE

All the different forms of gellan gum are produced from the same basic fermentation process (see Fig. 8.1). Large sterile fermentation vessels are used to allow the bacteria to convert simple sugars and other nutrients into the gellan gum polysaccharide. Once the cells have been killed, several different process steps can be followed to produce to four different product types. At present, the low-acyl, unclarified version is not produced commercially.

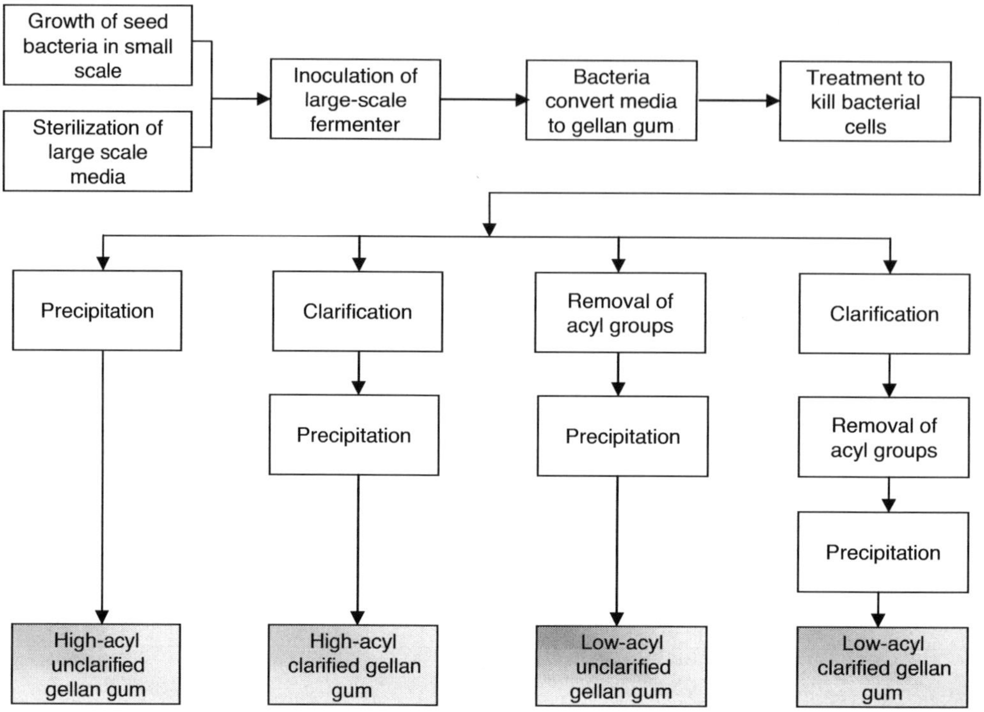

Fig. 8.1 Production of various gellan products. (Reproduced with permission from C.P. Kelco.)

Even within a defined product type, such as clarified, low-acyl gellan gum, many different commercial grades can exist depending upon the intended use of the product. Properties such as particle size and gel strength can vary between different grades. Additionally, gellan gum is often blended with other ingredients to make products with properties that are targeted to the needs of specific applications.

8.3 CHEMICAL COMPOSITION

The molecular structure of gellan gum is a straight chain based on repeating glucose, rhamnose and glucuronic acid units (see Figs 8.2a and 8.2b). In its native or high-acyl form, two acyl substituents – acetate and glycerate – are present. Both substituents are located on the same glucose residue and, on average, there is one glycerate per repeat and one acetate per every two repeating unit (Kuo *et al.*, 1986). In low-acyl gellan gum, the acyl groups are absent.

The structure of gellan gum has been studied extensively using techniques such as X-ray diffraction and molecular modeling (Chandrasekaran *et al.*, 1988a,b, 1992; Chandrasekaran and Thailambal, 1990). The studies suggest that gellan gum polymer adopts a double helical structure after heating and cooling. If the cooling is done in the presence of cations, the double helices aggregate into long-range networks. When the gum concentration is high enough, this network structure becomes a demoldable gel. The acyl groups on the high-acyl gellan

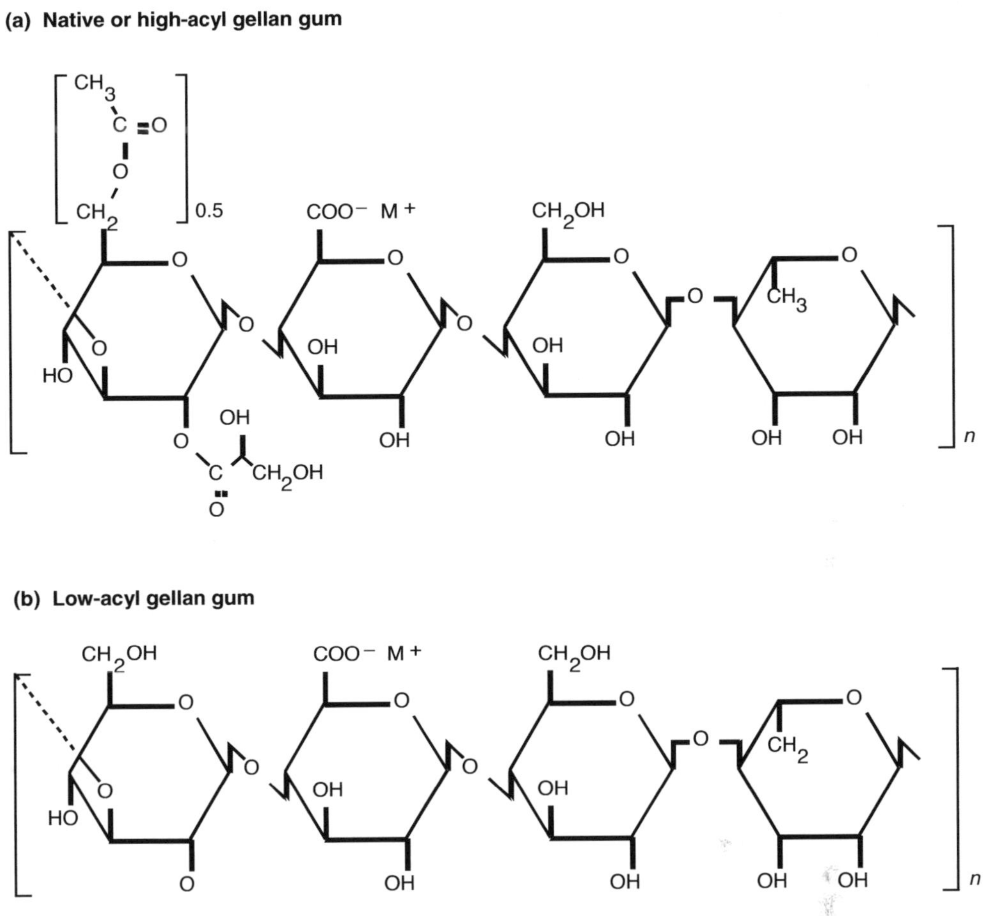

Fig. 8.2 Primary structure of gellan gum. (Reproduced with permission from C.P. Kelco.)

gum help stabilize the double helix, but they interfere with the double helix aggregation. The resulting high-acyl gellan gum gel is soft and elastic. The low-acyl form of gellan gum, without the interfering acyl groups, forms gels that are very firm and brittle.

8.4 FUNCTIONAL PROPERTIES

The presence or absence of the acyl groups on the gellan gum polysaccharide backbone has a profound effect on its functional properties. Therefore, when discussing hydration or gel properties, it is important to distinguish between low- and high-acyl gellan gum types.

8.4.1 Hydration of low-acyl gellan gum

The hydration temperature of low-acyl gellan gum is sensitive to the ionic environment and particularly sensitive to divalent cations (see Fig. 8.3). Low-acyl gellan gum itself contains

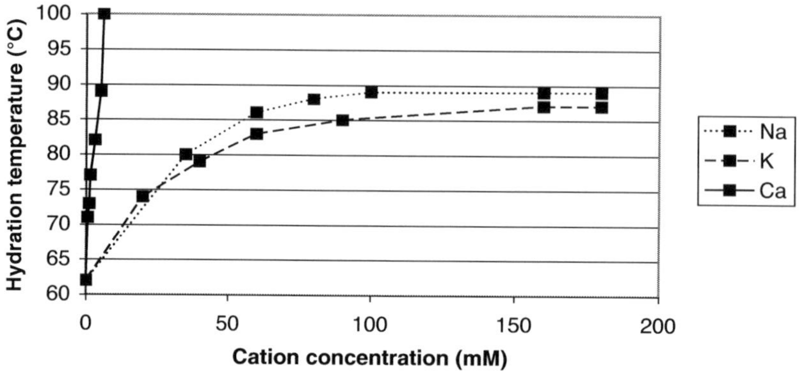

Fig. 8.3 Cation effect on low-acyl gellan gum hydration temperature. (Reproduced with permission from C.P. Kelco.)

divalent cations and will only partially hydrate in cold, deionized water. Hydration is further inhibited by the divalent ions present in most water supplies. This effect makes low-acyl gellan gum easy to disperse in cold water without forming lumps. Subsequently, the gum can be hydrated by adding sequestrants or chelators, such as citrates and phosphates, to control the divalent ions, heat or a combination of both. Hence, the hydration temperature of low-acyl gellan gum can be effectively controlled (see Fig. 8.4). Without sequestrants, low-acyl gellan gum requires a temperature above about 75°C (167°F) to fully hydrate in soft water. However, it can be hydrated in cold, soft water using 0.12% sodium citrate, as given in Table 8.1.

The solution pH also affects the hydration characteristics of low-acyl gellan gum (see Fig. 8.5). At pH values above the pK_a of gellan gum, about pH 3.6, the gum is in a form that allows easy dissolution. If the solution pH is below 3.6, however, the gum will exist in a predominately acid form which is not completely soluble. Therefore, when formulating acidic products, the acid should be added after the gum has been hydrated.

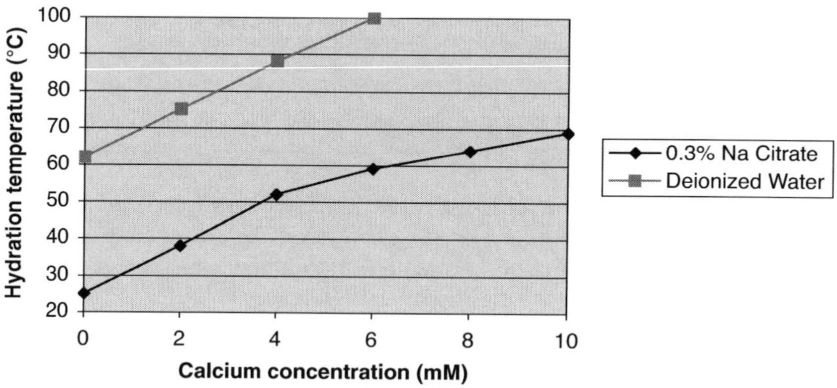

Fig. 8.4 Sodium citrate effect on low-acyl gellan gum hydration temperature. (Reproduced with permission from C.P. Kelco.)

Table 8.1 Hydration guide for low-acyl products.

Water quality	Hardness as ppm of $CaCO_3$	Na Citrate level (%)	Hydration temperature, °C or (°F)	Notes
Deionized	0	0.07	22 (72)	Hydration time <2 min with good dispersion and mixing
Soft	1–160	0.12	22 (72)	Hydration time <2 min with good dispersion and mixing
Slightly hard	161–320	0.16	22 (72)	High shear mixing improves hydration rate
Hard	321–460	0.20	28 (82)	Nearly 95% hydrated at 22°C with high shear mixing, slight heat needed for full hydration
Very hard	460–920	0.34	58 (136)	Heat required for full hydration. Forms a gel (sodium gel) upon cooling

Source: Reproduced with kind permission from C.P. Kelco.

Dissolved sugars have an effect on gellan gum hydration (see Fig. 8.6). Up to dissolved sugar levels of 25% or so, hydration proceeds in a normal manner. As the sugar level increases above this point, more heat needs to be applied to fully solubilize the gellan gum. When using gellan gum, it is always better to keep the level of dissolved sugars low until after the gum has been hydrated. After gum hydration, the remaining sugar solids can be added.

8.4.2 Hydration of high-acyl gellan gum

High-acyl gellan gum swells in deionized water, creating a consistency like a swollen starch paste. Low levels of sodium ions inhibit this swelling behavior so the addition of sodium salts is a useful strategy for improving gum dispersion and for minimizing viscosity during processing. Heat is required to fully hydrate high-acyl gellan gum. While ions affect dispersion and particle-swelling behavior, the full hydration of high-acyl gellan gum is relatively insensitive to ions. High-acyl gellan gum hydrates between 70°C and 80°C (158°F and 176°F), even in relatively high ion concentrations (see Fig. 8.7). In contrast

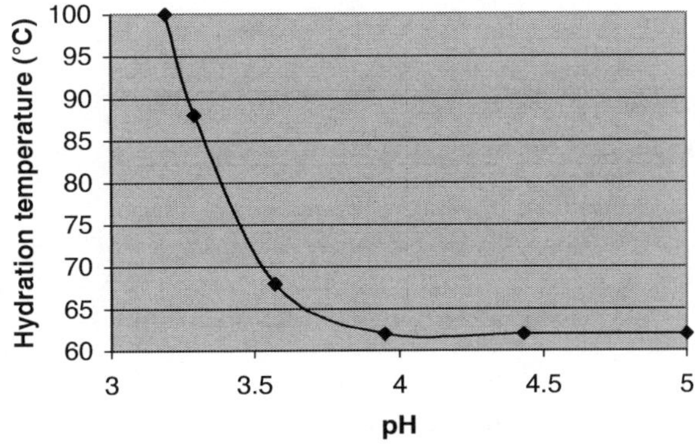

Fig. 8.5 Effect of pH on low-acyl gellan gum hydration temperature. (Reproduced with permission from C.P. Kelco.)

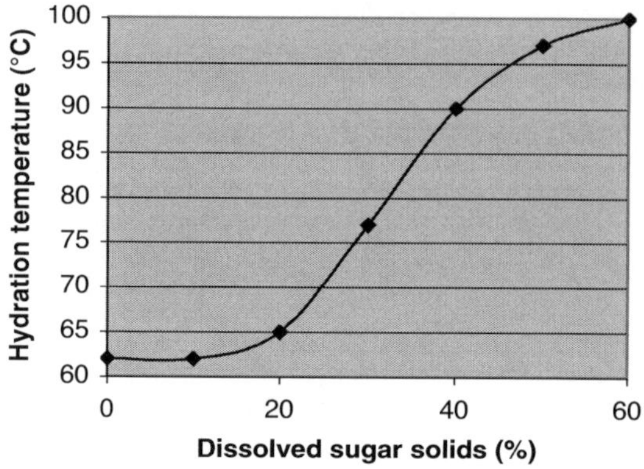

Fig. 8.6 Effect of dissolved sugar solids on low-acyl gellan gum hydration temperature. (Reproduced with permission from C.P. Kelco.)

to low-acyl gellan gum, the calcium effect on high-acyl gellan gum is small. Because calcium has little influence on the hydration temperature, sequestrants do not facilitate hydration.

Both low- and high-acyl forms of gellan gum can be dispersed directly in milk and will hydrate during normal heat processing. As mentioned already, dissolved solids and low pH inhibit gum hydration for both forms of gellan gum. In acidic environments, the pH must be above 4 for good hydration. In high-solids systems, extra care must be taken to ensure that the gellan gum hydrates. A fine mesh product is needed to facilitate hydration in the presence of sugars. Gellan gum's sensitivity to monovalent ions increases in high-solids systems, so

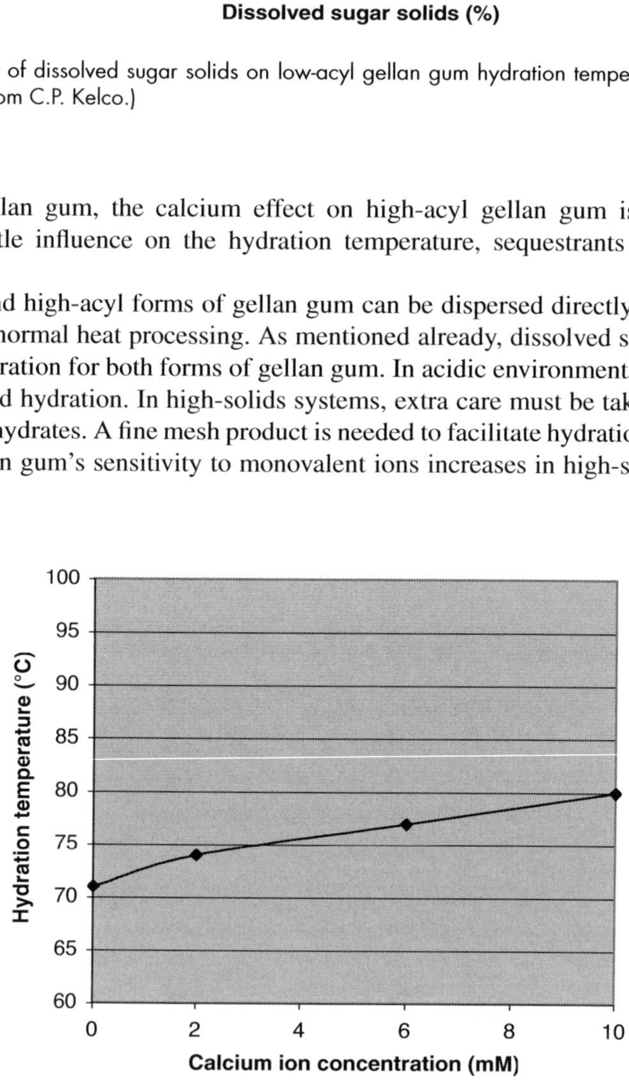

Fig. 8.7 Calcium ion effect on low-acyl gellan gum hydration temperature. (Reproduced with permission from C.P. Kelco.)

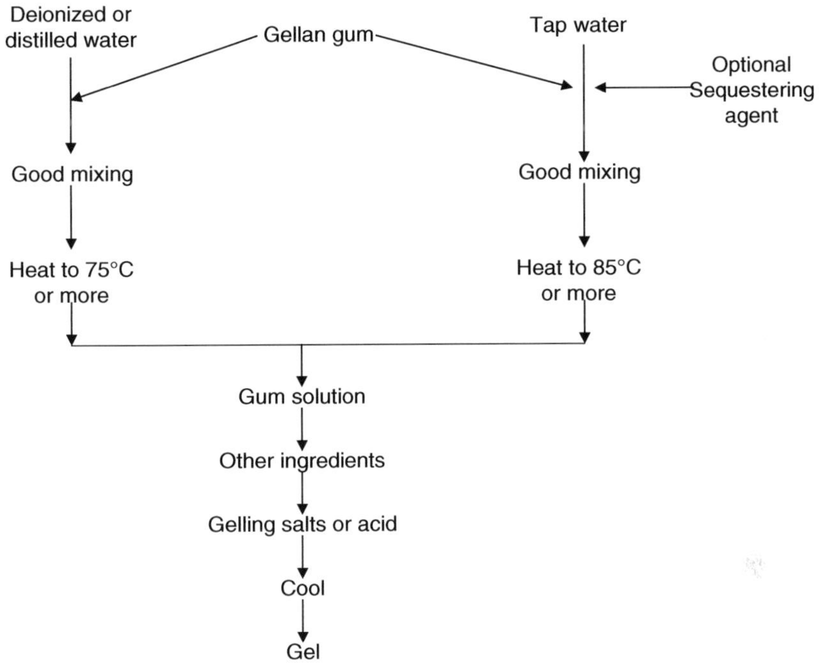

Fig. 8.8 Suggested hydration procedures for gellan gum. (Reproduced with permission from C.P. Kelco.)

high sequestrant levels will inhibit, rather than aid, gum hydration. The inhibition effect can be avoided by keeping sodium citrate levels less than 0.2%.

An overall strategy for gellan gum hydration is given in Fig. 8.8. While designed for the low-acyl form of the gum, these conditions will also permit full hydration of the high-acyl form. As a rule, the gellan gum is added to solutions with low levels of calcium, moderate pH and dissolved sugar levels of less than 25%. Once hydrated, salts, acid and additional sugars can be added to form the final food.

8.4.3 Gel properties of low-acyl gellan gum

Gellan gels can be formed over a broad range of conditions. The matrix shown in Fig. 8.9 describes the gel properties that can be expected in each of the four quadrants. Figures 8.9–8.11 show how pH and dissolved sugar solids interact.

Quadrant 1 represents a neutral pH value with a low level of dissolved sugar. Much of the literature published on gellan gum has focused on gels in this domain. Because the pH is not acidic, the carboxyl groups on the gellan gum molecule interact with sodium, potassium, calcium or magnesium counter-ions. This interaction results in a high degree of molecular association and the formation of a long-range network or gel. Gellan gum is an effective gel former in this quadrant, creating demoldable gels at gum concentrations as low as 0.05%.

With only a small amount of dissolved sugar, gellan gum is in a low-viscosity, aqueous environment. With high molecular mobility, a gel forms quickly upon cooling. The exact

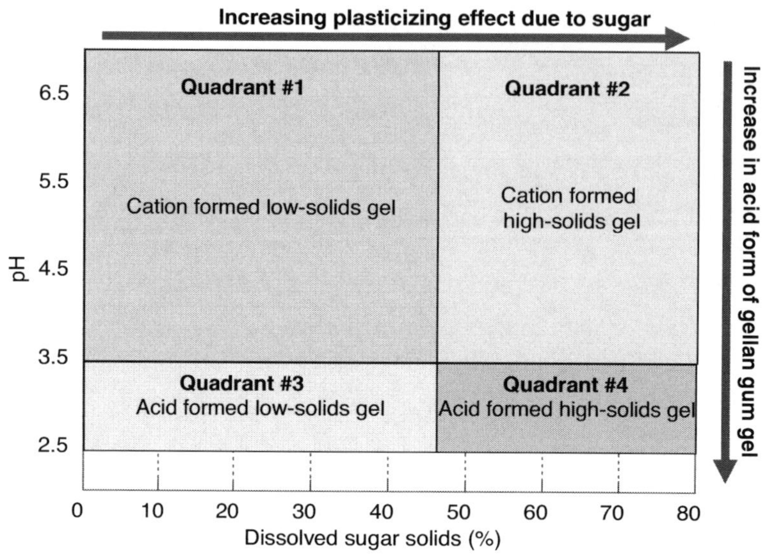

Fig. 8.9 Low-acyl gellan gum texture matrix. (Reproduced with permission from C.P. Kelco.)

temperature of gelation depends upon the cations present and, to a certain extent, upon the cooling rate. The average temperature for gelation is 30–45°C (86–113°F) (see Fig. 8.10).

Complete gelation occurs almost instantly at the gelling temperature, so there is little change in gel properties with storage time. Calcium tolerance is good because gel properties are stable over a range of calcium concentration from 5 to 15 mM, as shown in the neutral pH curve in Fig. 8.11. Unless low levels of gelling ions are used, these gels do not remelt.

Quadrant 2 in Fig. 8.9 covers gels formed at neutral pH and higher sugar levels. Gels prepared under high-solids conditions include confections, icings and the like. Figure 8.12

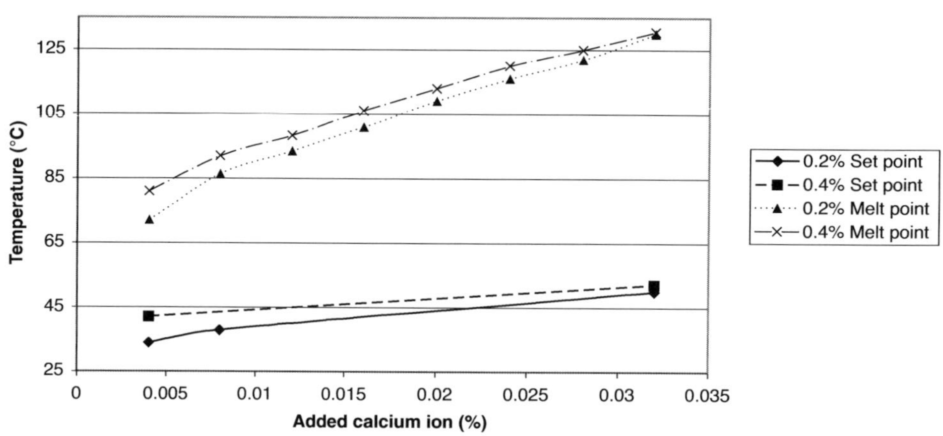

Fig. 8.10 Calcium effects on low-acyl gellan gum set and melt temperatures. (Reproduced with permission from C.P. Kelco.)

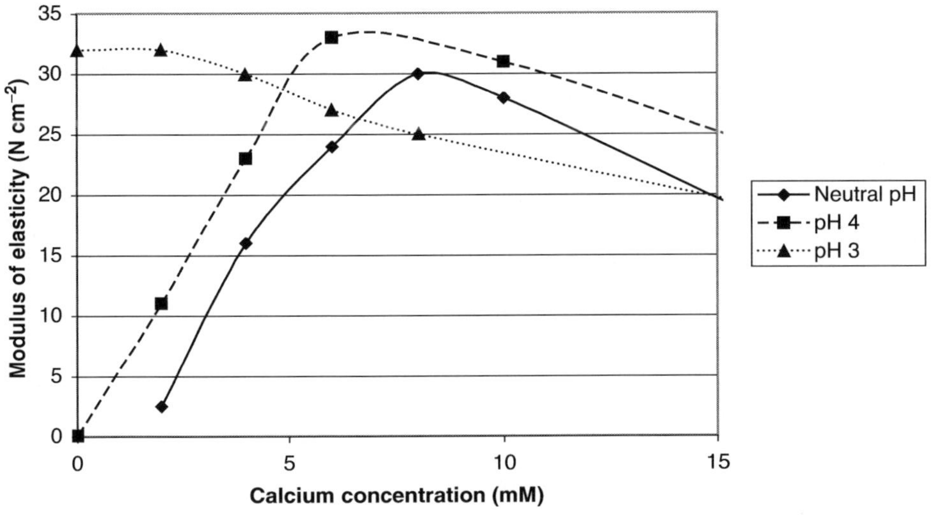

Fig. 8.11 Effect of calcium concentration on modulus of elasticity of low-acyl gellan gum gels at different pH values. (Reproduced with permission from C.P. Kelco.)

shows how the strength of these gels varies with calcium level. Compared to the low solids gels in Quadrant 1, these gels are softer and much less calcium sensitive. Because gellan gum is less effective at forming demoldable gels in Quadrant 2, higher gum levels, typically 0.3% or higher, are required.

At higher dissolved solids, the gellan gum molecules are dispersed in a high-viscosity environment. In general, sugars have a plasticizing effect on the gel. Associations between

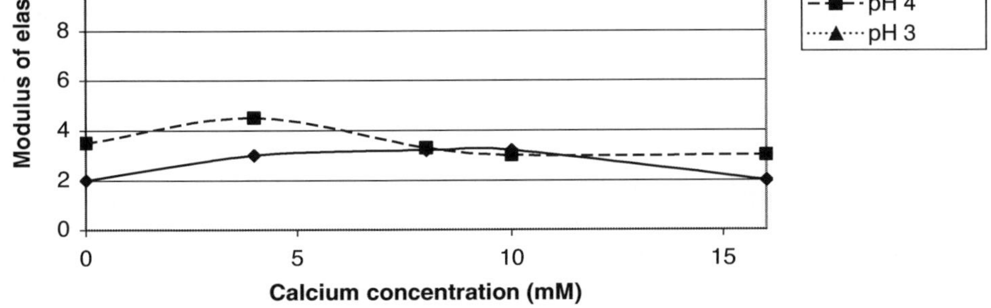

Fig. 8.12 Effect of calcium concentration on modulus of elasticity of high-solids, low-acyl gellan gum gels at different pH values. (Reproduced with permission from C.P. Kelco.)

molecular chains of gellan gum are slower to form and they are not as extensive once they do form. Like Quadrant 1 gels, the Quadrant 2 gels form with cooling. Unlike the Quadrant 1 gels, however, gel formation is not always fast in high-solids systems. Even at low temperatures, it can take weeks for gellan gum to develop its full structure. Once formed, high-solids gels are softer and more elastic and they have a greater ability to re-heal after being sheared.

Different sources of sugar solids, such as sucrose, glucose and fructose, affect both viscosity and molecular mobility of the gellan gum molecules, so the exact nature of the dissolved solids has a strong influence on the properties of the gellan gum gel. Sucrose has a more pronounced influence on the set time and gel texture than does fructose or glucose. Sucrose lengthens the set time and softens the gel and it also increases the gel's ability to re-heal after shearing.

Quadrant 3 represents gels with fairly low dissolved sugar levels and acidic pH values. When the pH goes below 3.7 or so, most of the carboxyl groups on the gellan gum molecule are protonated, so they cannot interact with cations. The curves for pH 3 and 4 in Fig. 8.11 show how calcium ions have far less effect on these gels, especially as the pH is lowered. Acid-induced gellan gels do not need calcium to set. Demoldable gels can be made at gum concentrations of 0.05% or less in this quadrant.

Compared to Quadrant 1 gels formed with ions at low solids, low-solids, acidic, gellan gels are firmer but they are quite brittle. The high brittleness of these gels is often perceived as soft and mushy because the gels break under low levels of strain. The gels have slightly more tendency to synerise than do gels at a more neutral pH. Like Quadrant 1 gels, gel formation in Quadrant 3 occurs quickly with cooling. However, acid-induced gelation generally occurs at a lower temperature than ion-induced gelation, and the resulting gels are not as thermostable. In this quadrant, the gels will not re-heal after shearing.

Quadrant 4 is for acidic gels formed with high levels of dissolved sugar solids. Figure 8.12 shows that the texture of these gels is relatively insensitive to calcium. As the pH drops below the pK_a of the gellan gum, these gels get more and more firm because the gelation mechanism changes from ion induced to acid induced. With the plasticizing effect of sugars, however, the gel textures do not become overly brittle like the Quadrant 3 gels. Because gellan gum is less effective at forming demoldable gels in Quadrant 4, higher gum use levels, typically 0.3% or higher, are required.

The gelation temperature in Quadrant 4 depends on the level of the sugar solids, as well as the pH and the ion levels. High levels of solids and ions increase the set temperature, while higher pH lowers the set temperature. Like Quadrant 2 gels, the composition of the sugar solids affects both the setting temperature and time. The setting temperature and time are reduced when sugar solids come from fructose, dextrose or sugar alcohols. Sucrose, however, increases the gel temperature and extends the gel set time. In most cases, these acid-induced, high-solids gels do not remelt. As the gels get firmer with lower pH, they also lose the ability to re-heal after shearing.

8.4.4 Gel properties of high-acyl gellan gum

Overall, the high-acyl form of gellan gum is not as sensitive to cation levels as the low-acyl gellan products. Figure 8.13 shows that large changes in the calcium level contribute only small changes in gel properties. Slightly firmer gels result as the calcium level is increased. Similar effects are seen for magnesium and, to a lesser extent, for sodium and potassium ions.

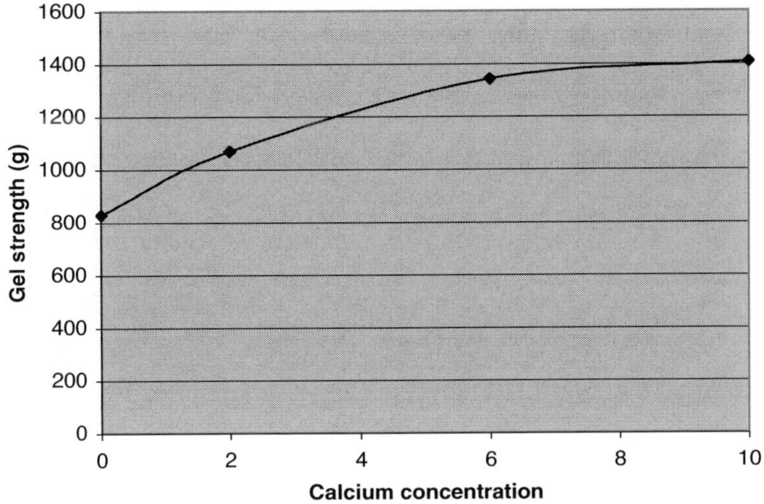

Fig. 8.13 Effect of calcium concentration on gel strength of high-acyl gellan gum gels. (Reproduced with permission from C.P. Kelco.)

Adding more dissolved sugar solids to the high-acyl form of gellan gum tends to strengthen the gel slightly (see Fig. 8.14). Again, the effect is less for this form of gellan gum than for the low-acyl form. At solids levels greater than 50%, however, the sugars raise the set temperature, interfere with gelation and plasticize the gel. Higher levels of high-acyl gellan gum are needed to form a demoldable gel in high-solids environments.

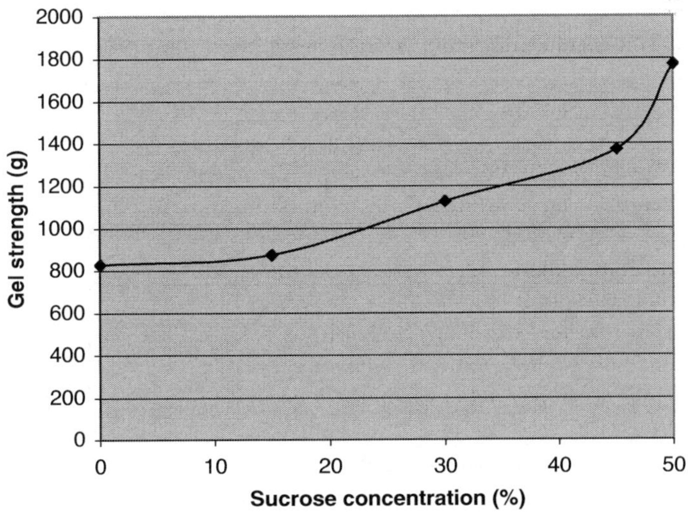

Fig. 8.14 Effect of dissolved sugar on gel strength of high-acyl gellan gum gels. (Reproduced with permission from C.P. Kelco.)

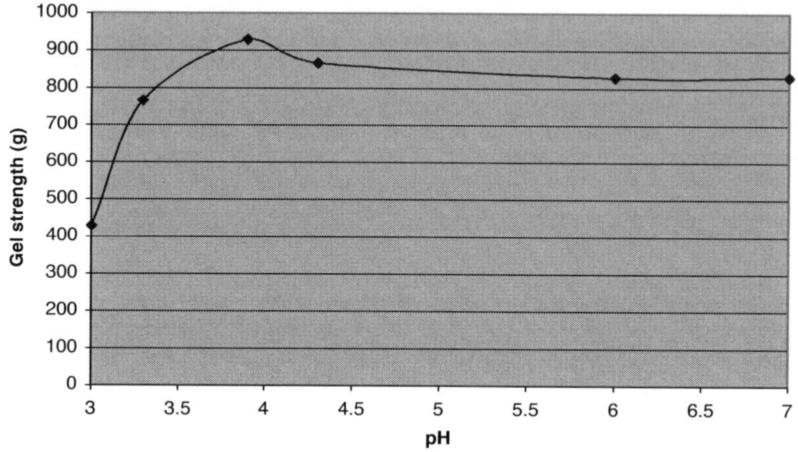

Fig. 8.15 Effect of pH on gel strength of high-acyl gellan gum gels. (Reproduced with permission from C.P. Kelco.)

Very low pH values below 3.5 make the high-acyl gellan gum gels soften somewhat as shown in Fig. 8.15. Unlike low-acyl gellan gum, there is no evidence of an acid-induced gel mechanism for high-acyl gellan gum. The gum simply loses the ability to form a gel network as the pH drops. However, over most of the pH range for foods, high-acyl gellan gum gels are not influenced by pH.

Gels made with the high-acyl form of gellan gum generally show little thermal hysteresis as shown in Fig. 8.16. The setting and melting temperatures increase as the calcium ion concentration increases. With low ionic strength gels, these two temperatures are within 3°C

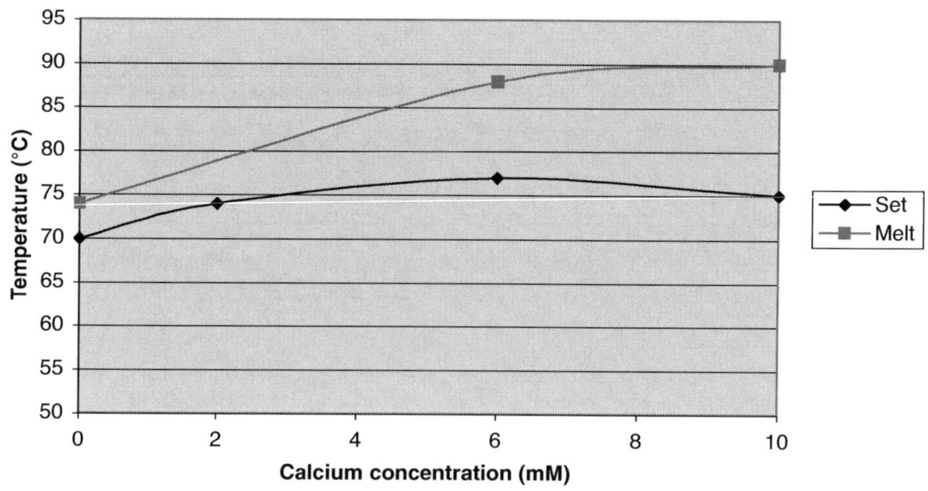

Fig. 8.16 Effect of calcium level on set/melt temperatures of high-acyl gellan gum. (Reproduced with permission from C.P. Kelco.)

of each other. With high ion levels, the difference increases to 8°C. The low ionic strength dependence and the low thermal hysteresis reveal that the high-acyl form of gellan gum forms gels as a result of interchain association, lacking the significant cooperative ion cross-linking of the low-acyl form of the gum.

8.4.5 Nongelling network properties

Both low- and high-acyl gellan gums can be used at concentrations that are too low to form a demoldable gel. While the gellan gum molecules still associate and form a long-range network, the system remains very fluid. These systems are often called 'fluid gels'.

Fluid gels exhibit a highly pseudoplastic flow property – the viscosity decreases with increasing shear. Therefore, fluid gels are perceived as thin in terms of mouthfeel, because the act of swallowing during consumption creates shear. While low in viscosity, however, fluid gels exhibit a high elastic modulus, which imparts suspension properties to the system. In fact, fluid gels made with gellan gum have a yield stress. A particle trapped in a fluid gel system will not move unless its weight or buoyancy can overcome the yield stress of the network (Fig. 8.17).

8.4.6 Gum combinations

Earlier in this chapter, it was mentioned that the two forms of gellan gum could be blended to form intermediate texture values. This feature is one of the primary strengths of gellan gum. Gellan gum blends appear to be composed of interpenetrating rather than cooperative networks of polymers (Manson and Sperling, 1976; Mao *et al.*, 2000). This structure means

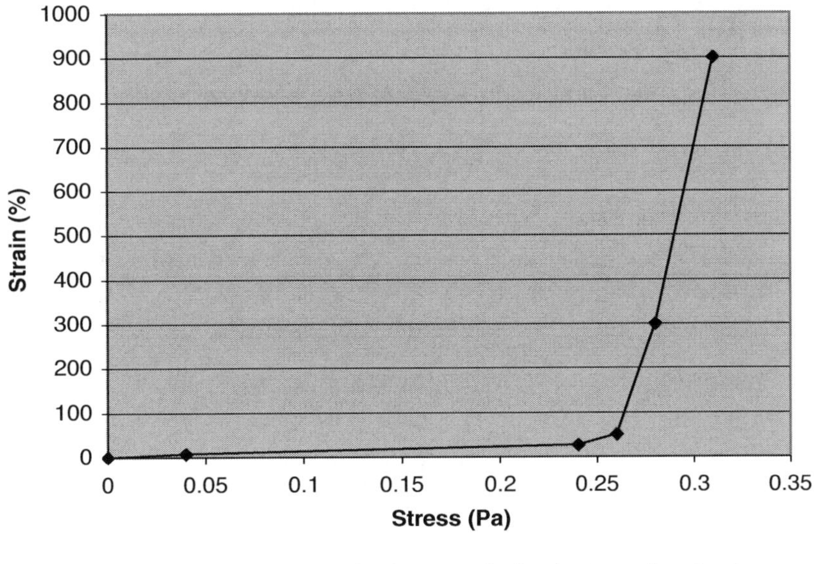

Fig. 8.17 Effect of strain on yield stress of gellan gum fluid gels. (Reproduced with permission from C.P. Kelco.)

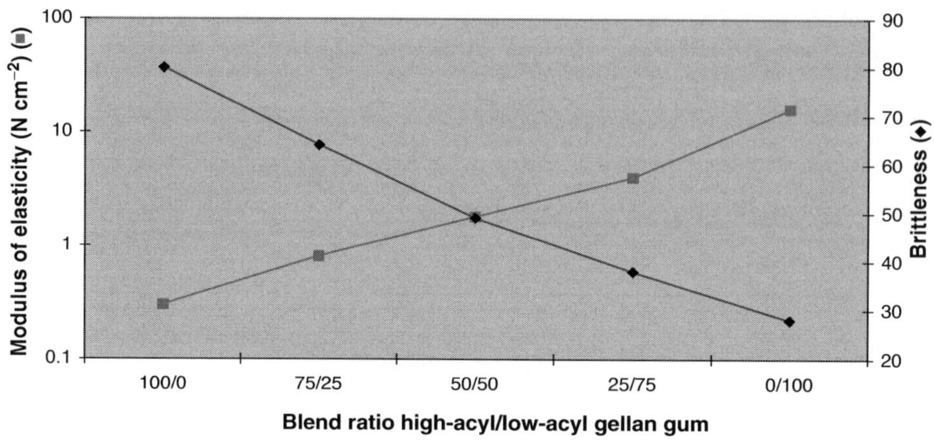

Fig. 8.18 Texture effect of blending high- and low-acyl gellan gum forms. (Reproduced with permission from C.P. Kelco.)

that, in most cases, the high-acyl form will set at a higher temperature as shown in previous figures. With further cooling, the low-acyl form sets up within the high-acyl network. This 'gel within a gel' allows for one form to smoothly transition into the other form as the ratio of low- to high-acyl gellan gum changes. This smooth transition results in the wide range of textures illustrated in Fig. 8.18. The flexibility of this blending allows formulators to produce exactly the texture that is needed for an application, while keeping most of the remaining formulation constant.

Blends with other gums are possible and useful. In general, nongelling hydrocolloids do not appreciably change the functionality or texture of gellan gum. High levels of viscosifying gums, such as xanthan gum or cellulose gum, however, can interfere with gellan gum mobility and cross-linking, resulting in softer gels. The antagonistic effect can be used advantageously to maintain a smooth texture when making fluid gels with low-acyl gellan gum.

Blending gellan gum with other gelling hydrocolloids typically results in gels with intermediate properties. For example, low-acyl gellan gum can be used to increase the firmness of soft, elastic gels such as iota carrageenan, gelatin or combinations of xanthan gum and locust bean gum. Adding gellan gum to other gelling agents can be particularly advantageous for altering the setting and melting properties of the other gum systems. For example, gellan gum can be used to raise the set temperature of gelatin-based confections and dessert gels.

Synergies with gelatin, carrageenan and xyloglucans have been reported in the literature (Blecker *et al.*, 2000; Loh *et al.*, 2002; Ikeda *et al.*, 2004). However, these synergies are likely based on ionic effects and volume exclusion rather than on direct gum interactions.

In other cases, it may be desirable to mix gellan gums with other gums to take advantage of their functional properties. For example, gellan gum can be mixed with gelatin to improve the heat stability of gelatin gels. Additionally, confectionery products, such as gummy candies, can be made without gelatin by using mixtures of carrageenan with gellan gum.

8.5 REGULATORY STATUS

Gellan gum first received approval for food use in Japan in 1988. It is now approved for food, nonfood, cosmetic and pharmaceutical use in the USA, Canada, Australia and many other countries in Latin America, South America, Asia and the European Union.

Gellan gum first appeared in the USP/NF first Supplement 1 April 2004 USP 27/NF 22. The NF monograph appears in the current USP/NF. Gellan gum is an approved FDA food additive under 21 CFR 172.665. Gellan gum appears as E418 in the European Community Directive EC/95/2 in Annex 1. Gellan gum is listed in the Food Chemicals Codex, Canada Food and Drug Act (Division 16, Table IV, G.2) and the Japanese Specifications and Standards for Food Additives (JSSFA).

Both the Joint FAO/WHO Expert Committee on Food Additives (JECFA) and the European Community Scientific Committee for Food have given gellan gum an acceptable daily intake (ADI) of 'not specified'.

On food labels, high-acyl and low-acyl gellan gum or any combination of the two may be declared under the name of 'Stabiliser: E418' or 'Stabiliser: Gellan Gum' in the EU.

8.6 APPLICATIONS

With its unique and versatile properties, gellan gum is used commercially in a wide range of food applications. These applications can be generally categorized into six application areas:

- Water-based gels
- Bakery applications
- Dairy foods
- Beverages
- Confections
- Fruit applications

Within these broad categories, a number of specific applications of gellan gum are discussed in greater detail.

8.6.1 Water-based gels

The most common commercial use of gellan gum is creating water-based gels.

8.6.1.1 *Dessert gels*

In dessert gels, low-acyl gellan gum creates a firm, brittle texture with excellent clarity and heat stability. Because it is highly efficient at forming gels, the gellan gum use level is very low, with typical use levels for dessert gels ranging from 0.15 to 0.35%. Gellan gum also provides particularly good flavor release compared to other gums (Morris, 1994; Costell *et al.*, 2001). However, the brittle texture of low-acyl gels is not typical of common dessert gels. Therefore, low-acyl gellan gum is often blended with other gelling agents to provide softer, more elastic textures. A blend of low-acyl gellan gum with xanthan gum and locust bean gum, targeted for a dessert gel texture, is commercially available.

The recent commercialization of clarified, high-acyl gellan gum also makes it possible to create a range of textures using blends of low- and high-acyl gellan gum. The ingredient declaration of the resulting product is simplified because both types of gum are still labeled as gellan gum. A ratio of 80% high-acyl gellan gum to 20% low-acyl gellan gum results in a gel texture that is similar to the texture of a gelatin dessert gel.

Sensitivity to heat and acid limits both the batch size and the pasteurization conditions for most gums that are used in dessert gels. Gellan gum's exceptional stability to heat and acid, however, allows larger batch sizes and more flexibility with pasteurization conditions. For example, unpasteurized fruit can be added to the cup before filling with the gel and the pasteurization conditions can be extended, so that the fruit is fully pasteurized in the sealed cup. Gellan gum's acid stability also allows dessert gels to be formulated at lower pH to enhance the flavor impact of the gels.

Multi-layer RTE (ready-to-eat) dessert gels are popular in Asia. The different layers can contain different colors, textures and flavors. Because gellan gum has a high melting temperature, hot gum solutions can be poured on top of a gelled gellan gum layer without melting it. Gellan gum can also be used to make heat-stable colored beads or shaped novelties that can be used as inclusions in a clear gel matrix.

8.6.1.2 *Drinking jellies*

Drinking jellies are formulated in a similar way to dessert gels, but lower levels of gelling agents are used to create a very soft gel. There are different types and textures of drinking jellies, depending on how they will be consumed. Some drinking jellies are soft enough to be sucked through a straw. Others will be crushed and squeezed out of a pouch.

In many Asian countries, drinking jellies are consumed simply for their interesting textures. Increasingly, however, gellan gum drinking gels are being formulated to provide enhanced nutrition or nutraceutical benefits. Sports gels are an example of such a product. They are typically packaged in pouches with tear-off tops for consumption while exercising. The gelled contents provide water and nutrients without spilling and the juicy gel bits slide down the throat without leaving a sticky coating.

Low-acyl gellan gum is typically preferred in these applications because the brittle gels are easy to crush and break. Once broken, the gels quickly exude water, enhancing flavor release and increasing the lubricity of the gel. Typical gellan gum use levels range from 0.05% to 0.1%. However, drinking jellies with different textures can be formulated by adding other gums, including high-acyl gellan gum.

8.6.1.3 *Asian foods*

In many Asian countries, water-based gels are a traditional food. In Japan, for example, mitsumame is a popular dessert, made of firmly gelled cubes. Tokoroten, a clear jelly noodle, is another example of a popular Asian water-based jelly. Traditionally, these dishes were made with agar-agar, but the firm, brittle texture of low-acyl gellan gum is well suited for these gels.

In Asia, gellan gum is often used to simulate the texture of some traditional foods such as bird's nest soup. Authentic bird's nest soup is made with the dried saliva from the nests of swiftlet birds, but it is expensive due to its limited supply. Gellan gum is often used to create an inexpensive imitation bird's nest. Low-acyl gellan provides a fine, brittle texture and provides heat stability for pasteurizing or retorting these products.

8.6.2 Bakery applications

Gellan gum is used in the commercial bakery applications almost as frequently as it is used in water gels. Many different applications are found in the bakery segment.

8.6.2.1 *Bakery mixes*

Gellan gum is highly functional in bakery dry mixes. Unlike some other gums commonly found in such products, low-acyl gellan gum does not readily hydrate in cold water. Using low-acyl gellan gum in a bakery formulation does not increase the batter viscosity because the gum does not hydrate immediately. However, gellan gum hydrates in the oven during baking, so it provides moisture retention and shelf life extension in the final baked goods. Gellan gum finds most of its use in dry mixes formulated to produce thick batters, like brownies, where excessive viscosity would impede batter spread in a baking pan. The textural effects of gellan gum are obtained at low use levels, typically ranging from 0.08% to 1.0% gum.

8.6.2.2 *Bakery fillings*

When considering gellan gum in bakery filling applications, it is convenient to discuss bakery fillings in terms of low-solids, medium-solids and high-solids fillings. Low-solids fillings have up to 45% soluble solids, medium solids fillings have 45–65% soluble solids and high-solids fillings have in excess of 65% soluble solids.

Low-solids fillings

Because of its very high melting temperature, low-acyl gellan gum is used to improve the bake stability of low-solids fillings. High melting temperatures are obtained when the gels are set by calcium cross-linking. Used alone, low-acyl gellan gum creates a firm brittle texture that is difficult to pump or spread. However, when used with modified starch, a smooth pumpable texture with good flavor release and heat stability can be obtained. Typical use levels for gellan gum range from 0.08% to 0.3%.

Texture and the pumpability of the filling can be adjusted by changing the levels and ratio of gellan gum and starch. Divalent salts, like calcium, may be added to increase the firmness and heat stability of the filling.

A combination of low-acyl and high-acyl gellan gum is appropriate when the filling is applied post-baking.

Medium-solids fillings

In medium-solids fillings, gellan gum can be used without starch when sucrose is the major source of solids. Calcium cross-linked gels can be sheared and pumped with a smooth texture that will re-heal with time. However, the softening effects of sucrose on the gel network generally call for higher levels of gum to maintain heat stability. Use levels typically range from 0.25% to 0.40% gum.

Special care must be taken when the pH of the filling is below 3.5. The combination of calcium cross-linking and acid gelation with higher solids can result in very high setting temperatures. Formulating gellan gum fillings with high levels of calcium at low pH can result in pregelation. Because of this, most medium solids fillings with gellan gum are formulated with high fructose corn syrup because fructose tends to lower the setting temperature. With higher levels of fructose, the gels are firmer, but starch is usually needed to provide a smooth texture.

High solids fillings

In high solids, water mobility is severely limited. Gellan gum gel does not form a gel network using ionic cross-links in this environment. Control of gellan gum properties in this environment is only available through the manipulation of pH. As the pH approaches 3.0, a very firm, heat-stable gel can be formed. Using starches to modify the gel to give a smooth texture, this system can provide an excellent, high-solids bakery filling. Because sugars interfere with the gel network, higher levels of gellan gum are needed to provide good heat stability compared to lower solids fillings and typical use levels range from 0.3% to 0.7% gum.

8.6.2.3 *Binding systems*

Gellan gum is an effective binder and film former. Binding is particularly important for producers of nutritional bars and high protein bars. These products typically contain protein nuggets that are bound together with a mixture of corn syrup and stabilizer. If the binding system contains too much moisture, the bar will become soggy. The binding system should also be short textured and nonsticky. Low-acyl gellan gum is commonly used in nutritional bars because it can be hydrated directly in corn syrup or glycerin.

8.6.2.4 *Glazes, icings and frostings*

The bulk composition of glazes, icings and frostings is undissolved powdered sugar. Stabilizers are effective only in the saturated sugar syrup that glues the particles of sugar together. Because the soluble solids in the syrup are mostly sugar, working with gellan gum in this system can be challenging.

To be effective, gellan gum must form a network or a gel, but the high level of sucrose interferes with calcium cross-linking. The secret to using gellan gum successfully in these systems is to create a sodium-crosslinked gel. Without any sodium ions, no network will be formed. However, too much sodium weakens the network. Sodium control is typically achieved through the use of sodium citrate, which is commonly used as a sequestrant and hydration aid. The optimum level of sodium citrate depends on the amount of water in the formulation, but the saturated syrup should not contain more than 0.1% sodium citrate. The gel network sets at approximately 37°C. The sucrose gel is thixotropic, so shearing through the set point is not a problem. However, there is an increase in the icing viscosity at the set temperature, so, for glazes and flat icings, it is advantageous to apply the icing before this viscosity develops. For frostings, it is best to shear though the set point.

8.6.3 Cultured dairy foods

Cultured dairy foods are based on the fermentation of milk. Typical products include yogurts, sour creams and cheeses. Gelatin is commonly used to stabilize many of these products, but it is not acceptable for consumers following vegan, Kosher or Halal diets.

Yogurts can be produced using two methods: (1) cup set and (2) stirred. Cup set yogurt is cultured directly in a serving-sized cup, which may have fruit on the bottom. Stirred yogurt is cultured in a large vat and, after fermentation, the curd is stirred with a fruit preparation and deposited into a cup. Both low- and high-acyl gellan gums can be used in a stirred yogurt, but cup-set yogurts are made only with high-acyl gellan gum. Low-acyl gellan gum creates a lumpy texture after culturing that requires mixing to create a smooth texture.

Gellan gum is typically added to the raw milk prior to homogenization and pasteurization. These processes, which are typical in yogurt production, will hydrate the gum. The use of low-acyl gellan gum must be limited to less than 0.06% or a grainy texture will result. High-acyl gellan gum can be used at up to 0.1% gum before excessive graininess develops. At these use levels, gellan gum adds a light texture and significantly reduces whey-off. Pectin or starch can be added to build a heavier body if desired.

Sour cream production is analogous to yogurt, but the fermentation includes butterfat. Like yogurt, sour cream is commonly stabilized with gelatin. Both low- and high-acyl gellan gums are used to stabilize this product. Low-acyl gellan gum is used to add heat stability to the sour cream, so that it retains structure after being added to hot foods. High-acyl gellan gum is used to provide a thixotropic rheology to the sour cream, allowing the structure to reform after stirring.

8.6.4 Beverages

Shelf-stable, ready-to-drink (RTD) juice beverages are steadily growing in popularity. However, one drawback is that many juices suffer from both juice cloud settling and pulp separation during storage.

Gellan gum can be used to overcome cloud and pulp settling in juices while providing a mouthfeel that is light and refreshing when compared to other stabilizers. Gellan gum can stabilize beverages with a pH as low as 3.0. Gellan gum has low protein reactivity which makes it compatible with a wide variety of juices. Gellan gum is easy to disperse and hydrate and so it can be used in most juice processing plants without having to add special mixing equipment.

Combinations of gellan gum and pectin are particularly advantageous in this application. A special blend high-acyl gellan gum and pectin is commercially available for this application. The gum blend is added to the juice prior to sterilization. Hydration is achieved simply by heating the juice to 85°C (185°F) for 30 s. These processing conditions are commonly used to sterilize juices in hot-fill bottling operations. For optimum stability of juice clouds, pulp and calcium salts with minimal increase in mouthfeel, the high-acyl gellan gum and pectin blend should be used at levels from 0.25% to 0.30%.

Neutral pH dairy beverages, including chocolate and other flavored milks, ready-to-drink coffee or tea and nutritional beverages, can be stabilized with gellan gum. Carrageenan is often used in these systems because of its synergistic interaction with milk caseins. Gellan gum does not have the same synergy with milk, but its independence of milk proteins allows it to be used in products with low milk protein and in systems with low-quality or heat-damaged proteins as will be found in spray-dried milk powders.

Most grades of high-acyl gellan gum are standardized on gel strength but this is not related to its suspension properties. Standard grades of high-acyl gellan gum may also develop off-flavors in neutral pH, long shelf life milk systems. Consequently, Kelcogel® HM-B gellan gum has been specially developed and standardized as a beverage stabilizer for use in neutral pH, long shelf life milk systems without developing off-flavors with storage time.

Soy milk stabilization can be more difficult to stabilize than dairy milk because soy sources usually contain large amounts of insoluble material. Carrageenan does not have a strong synergistic interaction with soy protein, so higher levels of carrageenan and higher viscosities are needed to fully stabilize the beverage. High-acyl gellan gum functions independently of soy protein type or concentration. This independence is particularly advantageous because the types of soy protein vary from ground whole beans or concentrates and isolates. A special grade of gellan gum has been specially developed to work as a beverage stabilizer in soy protein systems.

8.6.5 Confections

Low-acyl gellan gum is used commercially in high-solids, gelled confections. It produces jellies with short texture and good flavor release. Because of the high melting temperature, low-acyl gellan gum is also used to increase the melting temperature of gelatin gummy confections. The gum use level depends on the structure needed. A soft textured jelly can be made with 0.35% gum, but use levels around 0.75% are needed to produce a firm jelly.

To make smooth-textured confections, low-acyl gellan gum must be hydrated in the absence of divalent ions. Unsequestered divalent ions will create a grainy texture during cook-up. It is also difficult to add divalent ions to a high-solids gellan gum solution and control the gelation. Therefore, it is typical to hydrate the gum using low levels of sequestrants, such as sodium citrate, and then to use acids to set the gel after cooking. The gel texture is largely controlled by the pH, but calcium released from the sequestrant at low pH will raise the set temperature and shorten the set time. By controlling both pH and calcium, it is possible to obtain a variety of textures and setting profiles.

Gellan gum can be used to improve the functionality of other hydrocolloids in confections. Confections made from starch can take several days to develop a demoldable texture. Adding a small amount of gellan gum to a starch formulation significantly reduces the set time, so the gels can be demolded on the same day. Gummy confections made with gelatin also benefit from the addition of gellan gum. Gellan gum increases the heat stability so that the individual candy pieces do not melt together when exposed to a warm environment.

8.6.6 Fruit applications

Fruit bakery fillings were discussed in Section 8.6.2. However, gellan gum is used in a few additional fruit applications, including low-solids jam and yogurt fruit preparations.

With increased consumer interest in nutrition and diet, food manufacturers are continually challenged to create high-quality, low-sugar food products. This is particularly true in the development of imitation jams or spreads which cannot rely upon sugar and corn syrups to provide body and texture in the final product. The challenge for manufacturers of these products is to develop the necessary physical properties of set, spread and stability, while maintaining natural fruit flavors and colors. Pectin has long been used in the manufacture of jams and jellies, but its gelling mechanism requires some sugar and a relatively low pH. Gellan gum is used to work around these key formulation requirements, allowing low calorie jams to be formulated without sugar.

Yogurt fruit preparations use gelling agents to provide viscosity and suspension of fruit pieces. High-acyl gellan gum has ideal properties in this application. With low calcium and protein sensitivity, the powder can be added directly to the fruit and it can be heated just like starch. With a high setting temperature, high-acyl gellan gum suspends fruit at a use levels from 0.08% to 1.0% gum, making it one of the most cost-effective stabilizing systems available.

8.6.7 Miscellaneous applications

Gellan gum can also be in sauces, films and adhesion systems.

Gellan gum fluid gels are used to suspend herbs in no-fat salad dressings. It is also used in full fat dressings that are sold with separating oil and aqueous layers. The fluid gel is used to suspend herbs in the aqueous layer.

Fluid gels are also used in some hot sauces to suspend particulates of chili peppers. Gellan gum is able to tolerate the harsh environment while minimizing viscosity without masking flavors.

Gellan gum solutions can be dried to form a film. The film is a poor moisture barrier, but it is an excellent oil barrier. Gellan gum can be used to coat substrates such as french fries or chicken. It will then act as an oil barrier during subsequent frying, resulting in a significant reduction of fat in the final product.

8.7 FUTURE DEVELOPMENTS

Currently, gellan gum is available in its high or low-acyl forms. However, it is theoretically possible to produce gellan gum with intermediate levels of acyl content. The acyl content will dramatically affect the gum functionality so it may be possible to target very specific gel textures and properties by making products with a different acyl content.

With improvements in manufacturing processes, it may also be possible to produce gellan gum with significantly greater molecular weight. Gellan gum with exceptionally high gel strength would result from the higher molecular weight. This material would make gellan gum more cost-effective in many applications and it would also be expected to expand gellan gum's functionality into new applications.

References

Blecker, C., Razafindralambo, H. and Paquot, M. (2000) *Chemical Reactions in Foods IV*, September 2000 Conference Proceeding, Institute of Agricultural and Food Information, Federation of European Chemical Societies, Prague, **2000**, 83–85.

Chandrasekaran, R., Millane, R.P., Arnott, S. and Aktins, E.D.T. (1988a) The crystal structure of gellan. *Carbohydrate Research*, **175**, 1–15.

Chandrasekaran, R., Puigjaner, L.C., Joyce, K.L. and Arnott, S. (1988b) Cation interactions in gellan: an X-ray study of the potassium salt. *Carbohydrate Research*, **181**, 23–40.

Chandrasekaran, R., Radha, A. and Giacometti, A. (1992) Roles of potassium ions, acetyl and L-glycerate groups in native gellan double helix: an X-ray study. *Carbohydrate Research*, **224**, 1–17.

Chandrasekaran, R. and Thailambal, V.G. (1990) The influence of calcium ions, acetate and L-glycerate groups on the gellan double helix. *Carbohydrate Polymers*, **12**, 431–442.

Costell, E., Peyrolon, M. and Duran, L. (2001) Influence of texture and type of hydrocolloid on perception of basic tastes in carrageenan and gellan gels. *Food Science and Technology International*, **6**, 495–499.

Ikeda, S., Nitta, Y., Kim, B.S., Temsiripong, T., Pongsawatmanit, R. and Nishinari, K. (2004) Single-phase mixed gels of xyloglucan and gellan. *Food Hydrocolloids*, **18**(4), 669–675.

Kuo, M.-S., Mort, A.J. and Dell, A. (1986) Identification and location of L-glycerate, an unusual acyl substituent in gellan gum. *Carbohydrate Research*, **156**, 173–187.

Loh, J., Dulin, D.A. and Cha, A.S. (2002) Anionic hydrocolloid gums as gelatin extenders. US Patent Application 20020164412.

Manson, J.A. and Sperling, L.H. (1976) *Polymer Blends and Composites*. Plenum Press, New York and London.

Mao, R., Tang, J. and Swanson, B. G. (2000) Texture properties of high and low acyl mixed gellan gels. *Carbohydrate Polymers*, **41**, 331–338.

Morris, M. (1994) Rheological and organoleptic properties of food hydrocolloids. In: K. Nishinari and E. Doi (eds), *Food Hydrocolloids: Structures, Properties and Functions*. Plenum Press, New York, pp. 201–210.

9 Gum Tragacanth and Karaya

Jenny M. Mayes

ABSTRACT

Gum tragacanth and gum karaya are powdered exudate gums originating from Asia which can be used in a wide range of food and pharmaceutical applications. Gum tragacanth has a long history of safe use as a thickener, binder and emulsifier. The excellent acid stability of gum tragacanth makes it a particularly suitable stabiliser for acidic sauces and dressings. The price of both gums has risen over recent years due to supply shortages and competition from xanthan gum has seen a reduction of many traditional food markets. Gum karaya is used mainly in pharmaceutical applications as an adhesive, thickener and binder. In spite of the restricted food use of karaya in Europe, this acid-stable gum can be used in a diverse range of food products as a thickener and stabiliser.

9.1 GUM TRAGACANTH

9.1.1 Introduction

The natural plant exudate gum tragacanth has been used for centuries as an emulsifier and thickener in foods. The name is derived from the Greek *tragos* (goat) and *akantha* (horn) and reflects the shape of the curved exudate. In spite of the availability of alternative materials, the continued use of the gum is the result of its unique functional properties combined with a high degree of stability in a range of conditions.

In the *European Pharmacopoeia* (6th edition, 2007), *gum tragacanth* is defined as 'the air-hardened gummy exudates, flowing naturally or obtained by incision from the trunk and branches of *Astragalus gummifer* Labillardiere and certain other species of *Astragalus* from western Asia'. The *Food Chemical Codex* (6th edition, 2008) defines *gum tragacanth* as the 'dried gummy exudation obtained from the stems and branches of *Astragalus gummifer* Labillardiere or other Asiatic species of *Astragalus* (Family Leguminosae)'. The two main commercially exploited species are *Astragalus gummifer* Labillardiere and *Astragalus microcephalus* Willd which are grown predominantly in Iran.

Approximately 1000 tonnes per annum of gum tragacanth were used worldwide until the early 1980s when demand fell to approximately 300 tonnes per annum. The introduction of the newly approved xanthan gum replaced tragacanth in many thickening applications at this time and the Gulf War made tragacanth supplies erratic. Also, increasing labour costs combined with the Iranian government's attempt to fix gum prices led to higher export costs.

At the beginning of the twenty-first century, the availability of good-quality gum tragacanth has improved and price fluctuations have settled.

The US Food Chemical Codex classifies the *gum* as 'generally recognised as safe' (GRAS). The Joint WHO/FAO Expert Committee on Food Additives (JECFA) classifies the gum as 'acceptable daily intake (ADI) not specified' reflecting the safe use of gum tragacanth over the centuries. The Scientific Committee for Food of the European Community lists gum tragacanth as E413 in the Food Additives other than Colours and Sweeteners Directive 1995, and as gum tragacanth is listed in Annex I, it has unrestricted status. In Europe, therefore, the gum may be used *quantum satis* in foods other than those foods listed in Article II of the legislation. Although allergenic responses to gum tragacanth have been reported (Stauffer, 1980), tragacanth is not listed in Annex IIIa of the European Directive 2003/89/EC (as amended) and is not subject to allergen labelling in Europe.

9.1.2 Raw materials

There has been very little change in the traditional agricultural practices over the years and, as yet, no known genetically modified species of *Astragalus* exist.

The gum is obtained from small, thorny shrubs of the commercial *Astragalus* species grown in Iran, Syria and Turkey. The perennials have a large taproot and small stem that are tapped for gum, which hardens on the stem in the dry, hot climate to form a tough opaque material. The gum is exuded as a pale yellow to off-white, bulky and irregularly shaped material, known as flake grade. However, the best quality material comes from gum exuded as flat, white to off-white or translucent ribbons, sold as ribbon grades.

Using a knife, longitudinal incisions are made in the lower stem of the shrubs in May or June. Collection of the exudate occurs in the hot summer months of August and September for the high-quality ribbon grades and in August to November for flake grades. Wet and windy conditions during the exudation period result in a poorer quality, discoloured gum. Collected material is hand-sorted into the various grades of flake or ribbon. The commercial producers in Iran and Turkey have different grading systems. The less complex Iranian system sorts the ribbon into grades numbered 1, 2, 3, 4 and 5. The flake grades are numbered 26, 27, 28 and 55. All remaining gum is classified as nubs and siftings. The gum grades and their viscosity properties are given in Table 9.1.

Table 9.1 Commercial grades and viscosity of gum tragacanth.

			Approximate viscosity range	
Iranian grade		**Turkish grade**	**Redwood (s) (0.44%, 20°C)**	**Brookfield (mPa s) (1.0% 25°C, 20 rpm)**
Ribbon	1	Fior Extra	350–600	2200–3400
	2		250–400	1800–2500
	3		200–350	1400–2000
	4	Fior	120–170	1000–1600
	5		80–100	800–1000
Flake	26		70–85	600–800
	27	Bianca	65–75	400–700
	28		45–60	300–500
	55	Pianto	40–50	200–400

Source: Reproduced from Wareing (1997), with kind permission of Springer Science and Business Media, copyright 1997.

9.1.3 Processing

9.1.3.1 *Grinding and blending*

Gum purchase is based on the approval of pre-shipment samples. The majority of the gum is purchased by customers in Europe and the USA where further processing is then carried out. The exudate is mechanically ground to a fine mesh powder, usually below 150 μm. Throughout the grinding process, foreign matter is removed by selective sieving, aspirating and density-table separation. The gum is conveyed by clean, filtered air which also acts to cool the gum during processing preventing detrimental viscosity losses due to overheating (Wang, 2000). Powdered grades may be blended together in order to achieve a product with consistent colour and viscosity properties.

9.1.3.2 *Hygiene*

Whilst efforts are made to minimise the microbial load of the gum when collecting and sorting, contamination from the soil and air can occur. Superior ribbon grades tend to have lower total viable counts (TVCs) compared to flake grades, but the use of high-pressure steam heat treatment on the collected gum reduces the TVC considerably. Fumigant use with food ingredients is now severely restricted and steam heat treatment remains the safest method of reducing microbial load.

9.1.4 Composition and chemistry

9.1.4.1 *Composition*

Gum tragacanth can be described as a complex, acidic, highly branched, heterogeneous hydrophilic polysaccharide. The molecular weight is approximately 840 000 daltons and the molecular shape is believed to exist as an elongated polymer of 4500 nm by 19 nm in flake gum (Gralen and Karrholm, 1950) thus accounting for the high viscosities obtained. Heterogeneous gum tragacanth is a mixture of two components present as a mixed calcium, magnesium and potassium salt. The first component is tragacanthic acid, commonly known as bassorin, a water-swellable polymer. The second component is a water-soluble arabinogalactan polysaccharide known as tragacanthin. Generally, tragacanthic acid is present between 60% and 70% in commercially exploited gum. Species type and geographical location, however, appear to determine the percentage of the two components.

The water-swellable tragacanthic acid yields D-xylose, L-fucose, D-galactose, L-rhamnose and D-galacturonic acid on hydrolysis. The viscous property of gum tragacanth is largely due to this component having a high molecular weight and rod-like molecular shape as it has a 1,4-linked D-galactose backbone with short side chains of D-xylose, L-fucose or D-galactose connected by 1,3-linkages. Tragacanthin is a highly branched arabinogalactan in which L-arabinose is the main sugar component. The structure is believed to consist of a core of repeating 1,6-linked D-galactose units to which highly branched chains of L-arabinofuranose residues are attached (Wang, 2000). Traces of starch and cellulosic materials are present and the 3–4% protein content is thought to play a part in the emulsification properties of the gum. Chemical variability between gums derived from different *Astragalus* species has been reported (Anderson and Bridgeman, 1985). Studies conducted with Iranian gum samples indicate that the more viscous gum grades contain higher proportions of L-fucose, D-xylose, D-galacturonic acid and methyl ester groups and low proportions of L-arabinose and nitrogenous fractions. The lower viscosity gum grades contain higher proportions of L-arabinose and nitrogenous fractions and lower proportions of D-galacturonic acid and methyl ester groups (Anderson and Grant, 1989).

9.1.4.2 *Solubility, viscosity and rheology*

Gum tragacanth hydrates in hot or cold water to form viscous solutions. The viscosity depends on the grade of gum used and is regarded as the main indicator of gum quality. For example, a commercial ribbon No. 1 grade tragacanth will give a 1% solution viscosity of approximately 3500 mPa s (1.0%, 25°C, 24 h at 20 rpm measured on a Brookfield Viscometer). Flake grades have viscosities of between 400 and 800 mPa s. At 25°C, the gum is fully hydrated and has attained maximum viscosity after 24 h. The hydration time is reduced as the initial water temperature increases.

If fine powdered gum is used, a good dispersion technique is required to avoid lumping. When using the gum in a formulation, it is best to blend the tragacanth together with a diluent, such as sugar or other dried ingredients, before adding to the aqueous phase using adequate stirring techniques to avoid lumping. Tragacanth solutions are pseudoplastic, displaying a reversible apparent viscosity decrease with increasing shear rate.

9.1.4.3 *Emulsification ability*

Low concentrations of gum tragacanth can rapidly reduce the surface tension of water. Levels of less than 0.25% in oil–water mixtures can lower the interfacial tension to 190–230 μN/cm and facilitate emulsification. The aqueous viscous phase will also help to stabilise the emulsion making gum tragacanth a very effective bifunctional emulsifying agent. Figure 9.1 demonstrates that the low viscosity flake grades have superior surface activity properties compared to the higher viscosity ribbons. The lower viscosity grades of tragacanth contain higher nitrogen fractions compared to the higher viscosity grades. Dickinson *et al.*

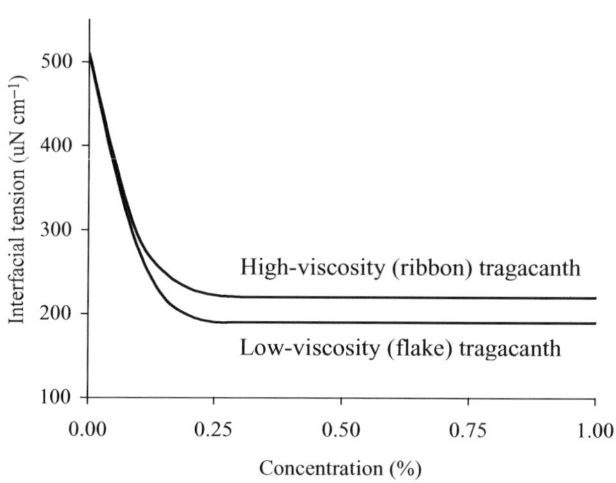

Fig. 9.1 Effect of gum concentration on interfacial tension of oil–water emulsions (Stauffer, 1980). (Reproduced with kind permission of Robert L. Davidson III.)

(1988) working with gum arabic have shown that polypeptide present is involved with the surface activity and emulsification properties.

9.1.4.4 *pH stability*

The pH of 1% tragacanth solutions is naturally acidic and ranges from 4.5 to 6.0. The gum is stable over the pH range 2.0–8.0 but it is most stable between 4.0 and 8.0. The stability of tragacanth in acidic conditions makes it suitable for use in long shelf life, low pH food systems, such as salad dressings.

9.1.4.5 *Heat stability*

As with many hydrocolloids, high temperatures will temporarily lower the viscosity of the gum solution. On cooling, the solution reverts to the original viscosity. Permanent viscosity loss can occur during prolonged heating.

9.1.4.6 *Synergies and compatibility*

Most hydrocolloids are compatible with gum tragacanth, as are most proteins and fats used in food and pharmaceutical production. However, interaction with gum arabic results in a viscosity loss. Such a reaction is commercially exploited to produce good-quality, thin, pourable emulsions with long shelf life.

9.1.4.7 *Microbiological stability*

Powdered gum has a shelf life of at least 12 months providing good hygiene practices have been employed during processing and the gum is suitably packaged and stored in a cool, dry environment. Gum solutions degrade under microbial action unless preventative measures such as heat treatment, refrigeration, freezing, pH reduction and preservative addition are employed. Depending on the food formulation, sorbic and benzoic acids and their salts are effective preservatives in acidic conditions below pH 6.0. For food and pharmaceutical systems at pH 3.0–9.0, the combined use of 0.17% methyl and 0.3% propyl parahydroxybenzoate is reported to be effective (Wang, 2000).

9.1.5 Applications

Many of the traditional tragacanth markets now use cheaper alternatives but tragacanth gum is still used where the viscosity properties, emulsification ability and acid stability are superior to other food hydrocolloids.

9.1.5.1 *Solution preparation*

Finely powdered gum tragacanth will form lumps if added directly to poorly agitated water. Improved dispersion using high shear will reduce gum particle aggregation and ensure that all the gum particles are able to hydrate. Coarse powders, on the other hand, will disperse easily but will take longer to hydrate. Kibbled gum of 1–4 mm disperses easily but will take up to 24 h soaking at ambient temperature for full hydration to be attained. The method of ensuring a homogenous dispersion depends upon the formulation. Common methods used in the food industry are outlined:

a. Thoroughly blend the gum with at least five to ten times its own weight of dried ingredients that may also be present in the formulation. Pre-blending with starches, maltodextrin or sugar, for example, before addition to the water phase will aid dispersion.
b. Solutions may be prepared by adding the gum into the water vortex created by a high-shear or high-speed mixer. This method ensures that lump-free, rapidly hydrated dispersions are produced and is probably the fastest method for preparing solutions. Large solution volumes may be prepared by using in-line mixers or by controlled powder addition using a Venturi suction valve into an agitated water tank.
c. Pre-mixing the gum into a quantity of alcohol, oil, glycerine, or propylene glycol will disperse but not swell the gum. When water is added to this mixture using high shear, controlled hydration is achieved.

9.1.5.2 *Icings*

Tragacanth gum will hydrate in high sugar solids systems and effectively bind water. Thus, the gum is used widely in sugar confectionery systems with great success. The hydration rate is appreciably slower in these high-solid systems, and for this reason, fine powder of 75 μm or less is used to help speed the hydration process. Gum tragacanth is also a very effective emulsifier and is used extensively in fat-containing 'ready-to-roll' icings resulting in a plastic, pliable sugar paste. As well as emulsifying the fat, the gum aids moisture retention preventing drying, flaking and cracking of the finished rolled or moulded icing. The pliability of the icing is maintained throughout the shelf life of the ready-to-roll product and, organoleptically, the gum contributes a smooth texture, body and creamy taste.

Commercial icing producers frequently use a pre-hydrated gum stock solution at a gum use level of approximately 0.3% in the final product. Adding aqueous colourants to the icing before the gum has fully hydrated will result in the gum selectively absorbing the colour giving an unacceptable speckled appearance to the final product. Therefore, it is important to ensure complete gum hydration before colour addition. Sugarcraft is now a hobby on the increase in Europe and gum tragacanth is available to the domestic user. Coloured icing blocks are also available and the gum is used extensively in these products. The success of ready-iced cakes and 'celebration-style' supermarket cakes has seen increased gum usage (Anon., 2005). Chinese and Russian 'premium' western-style cake markets are also predicted to grow, potentially increasing the use of gum tragacanth.

9.1.5.3 *Confectionery*

Candy cream centres can be manufactured using gum tragacanth as a thickener, to produce a smooth, non-stringy texture and emulsify any fats and flavour oils present. The gum is also resistant to hydrolysis by the fruit acids present and helps maintain a long shelf life. Tragacanth has also been used as a binder and emulsifier in cold-process lozenge production where it can be added as a solution directly to the icing sugar prior to flavour addition and cold pressing. Dry-tableted confectionery produced by direct compression can be made using tragacanth as a binder. In combination with gelatin, tragacanth may also be used to produce smooth textures and good flavour release in gum drops, pastilles and fruit chews.

9.1.5.4 *Dressings and sauces*

Pourable, creamy salad dressings and dips can be produced using gum tragacanth to emulsify the oil and thicken the water phase preventing coalescence of oil droplets. The gum can be used in condiment sauces, horseradish, mustards and relishes and will remain stable at the low pH values required for these food systems without viscosity loss. Low-calorie dressings containing less than 5% oil can be produced with 0.5–1.2% gum tragacanth. At this concentration, the gum is contributing to the mouthfeel and body of the product as well as emulsifying any oil present.

9.1.5.5 *Ice creams, ices and sherbets*

Tragacanth serves mainly as a thickener and emulsifier in frozen confectionery. Used at 0.2–0.5% in ice cream, the gum emulsifies the cream or butter, controls ice crystal growth and reduces water migration and crystal growth during storage. When used in ice lollies and sherbets, tragacanth thickens the water phase and controls crystal growth.

9.1.5.6 *Baked goods*

In fruit pie fillings, pastes and purees, the acid stability of gum tragacanth enables it to be used as a thickener to provide a creamy mouthfeel and give gloss and clarity to the filling. The gum will also suspend the fruit pieces in deep-filled pies and pastries. Tragacanth has also been used in meringues as an emulsification agent in the liquid phase in the air-in-liquid emulsion.

9.1.5.7 *Oil emulsions*

Fish oil and fruit oil emulsions can be stabilised by tragacanth acting both as an emulsifier and as a water-phase thickener. In the presence of gum arabic, however, superior emulsions are obtained which are thinner than those produced using gum tragacanth alone. The increase in demand in Asian markets for vitamin-fortified oils and related nutritional supplements has meant an increase in the use of gum arabic and tragacanth for this purpose. A use level of 0.8–1.2% is normally required.

9.1.5.8 *Miscellaneous*

Tragacanth has found uses in the soft drinks industry as a thickener and pulp-suspending agent. The acid-resistant gum ensures good shelf life stability as well as preventing settling of comminuted fruit particles. The pharmaceutical industry uses a substantial amount of gum tragacanth in tablets, emulsions, emollients, ointments, lubricants, syrups and lotions.

9.1.6 Future developments

The continued use of gum tragacanth is mainly due to the useful bifunctional properties of the gum as an emulsifier and thickener in spite of the competition from xanthan gum. If species of *Astragalus* could be further cultivated, prices may fall making the gum competitive once more. To this end, Iranian plants were planted in arid regions of Arizona and California, USA, in an attempt to produce a cost-effective product and regain some markets (Whistler and BeMiller,

1997). Even though some efforts are being made, more botanical research is required to investigate beneficial propagation and cultural techniques. A clearer understanding of the *Astragalus* species as a whole will achieve higher, better quality yields which may result in increased gum consumption.

9.2 GUM KARAYA

9.2.1 Introduction

Gum karaya, also known as sterculia gum, is a dried tree exudate obtained from certain species of *Sterculia*. The majority of the gum is used in pharmaceutical applications but karaya is still used in food applications where the individual qualities of the gum are difficult to replace with more economical alternatives.

The *Food Chemicals Codex* (6th edition, 2008) defines *gum karaya* as the 'dried exudation from *Sterculia urens* (Roxburgh) and other species of *Sterculia* (Family *Sterculiaceae*) or from *Cochlospermum gossypium* A. P. De Condolle or other species of *Cochlospermum* Kunth (Family *Bixaceae*)'. Commercial quantities of the tapped exudate gum are generally obtained from the tree *Sterculia urens* growing mainly in central and northern India. Other quantities of the gum can be obtained from *Sterculia setigera* in Mali, Senegal and the Sudan (Wang, 2000) and from *Sterculia villosa* in India and Pakistan (Wareing, 1997).

Demand for gum karaya has decreased over the years. In the early 1980s, approximately 6000 tonnes were used worldwide. The appearance of more cost-effective gums replaced gum karaya in a number of key applications causing the market to fall. The Indian government intervened and nationalised the industry to form the National Association for Export Development (NAFED). Private merchants, who had helped maintain price stability and product consistency, were excluded and Indian gum karaya quality became unreliable. Buyers turned to sources in the Sudan and Senegal, thus exacerbating the falling Indian market. In response to this decline, the Tribal Marketing and Development Federation of India (TRIFED) was formed returning some control to the merchants. It is now estimated that around half of the 3000 tonnes produced worldwide (1000–1500 tonnes) comes from North Africa but the competition this represents may help stabilise future prices.

Gum karaya is classified in the USA as 'generally recognised as safe' (GRAS). The Joint WHO/FAO Expert Committee on Food Additives (JECFA) has classified the gum as 'ADI not specified'. Within the EU, however, although listed as E416 by the Scientific Committee for Food of the European Community, the Food Additives other than Colours and Sweeteners Directive allocates gum karaya to Annex IV. This legislation, when applicable, restricts the use of gum karaya to the following food categories:

a. cereal and potato based snacks,
b. nut coatings,
c. bakery fillings, toppings and coatings,
d. desserts,
e. emulsified sauces,
f. egg-based liqueurs,
g. chewing gum,
h. dietary food supplements.

Few allergenic responses to gum karaya have been reported, thus the gum is not listed in Annex IIIa of the European Directive 2003/89/EC (as amended) and is not subject to allergen labelling in Europe.

9.2.2 Raw materials

Gum karaya is obtained by making incisions into the trunks of the *Sterculia* trees that can grow up to 10 m high depending on the species. The bulk of the gum exudes within the first day forming large, irregular lumps or 'tears' that dry in the hot, dry climate. Each tapping yields between 1 and 5 kg of gum and each tree can expect to be tapped approximately five times during its life time. In India, the best quality gum is collected in April, May and June, just prior to the monsoon season. The crop in Senegal is collected in September to January and from March to July. Gum is collected and delivered to collection points where bark and foreign matter (BFM) is removed by hand. Lumps are broken down into smaller pieces and sorting occurs on the basis of gum colour and residual BFM. The grades used in international trade are Superior No. 1, 2, 3 and siftings, and the maximum percentage BFM for these grades are 0.5%, 2.0%, 3.0% and 7.0%, respectively. Superior No. 1 karaya gives light-coloured solutions. As BFM content increases, solution colour becomes darker. Premium quality Superior No. 1 gum is used in food and pharmaceutical preparations requiring high viscosity, good solution colour and excellent moisture retention. The siftings exceed the BFM of 3% as laid down by the *Foods Chemical Codex* (6th edition 2008) and their use is therefore limited to technical applications.

9.2.3 Processing

9.2.3.1 *Grinding and blending*

The gum is purchased after evaluation of pre-delivery samples of the required grade. Further processing mainly occurs in Europe and the USA where the gum is mechanically ground, usually to a particle size below 106 μm. Fibres, sand, bark and other foreign material are removed from the gum by aspiration and density-table separation. Further blending may occur to ensure customer specifications for viscosity and colour are consistently met.

9.2.3.2 *Hygiene*

Microbial loads are similar to other exudate gums and similar treatments may be employed to reduce the TVC. The use of heat treatments should be avoided as this results in permanent viscosity loss.

9.2.4 Composition and chemistry

9.2.4.1 *Composition*

The chemical composition of gum karaya varies little between the exploited tree species. Anderson *et al.* (1982), working with *S. urens*, *S. villosa* and *S. setigera*, found few notable chemical differences. Gum karaya is a complex, branched, partially acetylated polysaccharide with a molecular weight of up to 16 million daltons (Le Cerf *et al.*, 1990). Hydrolysis yields D-glucuronic acid, D-galacturonic acid, D-galactose and L-rhamnose in varying proportions (Meer, 1980). Gum karaya has a much higher L-rhamnose content compared to other exudate gums and this can be a means of identification. The polysaccharide contains approximately 40% uronic acid residues and between 8% and 14% acetyl groups. Acetyl groups prevent the gum from fully dissolving in water but allow swelling (Wareing, 1997). Alkali treatment of the gum will chemically deacetylate the gum, thus modifying its characteristics and improving water solubility. The structure of gum karaya contains a central chain of galactose, rhamnose and galacturonic residues with side chains of glucuronic acid. The exudate occurs as a mixed calcium and magnesium salt. Approximately 1% proteinaceous material has been detected but the amino acid composition varies between *Sterculia* species.

9.2.4.2 *Solubility, viscosity and rheology*

Because of the presence of acetyl groups, gum karaya is the least soluble of the commercial exudates, yet it can absorb cold water very rapidly to form a viscous colloidal dispersion at low concentrations. At high concentrations of 3–5%, the gum forms thick, soft gel-like pastes which have a spreadable quality and are ideal for suspending food particulates.

Hydration rate is dependent on particle size with finer mesh material hydrating faster than coarse material to produce apparently homogenous dispersions. Application of high shear to grainy dispersions made with coarse-powdered gum results in smooth solutions having a reduced viscosity. Heating gum karaya mucilages increases solubility, due to the changes of the polymer conformation, but results in permanent viscosity loss. A maximum of up to 4% gum can be hydrated in cold water. Heating under pressure gives smooth, translucent, colloidal solutions at concentrations as high as 20% (Glicksman, 1983).

The solubility of pre-hydrated gum dispersions can be improved by increasing the pH. The polymer then behaves as a random coil producing higher viscosities having a ropy and cohesive texture (Le Cerf *et al.*, 1990). Gum karaya from African sources (*S. setigera*) has a slightly lower acetyl content from the Indian sources (*S. urens*) and therefore tends to give more ropy textures.

The viscosity of gum karaya dispersions ranges from 120–400 mPa s for 0.5% dispersions to about 10 000 mPa s for 3% dispersions depending on the grade. In powdered form, viscosity decreases with increasing age. Finely ground powder loses more viscosity compared to a coarse powder or the whole exudate and is most noticeable soon after grinding. High storage humidity or high storage temperature all contribute to viscosity loss. Boiling gum solutions for more than 2 minutes also reduces viscosity (Wang, 2000). Rheological studies have shown the gum solutions to be thixotropic and therefore the apparent viscosity depends on solution history.

9.2.4.3 *pH stability*

The pH of naturally acidic gum karaya ranges from 4.4 to 5.2 because of the acid residues present. Raising the pH causes deacetylation and the solution texture becomes ropy and cohesive. This phenomenon does not occur if the alkali is added to water prior to gum addition. Also, if electrolytes are present, maximum viscosity is retarded. Therefore, in food applications where fast development of maximum viscosity is required, it is always best to add the gum to the water before addition of salts and buffers. Gum karaya is acid stable and, therefore, it is most suitable for thickening acid-based sauces and dressings prepared with minimal heat processing.

9.2.4.4 *Heat stability*

Irreversible polymer conformation changes occur in heated gum karaya dispersions resulting in a permanent viscosity loss. However, under controlled temperature and pressure conditions, concentrations of up to 20% can be prepared giving very viscous solutions.

9.2.4.5 *Water-binding and adhesion properties*

In water, gum karaya can swell and absorb to more than 60 times of its original volume. At concentrations of 20–50% gum karaya yields pastes with good adhesion properties. These pastes were used primarily in dental adhesives, bulk laxatives and colostomy bag sealing rings. However, recent dental reports have shown that continued contact with acidic gum has a detrimental effect on natural teeth.

9.2.4.6 *Synergies and compatibility*

Most hydrocolloids are compatible with gum karaya, as are most proteins and fats used in food and pharmaceutical production. Interactions with sodium alginate have been reported to modify solution properties (Le Cerf and Muller, 1994) and gum karaya is also reported to reduce the gel strength of agar gels (Hoefler, 2004).

9.2.4.7 *Microbiological stability*

Unless preservatives are used, gum karaya dispersions are prone to microbial attack. Depending on the food formulation in use, benzoic or sorbic acids, methyl and propyl parahydroxybenzoate and propylene glycol are commonly used.

9.2.5 Applications

Gum karaya has been commercially exploited for the last 100 years. However, the solution colour, acidic odour and sour taste have limited the use of this material. Many of the applications traditionally using gum karaya now use xanthan gum. Also, the European Union has limited the use of the gum to the eight applications listed in Section 9.2.1. Aside from the European Union constraints, gum karaya has been reported in a number of applications.

9.2.5.1 *Solution preparation*

When gum karaya is dispersed in water, it swells very quickly. Powdered material may lump on initial mixing and behave in a similar manner to gum tragacanth unless suitable techniques are employed to ensure smooth homogenous dispersions (see Section 9.1.5).

9.2.5.2 *Dressings and sauces*

The acid stability of gum karaya makes it ideally suited for use in acid-based sauces and dressings. At low concentrations, gum karaya produces viscous solutions which can suspend particles which are stable and have long shelf life. The slight acid taste of the gum is masked by the overall flavour of the dressing or sauce. Gum karaya is best used in cold-processed dressings, as high temperatures reduce the viscosity. Use levels of between 0.6% and 1.0%

have been reported (Wareing, 1997), and combinations with gum arabic have been used to enhance emulsion stability by improving the viscosity of the water phase.

9.2.5.3 *Dairy products*

Gum karaya functions as an effective foam stabiliser maintaining foam integrity in products such as whipped creams and mousses. Gum karaya can also be used as a water binder in cheese spreads at an approximate use level of 0.8% to prevent whey separation.

9.2.5.4 *Meat products*

Acting as a water binder, gum karaya can be used at about 0.3% in sausages and meat products and improves adhesion between the meat ingredients during processing. The gum also stabilises the meat emulsion by thickening the water phase producing a smooth-textured meat product.

9.2.5.5 *Baked goods*

Gum karaya can be used in meringues to increase foam overrun. The moisture retention properties of the gum make it suitable for staling retardation and shelf life extension. Glazes thickened with karaya adhere to pastry give the baked goods an appealing sheen after baking.

9.2.5.6 *Frozen desserts*

Moisture migration, control of ice crystal size and reduction of suck out of flavour and colour, especially in ice lollies, can all be controlled by the addition of gum karaya. Traditional products using the gum include sorbets, sherbets, and ice lollies. The use of karaya in ice cream applications is now rare as it has largely been replaced by guar, cellulose or xanthan gums.

9.2.5.7 *Miscellaneous*

There is an expanding market for tablets with fibre for dietary supplements both in Australia and the USA. Gum karaya is widely used in this application as a source of fibre and is often used in combination with guar gum.

9.2.6 Future developments

Low-cost alternatives have largely replaced gum karaya in a number of key applications. The colour and flavour of the gum present the technologist with a limited number of uses, and cost and supply problems have meant a larger reliance on guar and xanthan gums. Markets for the gum still remain, however, and future market needs may be met by the African producers. For example, the suppliers in Sudan have the potential to export large quantities of good-quality gum as they now have large growing areas devoted to cultivating the *Sterculia* trees.

References

Anderson, D.M.V. and Bridgeman, M.M.E. (1985) The composition of the proteinaceous polysaccharides exuded by *Astragalus microcephalus, A. gummifer* and *A. kurdicus* – the sources of Turkish gum Tragacanth. *Phytochemistry*, **24**(10), 2301–2304.

Anderson, D.M.V. and Grant, D.A.A. (1989) Gum exudates from four *Astragalus* species. *Food Hydrocolloids*, **3**(3), 217–223.

Anderson, D.M.V., McNab, C.G.A, Anderson, C.G., Braown, P.M. and Pringuer, M.A. (1982) Gum exudates from the genus *Sterculia* (gum karaya). *International Tree Crops Journal*, **2**, 147–154.

Anon. (2005) The icing on the cake. *Bake and Take*, February 2005, pp. 16–17.

Dickinson, E., Murray, B.S., Stainsby, G. and Anderson, D.M.W. (1988) Surface activity and emulsifying behaviour of some Acacia gums. *Food Hydrocolloids*, **2**(6), 477–490.

Glicksman, M. (1983) Gum karaya (*Sterculia* Gum). In: M. Glicksman (ed), *Food Hydrocolloids*. Vol. II. CPC Press, USA, pp. 39–47.

Gralen, N. and Karrholm, M. (1950) The physicochemical properties of solutions of gum tragacanth. *Journal of Colloid Science*, **5**(1), 21–36.

Hoefler, A.C. (2004) *Hydrocolloids*. Eagan Press Handbook Series, Eagan Press, USA.

Le Cerf, D., Irinei, F. and Muller, G. (1990) Solution properties of gum exudates from *Sterculia urens* (Karaya gum). *Carbohydrate Polymers*, **13**(4), 375–386.

Le Cerf, D. and Muller, G. (1994) Mechanical spectroscopy of karaya-alginate mixed dispersions. *Carbohydrate Polymers*, **23**(4), 241–246.

Meer, W. (1980) Gum karaya. In: R.L. Davidson (ed), *Handbook of Water Soluble Gums and Resins*. McGraw-Hill, New York, pp. 10.1–10.14.

Stauffer, K.R. (1980) Gum tragacanth. In: R.L. Davidson (ed), *Handbook of Water Soluble Gums and Resins*. McGraw-Hill, New York, pp. 11.1–11.31.

Wang, W. (2000) Tragacanth and karaya. In: G.O. Phillips and P.A. Williams (eds), *Handbook of Hydrocolloids*. CPC Press LLC, USA, pp. 231–246.

Wareing, M.V. (1997) Exudate gums. In: A. Imeson (ed), *Thickening and Gelling Agents for Food*, 2nd edn. Blackie Academic & Professional, London, pp. 86–118.

Whistler, R.L. and BeMiller, J.N. (1997) *Carbohydrate Chemistry for Food Scientists*. AACC, USA.

10 Inulin

Rudy Wouters

ABSTRACT

Inulin is a storage carbohydrate found in many plants. The industrial production of inulin is from chicory using hot water extraction. Inulin is a polydisperse, linear carbohydrate consisting mainly of fructose units. Standard chicory inulin has an average chain length of 10–12 units and long-chain material contains 25 units on average. Hydrated inulin particles form a particle gel that resembles a network of fat crystals. Low levels of inulin significantly improve the mouthfeel and creaminess in low-fat dairy products, such as yogurt and cheese products, frozen desserts, meat spreads, sauces, soups and fat-continuous table spreads. Long chain inulin is also able to provide structure in water-continuous, spreadable products, such as table spreads and cream cheese, and it increases the stability of whipped dairy desserts. There is significant interest in inulin as a soluble fibre with prebiotic benefits of selectively promoting the growth and activity of beneficial bacteria, mainly bifidobacteria.

10.1 INTRODUCTION

Inulin is a storage carbohydrate found in many plants. It has been part of the daily diet of humans for some centuries as it naturally occurs in many vegetables, fruits and cereals. The average daily consumption has been estimated to be between 3 and 11 g in Europe and between 1 and 4 g in the USA (Van Loo *et al.*, 1995).

Table 10.1 shows the inulin content in some plant species. Commercially, inulin is obtained from chicory roots and used as a food ingredient offering interesting nutritional properties and important technological benefits as well. It significantly improves the organoleptic characteristics of food and drinks and shows a fat-like behaviour when used to form a gel in water.

Inulin was first isolated from *Inula helenium* by Rose, a German scientist, in 1804 (Rose, 1804), but it was Thomson who named this substance inulin (Thomson, 1818). One pioneer in fructan research was the German plant physiologist Julius Sachs (Sachs, 1864), who detected inulin in the tubers of dahlia, Jerusalem artichoke (*Helianthus tuberosus*) and *I. helenium* after ethanol precipitation. Already, from the beginning of the twentieth century, feeding diabetic patients with inulin was reported to be beneficial (Root and Baker, 1925), and especially since the 1990s, a spectacular increase in the number of publications dealing with the technological and nutritional benefits has been observed (Franck and De Leenheer, 2002).

Table 10.1 Inulin content (% of fresh weight) of plants that are commonly used in human nutrition.

Source	Edible parts	Dry solids content	Inulin content
Artichoke	Leaves, heart	14–16	3–10
Banana	Fruit	24–26	0.3–0.7
Barley	Cereal	NA	0.5–1.5[a]
Burdock	Root	21–25	3.5–4.0
Camas	Bulb	31–50	12–22
Chicory	Root	20–25	15–20
Dandelion	Leaves	50–55[a]	12–15
Garlic	Bulb	40–45[a]	9–16
Jerusalem artichoke	Tuber	19–25	14–19
Leek	Bulb	15–20[a]	3–10
Murnong	Root	25–28	8–13
Onion	Bulb	6–12	2–6
Rye	Cereal	88–90	0.5–1[a]
Salsify	Root	20–22	4–11
Yacon	Root	13–31	3–19

NA: figures not available.
[a] Estimated.

10.2 RESOURCES AND RAW MATERIALS

After starch, fructans are the most abundant non-structural polysaccharides found in nature. They are present in a wide variety of plants and in some bacteria. Fructan-producing plants are commonly present among the grasses (1200 species) and 15% of the flowering plants produce them in significant amounts. They are widely spread within the *Liliaceae* (3500 species) and most frequently among the *Compositae* (25 000 species) (Hendry and Wallace, 1993). Inulin-containing plants that are commonly used in the human diet belong mainly to the *Liliaceae* (leek, onion, garlic and asparagus) or the *Compositae* (Jerusalem artichoke, chicory and yacon). A plant with a special status is the *Agave Azul Tequila Weber (Liliaceae)*. Tequila is an alcoholic drink made by fermentation of the highly branched fructan present in this plant. Bacterial fructans are essentially of the levan type and are found among the *Pseudomonaceae*, *Enterobacteriaceae*, *Streptococcaceae*, *Actinomycetes* and *Bacillaceae*. Given their high inulin content of over 10%, dahlia, Jerusalem artichoke (*H. tuberosus*) and chicory (*Cichorium intybus*) have been considered as candidates for industrial production of inulin (Meyer *et al.*, 1993).

10.2.1 Dahlia

Many dahlia cultivars are available but they have all been selected for their flowers rather than for inulin production. The tuberous roots can be propagated only if attached to a piece of stem tissue. When propagated from seed, sowing has to be delayed until late spring, given the dahlia's extreme sensitivity to frost. Mechanical harvesting of the tubers is feasible only on sandy grounds. Although the degree of polymerisation (DP) of dahlia inulin is higher than the one for chicory, its yield is only half that of chicory (see Table 10.2). For all these reasons, dahlia does not appear to be an interesting crop for inulin production.

Table 10.2 Yields and composition of dahlia, Jerusalem artichoke and chicory.

	Dahlia	Jerusalem artichoke	Chicory
Roots or tubers (tonnes/hectare)			
Average	25	42	43.5
Variation		35–60	25–75
Dry matter (%)			
Average	18	22	22.5
Variation	15–22	19–25	20–25
Inulin (%)			
Average	11	16	16.5
Variation	10–12	14–18	15–18
Inulin	2.5–3	4.5–8.5	5–11
Mean DP	13–20	6–10	10–14

DP, degree of polymerisation.

10.2.2 Jerusalem artichoke

Jerusalem artichoke has a rather high inulin content of between 14% and 19%. It is not recommended to cultivate Jerusalem artichoke in clay soil as the tubers are small and irregular and, hence, a lot of soil is attached to them. Although Jerusalem artichoke is frost tolerant, its inulin metabolism is very sensitive to cold. As Jerusalem artichoke inulin has only 20% of chains with a DP longer than 10, it is unsuited for several food and non-food applications (Table 10.2).

10.2.3 Chicory

Chicory is a biennial plant. During the first season, the plants remain in the vegetative phase and make only leaves, tap roots and fibrous roots. The roots look like small oblong sugar beets. The inulin content is high (16–18%) and fairly constant from year to year for a given region. Yields in tonnes of roots per hectare show a wider variation but average around 45 tonnes/hectare (see Table 10.2). In order to keep the chicory culture healthy, a strict crop rotation is imposed with cultivation once every 5 years.

Industrial production of inulin is from chicory, *C. intybus*, which is of the same type as the one that can be roasted and used for the production of a coffee substitute.

10.3 PRODUCTION

The production of chicory inulin goes through two phases. The first stage includes the extraction and a primary purification step resulting in raw impure syrup. The second stage is the refining phase which results in a commercial end product that is more than 99.5% pure (De Leenheer, 1996; Franck and De Leenheer, 2002).

10.3.1 First stage

The first process phase is very similar to sugar beet processing. Roots are harvested and stored in piles on the field. To minimise losses, no more than 7 days should elapse between harvest and delivery. The roots are then transported to the factory by truck, weighed and carefully moved to a storage yard. From there, they are transported on a stream of water inside the factory, where they are washed and sliced. Raw inulin is extracted from the resulting 'chips' with hot water in a counter current diffuser. The leached 'chips' are dried and sold as animal feed. A first purification step is applied to the extraction juice by liming and carbonating at high pH. The resulting calcium carbonate sludge precipitates easily. Peptides, some anions, degraded proteins and colloids are trapped as a floc and this product is used by farmers to improve the soil structure, as it is rich in calcium and organic matter.

10.3.2 Second stage

The raw juice is further refined using cationic and anionic ion exchange resins for demineralisation and active carbon for decolourisation. This technology is comparable to the one used in starch processing. After demineralisation and decolourisation, the juice is passed through a 0.2-μm filter then sterilised, before being evaporated and spray dried. The resulting standard inulin has an average chain length which reflects the original DP distribution present in the chicory root of between 10 and 12 residues. A special long-chain grade inulin, with an average DP above 23, is also available. It is made by physical elimination of the small DP fraction (Smits *et al.*, 1994).

10.3.3 Environmental aspects

The chemicals used for regeneration of the ion exchange resins are not sodium hydroxide and hydrochloric acid, as usual, but ammonia and sulphuric acid. The advantage of this is that the effluents can be converted into reusable by-products. At high concentrations, easily crystallised salts such as ammonium and potassium sulphates precipitate. These are separated from the mother liquor and sold commercially as fertilisers. The mother liquor is further evaporated into a stable product that is sold as animal feed based on its high organic matter content. In this way, the circle between the processing industry and agriculture is fully closed.

10.4 CHEMICAL STRUCTURE

Inulin is a polydisperse, linear carbohydrate material consisting mainly, if not exclusively, of fructose units (F) linked by $\beta(2\rightarrow1)$ bonds. Fructan, a more general name, is used for any compound in which fructosyl–fructose bonds constitute the majority of linkages. A starting glucose unit (G) can be present in the chain but is not necessary. GF_n and F_n compounds are included under the same nomenclature and they are both a mixture of oligomers and polymers that are characterised by the average and the maximum DP. In chicory inulin, n, the number of fructose units can vary from 2 to 60 (De Leenheer and Hoebregs, 1994). Its molecular structure is shown in Fig. 10.1.

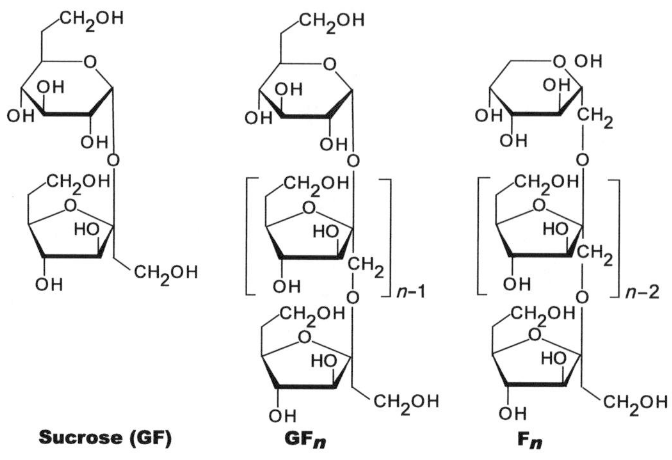

Fig. 10.1 Molecular structure of fructans.

The DP influences the functionality of inulin. Standard chicory inulin has an average DP of about 10–12. Long-chain chicory inulin, from which the lower DP fraction has been physically removed, has an average DP of about 25.

10.5 PHYSICAL AND CHEMICAL PROPERTIES

10.5.1 Basic properties

Chicory inulin is available as a white, odourless powder with a high purity and a well-known chemical composition. The taste is neutral, without any off-flavour or aftertaste. Standard chicory inulin is slightly sweet, about 10% of the sweetness of sucrose, whereas long-chain inulin is not sweet at all. Inulin behaves like a bulking agent, contributing to body and mouthfeel. It is often used in combination with high potency sweeteners, such as aspartame and acesulfame K, to provide a more-rounded mouthfeel and a better-sustained flavour with reduced aftertaste. Such combinations also exhibit a significant synergy in the sweetness level (Wiedmann and Jager, 1997).

Chicory inulin is moderately soluble in water, about 10% at room temperature, which allows it to be fully hydrated for incorporation in aqueous systems. To make a solution of inulin, the use of warm water between 50°C and 100°C is suggested. The viscosity of chicory inulin solutions is rather low, for example 1.65 mPa s at 10°C for a 5% dry matter (DM) solution and 100 mPa s for a 30% DM solution (Franck and De Leenheer, 2002). Inulin exerts a small effect on the freezing and boiling point of water; for example 15% chicory inulin decreases the freezing point by 0.5°C.

In very acid conditions, the $\beta(2 \rightarrow 1)$ bonds between the fructose units can be partially hydrolysed. Fructose is formed in this process which is more pronounced under low pH, high temperature and low dry substance conditions. Inulin is stable in applications with a pH higher than 4. Even at lower pH values, the hydrolysis of inulin is limited to less than 10% if the products either have a high dry substance content (>70%) or are stored at a low temperature (<10°C), or have a short shelf life. The physicochemical properties of chicory inulin are summarised in Table 10.3.

Table 10.3 Physicochemical properties of chicory inulin.

	Standard inulin	**Long-chain inulin**
Chemical structure	GF_n ($2 \leq n \leq 60$)	GF_n ($10 \leq n \leq 60$)
Average DP	12	25
Dry matter (%)	>95	>95
Inulin content (% on DM)	92	99.5
Sugar content (% on DM)	8	0.5
pH (10% w/w)	5–7	5–7
Sulphated ash (% on DM)	<0.2	<0.2
Heavy metals (ppm on DM)	<0.2	<0.2
Appearance	White powder	White powder
Taste	Neutral	Neutral
Sweetness (sucrose = 100%)	10%	None
Solubility in water at 25°C (g/L)	120	10
Viscosity in water (5%) at 10°C (mPa s)	1.6	2.4
Heat stability	Good	Good
Acid stability	Reasonable	Good
Functionality in foods	Fat replacement Body and mouthfeel Texture improvement Foam stabilization Emulsion stabilization Synergy with gelling agents	

F, fructosyl unit; G, glucosyl unit; DM, dry matter; DP, degree of polymerisation.

10.5.2 Gelling behaviour

10.5.2.1 *Particle gel*

At high concentrations in water, above 25% for standard chicory inulin and above 15% for long-chain material, inulin has gelling properties and forms a particle gel network after shearing. Inulin gels are formed by a network of small crystallites. Their rheological properties are quite distinct from that of non-crystallising polysaccharides and resemble more closely that of a network of fat crystals in oil (Bot *et al.*, 2004).

A white creamy structure results when the fructan is thoroughly mixed with water or another aqueous liquid, using a shearing device such as a rotor–stator mixer or a homogeniser, and this can easily be incorporated in foods to replace fat up to 100%. This process has been patented by Raffinerie Tirlemontoise/BENEO-Orafti (Frippiat and Smits, 1993). Such a gel provides a short spreadable texture and a smooth fatty mouthfeel, as well as a glossy appearance and a well-balanced flavour release.

Electron cryo-microscopy has shown that such an inulin gel is composed of a three-dimensional network of hydrated inulin particles in water having a diameter of 1–3 μm (see Fig. 10.2).

Figure 10.3 shows a sketch of a slice of an inulin gel (Bot *et al.*, 2004). The black objects are the primary non-spherical inulin crystals, having typical dimensions of 0.5–3 μm. The dashed lines indicate the more or less spherical aggregates that determine the effective volume that is observed in flow experiments. The aggregates, which include significant amounts of immobilised water, as determined by NMR experiments, ultimately interact to form a gel.

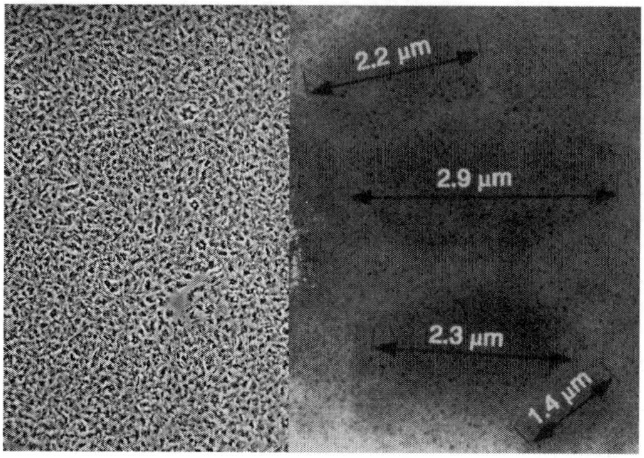

Fig. 10.2 Electron micrograph of hydrated inulin particles.

X-ray diffraction (Fig. 10.4) has confirmed the crystalline nature of the gel particles, whereas the starting inulin powder is essentially amorphous.

10.5.2.2 *Parameters affecting gel strength*

The gel strength depends on different parameters such as inulin concentration and total dry matter content, the inulin type, the shearing parameters and the type of shearing device used, but it is not influenced by pH between 4 and 9.

Fig. 10.3 Diagram of an inulin gel.

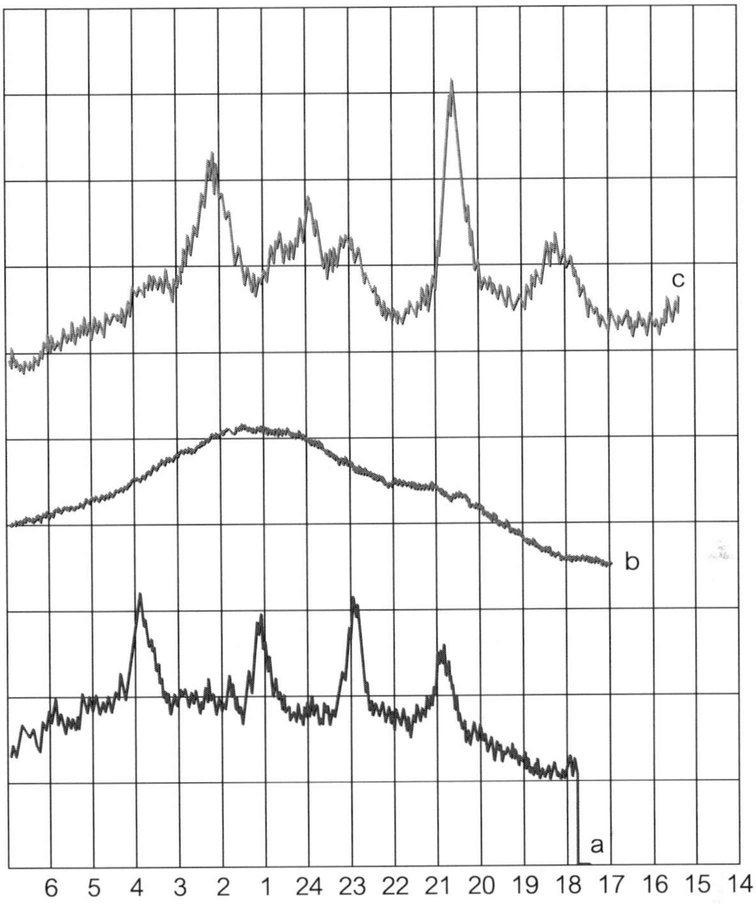

Fig. 10.4 X-ray diffraction of inulin.

Applying different shearing devices, an increase in gel strength is noticed with increasing mechanical pressure. For example, forming a dispersion with a colloid mill results in a lower gel strength compared to a rotor–stator mixer while the latter gives lower firmness values than a dispersion obtained with a high pressure homogeniser. The optimal gel strength is achieved after about 24 h as illustrated in Fig. 10.5.

A second parameter affecting gel strength is the dry solids content. Increasing the dry matter content of the system by using higher inulin dosages or adding other ingredients, like sugar or maltodextrins, results in higher gel strengths. A close relationship exists between the gel strength and the inulin content (Fig. 10.6). Approximately 15% long-chain inulin is the minimum concentration required to obtain a particle gel.

A very important parameter which strongly affects the final gel strength is the temperature during shearing. This parameter also demonstrates the difference in functionality between two different types of long-chain inulin, as shown in Fig. 10.7, and it determines the functionality of the inulin product in a food applications. For a 20% DM ORAFTI®HP gel, maximum gel strength is obtained when shearing is applied at moderate temperatures of 30–50°C but it loses its functionality at higher temperatures. A totally different behaviour is noticed for

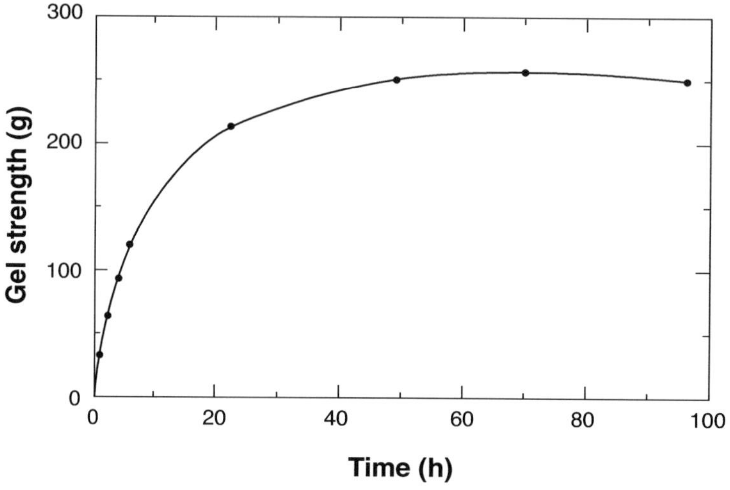

Fig. 10.5 Gel strength development with time for an inulin gel.

a 20% DM ORAFTI®HPX gel which shows an optimal gel strength at higher shearing temperatures between 65°C and 85°C. Consequently, the choice of which inulin type should be used to texturise a food product is determined by the shearing temperature.

10.5.2.3 *Gel characterisation*

An inulin gel exhibits the properties of a viscoelastic material and, as such, possesses gel strength and elasticity, characteristics of a solid, as well as a viscosity, a typical characteristic of a fluid. The rheological properties of such a particle gel can be reflected by its viscosity, yield stress, storage modulus and loss modulus. When a small pressure is applied to the gel, it behaves like a solid and shows some elasticity. A large pressure causes a loss of its gel-like structure and it then behaves like a fluid which is characterised by its viscosity. In between

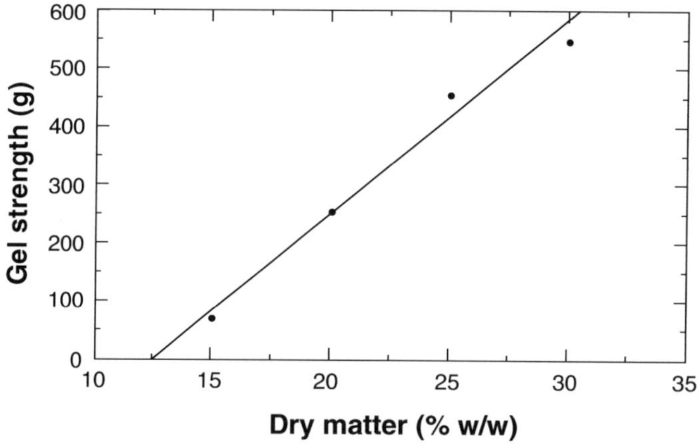

Fig. 10.6 Effect of dry matter content on gel strength of an inulin gel.

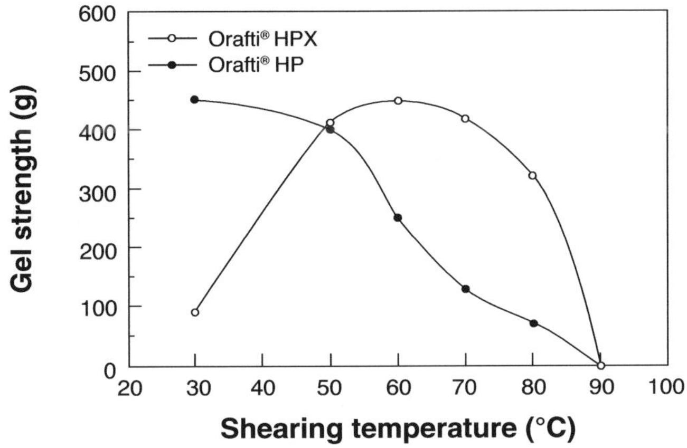

Fig. 10.7 Effect of shearing temperature on gel strength.

these two extremes, a state of transition can be defined when the viscous component (the fluid properties) begins to dominate over the elastic component (the solid properties). The pressure corresponding to this transition is called the yield stress. The storage modulus G' (or elasticity modulus) and the loss modulus G'' (or viscosity modulus) are two quantities expressing to which extent the gel exhibits the properties of a solid or a fluid. When the gel is subjected to an increasing deformation, the values of these two quantities change. From these changes, structural characteristics can be deduced.

An inulin gel is characterised by a relatively low yield stress; for example 1540 Pa for a gel of 30% standard inulin in water at 25°C. This results in an inverse dependency of the viscosity on the shear rate at low-shear rates. At high-shear rates above 100 per second, the viscosity becomes less dependent and starts to level off. The increasing and decreasing shear-rate regions show a different pattern which indicates a breakdown of the particle gel structure over time when shear is applied (see Fig. 10.8). This thixotropic breakdown starts when the yield stress is reached and is reversible.

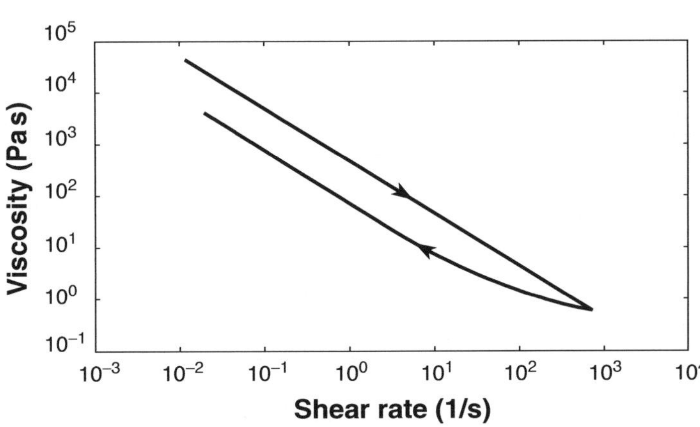

Fig. 10.8 Viscosity versus shear rate for an inulin gel.

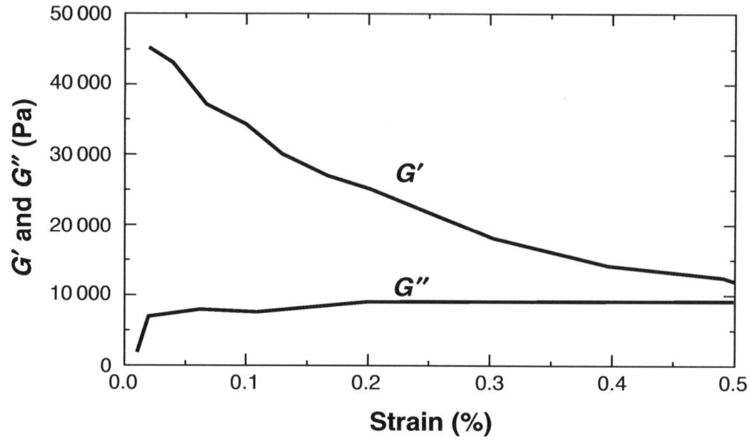

Fig. 10.9 Loss and storage moduli as a function of strain for an inulin gel.

Oscillatory experiments indicate that the inulin gel has non-linear behaviour: even at the lowest strain values, G' decreases with increasing strain, which means that the structure is broken down (see Fig. 10.9)

When the gel is subjected to an increasing deformation, it gradually loses its 'solid properties' (G' decreases) and the 'fluid properties' become more apparent. For larger deformations (>1%), G'' becomes even larger than G', which is typical for a suspension. Close to the linear region (0.05% deformation), frequency sweeps show a plateau value for G' and G'', with G' clearly larger than G''. This is a typical behaviour for a weak gel.

A partially broken inulin particle gel shows lower loss and storage moduli; however, the overall path of G' and G'' as a function of the strain shows a similar behaviour to that of an untreated gel. Thus, the breakdown of the structure does not change the qualitative rheological properties of the gel. The original structure is not irreversibly damaged by the shearing action. This is confirmed by the measurement of G' as a function of time after preshearing the sample. G' regains its original level after about 17 h (about 1000 min), which indicates a total recovery of the gel structure as shown in Fig. 10.10.

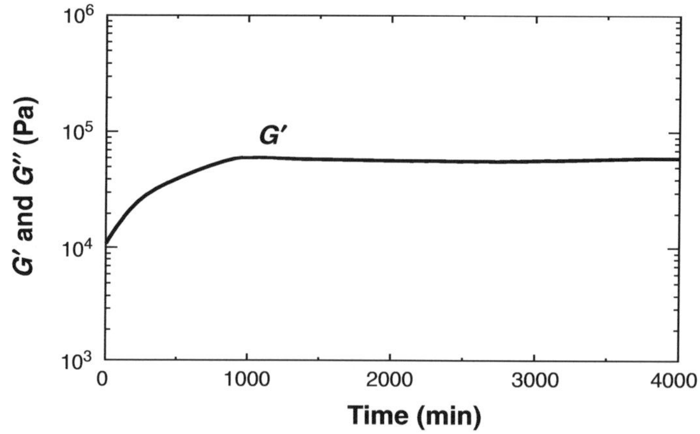

Fig. 10.10 Recovery of loss and storage moduli for a sheared inulin gel.

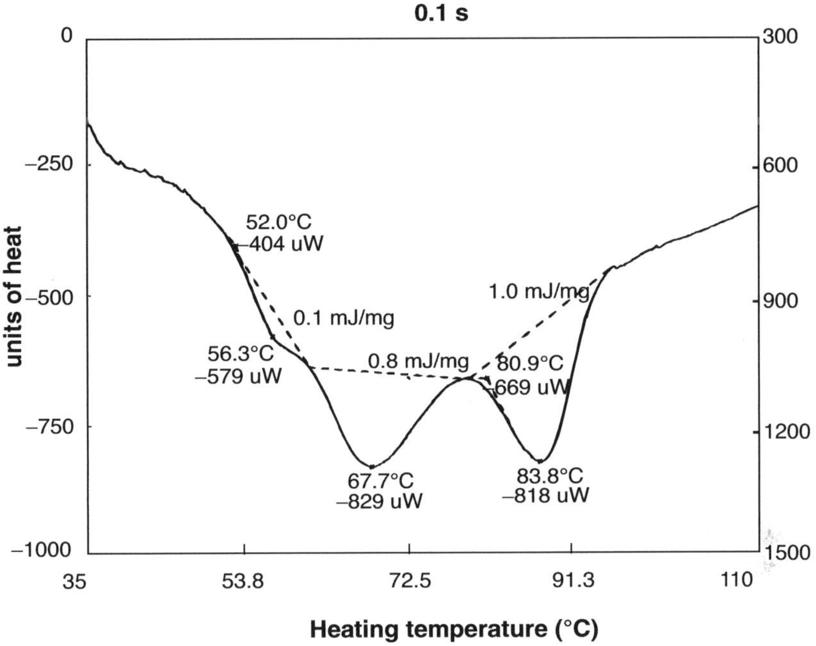

Fig. 10.11 Differential scanning calorimetry of an inulin dispersion.

The rheological behaviour is influenced by the inulin content. The concentration effects are, however, limited to changes in the absolute values of the rheological characteristics, without affecting the general rheological behaviour. Both viscosity and yield stress as well as G' and G'' increase with increasing inulin content.

Differential scanning calorimetry analysis of an inulin gel shows two characteristic peaks (Fig. 10.11). The first peak appears between 60°C and 70°C and correlates with the melting of crystalline particles. This transition is little influenced by the total amount of inulin. Increasing the amount of crystalline particles shows that this results in a larger endothermic transition. The second transition, which appears between 80°C and 90°C, correlates with the solution properties of inulin. This transition is dependent on the concentration and increases with increasing dry matter content.

Inulin also displays synergy with most hydrocolloids including gelatine, alginate, kappa and iota carrageenan, gellan gum and maltodextrin (Frippiat, 1998).

10.6 PRINCIPLE OF FAT REPLACEMENT

Inulin allows the development of low-fat foods while maintaining characteristics typical of fats. This fat-replacement ability is based on the particle gel properties as described above. Consequently, fat replacement with inulin is only possible in water-containing systems and, preferably, in food products where water is the continuous phase. The inulin particles are similar in size to fat droplets resulting in mouthcoating, mouthfeel and creaminess. These particles, formed by applying shear to a food product, have a size between 1 and 3 μm which is similar to fat droplets after homogenisation (see Fig. 10.12).

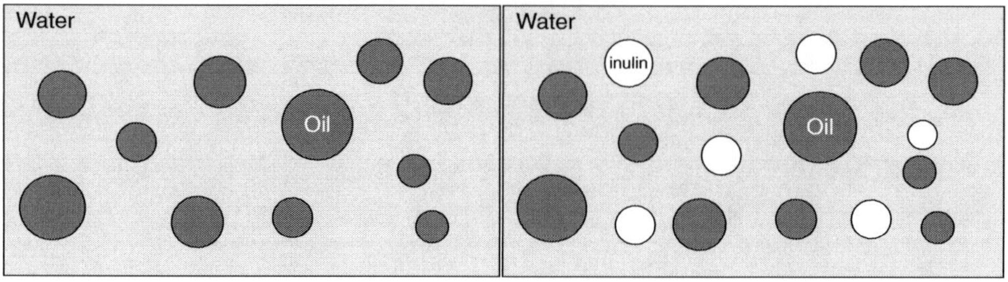

Fig. 10.12 Diagram showing inulin particles distributed within an oil in water dispersion.

To replace fat in food products, lower inulin dosages than required for gel formation can be applied as this property is related to the number of inulin particles in the food matrix. More inulin particles present in a food system will give a higher creaminess level. The number of these particles can be influenced by different parameters including shearing forces, type of inulin, level of inulin and so on. Increasing the inulin dosage in a food system obviously results in a higher number of particles and therefore a greater effect on mouthfeel.

A second parameter is the inulin type, and more precisely, its solubility (Fig. 10.13). A totally hydrated inulin molecule does not contribute to the formation of a particle and therefore it does not contribute to fat replacement. A long-chain inulin is less soluble than

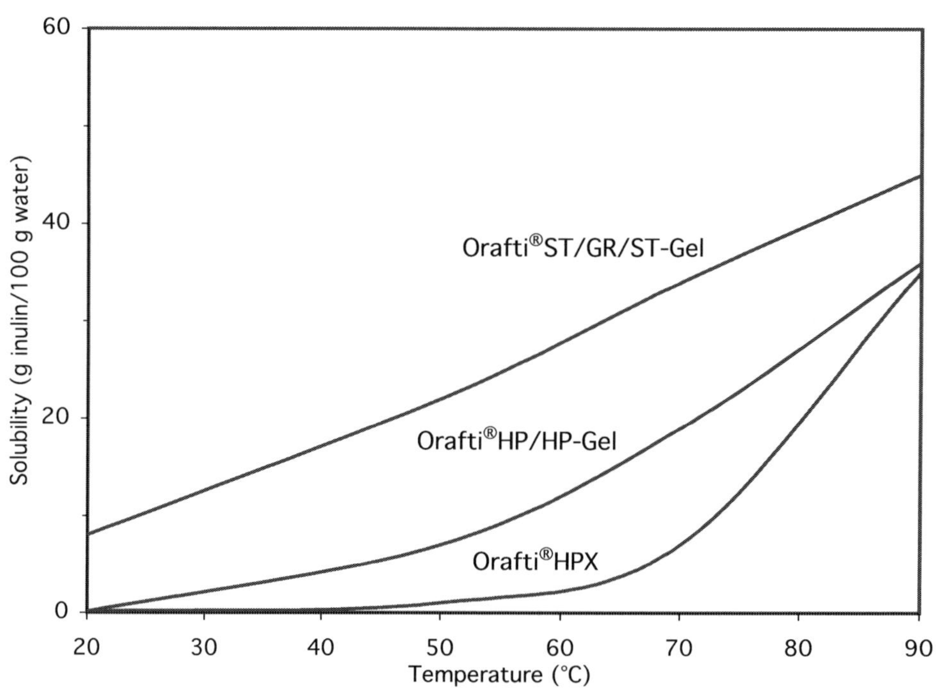

Fig. 10.13 The solubility of different inulins at different temperatures.

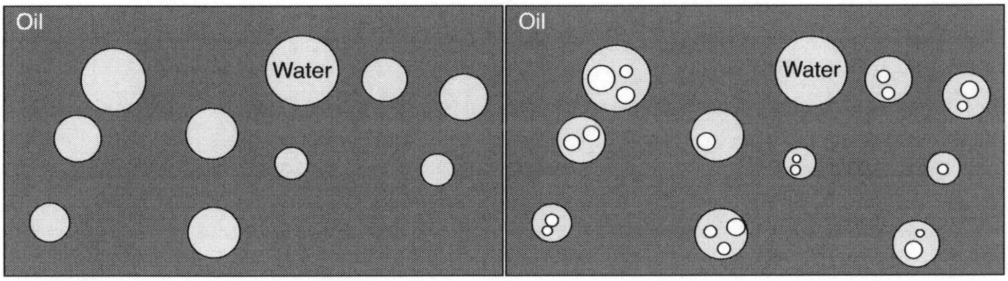

Fig. 10.14 Diagram showing inulin particles in water droplets distributed in a water-in-oil dispersion.

a standard inulin. This means that more inulin molecules can take part in the particle gel formation process and more particles will be formed which automatically results in more efficient fat replacement characteristics.

Finally, the mechanical energy applied to the mixture has an effect as well. A homogenised product will be creamier than if the product is produced with a high shear mixer.

In fat-continuous products, inulin functions in a different way (see Fig. 10.14). As inulin is not soluble in oil, it will be present in the water droplets surrounded by oil and contribute to the stability of the emulsion through an increase in the viscosity of the water phase. In this case, inulin merely functions as a stabiliser of such low-fat, oil-continuous products.

10.7 PHYSIOLOGICAL PROPERTIES

Inulin behaves like a bulk ingredient, contributing to body and mouthfeel. It is often used in combination with intense sweeteners to provide a more-rounded mouthfeel and a better-sustained flavour with reduced aftertaste. Inulin provides several interesting nutritional benefits to animals and humans (see Table 10.2) (Van Loo *et al.*, 1999). Because of its $\beta(2\rightarrow1)$ bonds, which the digestive enzymes of humans cannot hydrolyse, inulin passes through the mouth, stomach and small intestine without undergoing any significant change and without being absorbed (Ellegard *et al.*, 1997). Thus, inulin enters the colon almost quantitatively and there it is completely metabolised by the intestinal bacteria (Roberfroid *et al.*, 1993), mainly to short-chain fatty acids, bacterial biomass and gases. Only the short-chain fatty acids contribute to the host's energy metabolism, which explains the reduced caloric value of inulin of between 1.0 and 1.5 kcal/g. Inulin has no influence on blood glucose or insulin levels when ingested orally and it has been known as a food for diabetics since the beginning of the twentieth century (Beringer and Wenger, 1955).

It is a soluble dietary fibre (Prosky, 1999) which induces typical effects on the gut function, such as a reduction of the intestinal pH, a relief of constipation and an increase in stool weight and frequency. This faecal bulking effect is similar to that of other soluble fibres like pectin and guar gum (Roberfroid, 1997); each gram ingested increases the faecal wet weight by about 2 g. Inulin also has modulating effects on lipid metabolism, for example by reducing serum and liver triglycerides (Jackson *et al.*, 1999; Delzenne *et al.*, 2002).

In the colon, inulin selectively promotes the growth and metabolic activity of beneficial bacteria, mainly bifidobacteria, while repressing harmful ones, such as Clostridia. This is called the prebiotic or bifidogenic effect. The bifidogenicity of inulin has been demonstrated in in vitro models (Gibson and Wang, 1994) and in several in vivo human volunteer studies

(Gibson and Wang, 1994; Gibson *et al.*, 1995; Kleessen *et al.*, 1997). A daily intake of 5–10 g of inulin is sufficient to significantly enhance bifidobacteria in humans (Roberfroid *et al.*, 1998).

It has been demonstrated that inulin inhibits the development of cancer cells transplanted in the thigh or in the peritoneum of mice (Taper *et al.*, 1997). Recent research indicates that inulin also has a significant chemopreventive potential (Reddy *et al.*, 1997). Indeed, it can prevent the formation of pre-cancerous lesions and tumours in the colon of rats. Long-chain inulin is more effective than shorter-chain fructans. Synergistic effects were observed when inulin and bifidobacteria were administered together as a symbiotic mixture (Rowland *et al.*, 1998).

Inulin also increases the intestinal absorption of calcium, iron and magnesium as well as the bone mineral density in rats (Delzenne *et al.*, 1995; Roberfroid *et al.*, 2002). In an initial study with healthy adult volunteers who were given up to 40 g/day of inulin, an important increase in calcium absorption was observed (Coudray *et al.*, 1997). A significant increase in calcium absorption has now been confirmed in adolescent girls upon ingestion of only 8 g/day of a specific inulin product (ORAFTI® Synergy1, BENEO-Orafti) (Griffin *et al.*, 2002). All these results give promising evidence that inulin increases calcium absorption in humans and could therefore actively contribute to the reduced risk of osteoporosis (Franck, 1998).

10.8 APPLICATIONS

10.8.1 Food applications

In food and drink products, inulin can be used for either its nutritional advantages or its technological properties, but it is often applied to offer a double benefit: an improved organoleptic quality and a better-balanced nutritional composition. The use of inulin as a dietary fibre is easy and it is often used in all kind of food products, although more and more it is used in 'functional foods', especially in a whole range of dairy products, as a prebiotic ingredient to stimulate the growth of the beneficial gut microflora or to boost calcium absorption in the human body.

10.8.2 Fat replacement in water-based food products

Based on the particle gel characteristics of inulin described previously, it can be concluded that inulin functions as a fat replacer but only in water-based systems. Each inulin particle dispersed in the water phase of any food system will contribute to the creaminess of the finished food.

As far as fat replacement is concerned, long-chain inulin shows about twice the functionality compared to standard chicory inulin, thus allowing for lower dosage levels and ingredient costs. Special 'instant' products, which do not require shearing to give stable homogeneous gels, also have been developed using specific spray-drying technology and patented.

In dairy systems, 2–3% of inulin will have a significant effect on the mouthfeel characteristics of low-fat products. These dosages are typical for low-fat yogurts and all kind of dairy drinks. The incorporation of inulin in diet fruit yogurts, possibly through the fruit preparation, improves the mouthfeel and offers a synergistic sweetness effect in combination with aspartame and/or acesulfame K and with sucralose. Special, more-soluble inulin types have been developed which allow an easy incorporation of inulin in dairy products through the fruit preparation. In other low-fat dairy products, such as fresh cheese or processed cheese, the addition of a few percent of inulin gives a creamier mouthfeel and imparts a better-balanced round flavour.

Water-continuous, spreadable products do not only benefit from the fat replacement properties, but long-chain inulin is also able to provide structure, such as in a low-fat cream cheese. As discussed in Section 10.5, long-chain inulins show different gelling properties. ORAFTI®HP gives more texture when shearing is applied at moderate temperatures of 30–50°C, but it loses functionality at higher temperatures. ORAFTI®HPX works best at shearing temperatures of between 50°C and 85°C and it is ideally suited to processes where the product inlet temperature of the homogeniser or other shearing device is above 50°C. This means that ORAFTI®HPX is an ideal ingredient to provide texture in low-fat cream cheese products. Cream, butter, cheese, sour cream and yogurt all provide suitable bases. Combinations with fruit, chocolate and many other sweet ingredients are possible, as are savoury products containing, for example herbs and spices. In meat applications like pâté, the cutter process is applied at high temperatures and, therefore, ORAFTI®HPX acts as an excellent fat replacer resulting in smooth, spreadable textures with good mouthcoating properties and a neutral taste.

Inulin increases the stability of foams and mousses; its incorporation at 1–5% in dairy-based aerated desserts improves the processing characteristics and upgrades product quality. The resulting desserts retain their typical structure for longer and show a fat-like consistency.

Inulin is also destined to be used as a fat replacer in frozen desserts, as it processes easily to provide a fatty mouthfeel, excellent melting properties, as well as freeze–thaw stability, without any unwanted off-flavour.

Fat replacement can also be applied to meal replacers, meat products, sauces and soups. Fat-reduced meat products, such as sausages and pâtés, can be obtained with a creamier and juicier mouthfeel and improved stability due to better water immobilisation. The synergistic effect of inulin with other gelling agents constitutes an additional advantage in all these applications.

In fat-continuous table spreads, inulin functions as a stabiliser allowing the manufacturing of stable low-fat margarine or butter products. Inulin is not soluble in oil but only in water, and therefore it will be present in the water phase which is totally entrapped by fat. Inulin increases the viscosity of the water droplets and stabilises the emulsion. Long-chain inulin, incorporated at 2–10%, gives excellent results in water-in-oil spreads with fat contents ranging from 20% to 60%.

10.8.3 Other application areas

When used in bakery products and breakfast cereals, the use of inulin presents major progress in comparison with other fibres. Inulin gives more crispness and expansion to extruded snacks and cereals and it increases their bowl life. It also keeps breads and cakes moist and fresh for longer. Its solubility allows fibre incorporation in aqueous systems such as drinks, dairy products and table spreads.

Inulin has found an interesting application as a low-calorie bulk ingredient, often in combination with a polyol, in chocolate made with no added sugar. It is also used as a dietary fibre or sugar replacer in tablets.

So, inulin has become a key ingredient offering new opportunities to the food industry looking for well-balanced and yet better tasting products of the future.

References

Beringer, A. and Wenger, R. (1955) Inulin in der Ernährung des Diabetikers, *Deutsche Zeitschrift fur Verdauungs-und Stoffwechselkrankheit*, **15**, 268–272.

Bot, A., Erle, U., Vreeker, R. and Agterhof, W. (2004) Influence of crystallisation conditions on the large deformation rheology of inulin gels. *Food Hydrocolloids*, **18**, 547–556.

Coudray, C., Bellanger, J., Castiglia-Delavaud, C., Rémésy, C., Vermorel, M. and Rayssignuier, Y. (1997) Effects of soluble or partly soluble dietary fibres supplementation on absorption and balance of calcium, magnesium, iron and zinc in healthy young men. *European Journal of Clinical Nutrition*, **51**, 375–380.

De Leenheer, L. (1996) Production and use of inulin: industrial reality with a promising future. In: H. Van Bekkum, H. Röper and F. Voragen (eds), *Carbohydrates as Organic Raw Materials III*. VCH Publishers Inc., New York, pp. 67–92.

De Leenheer, L. and Hoebregs, H. (1994) Progress in the elucidation of the composition of chicory inulin. *Starch*, **46**, 193–196.

Delzenne, N., Aertssens, J., Verplaetse, N., Roccaro, M. and Roberfroid, M. (1995) Effect of fermentable fructo-oligosaccharides on energy and nutrients absorption in the rat. *Life Science*, **57**, 1579–1587.

Delzenne, N.M., Daubioul, C., Neyrinck, A., Lasa, M. and Taper, H.S. (2002) Inulin and oligofructose modulate lipid metabolism in animals: review of biochemical events and future prospects. *British Journal of Nutrition*, **87**(2), S225–S259.

Ellegard, L., Andersson, H. and Bosaeus, I. (1997) Inulin and oligofructose do not influence the absorption of cholesterol, and the excretion of cholesterol, Fe, Ca, Mg and bile acids but increases energy excretion in man – a blinded, controlled cross-over study in ileostomy subjects. *European Journal of Clinical Nutrition*, **51**, 1–5.

Franck, A. (1998) Prebiotics stimulate calcium absorption: a review. *Milchwissenschaft*, **53**, 427–429.

Franck, A. and De Leenheer, L. (2002) Inulin. In: A. Steinbuchel (ed), *Biopolymers*, Vol. 6. VCH Weinheim, Germany, pp. 439–479.

Frippiat, A. (1998) Tiense suikerraffinaderij, inulin based hydrocolloid compositions. Patent Application EP 867,470.

Frippiat, A. and Smits, G. (1993) Tiense Suikerraffinaderij, compositions having a creamy structure and containing fructan, preparation method therefore and uses thereof. Patent Application WO 93 06744.

Gibson, G.R., Beatty, E.R., Wang, X. and Cummings, J.H. (1995) Selective stimulation of bifidobacteria in the human colon by oligofructose and inulin. *Gastroenterology*, **108**, 975–982.

Gibson, G.R. and Wang, X. (1994) Bifidogenic properties of different types of fructo-oligosaccharides. *Food Microbiology*, **11**, 491–498.

Griffin, I.J., Davila, P.M. and Abrams, S.A. (2002) Non-digestible oligosaccharides and calcium absorption in girls with adequate calcium intakes. *British Journal of Nutrition*, **87**(2), S187–S191.

Hendry, G.A. and Wallace, R.K. (1993) The origin, distribution and evolutionary significancy of fructans. In: M. Suzuki and N.J. Chatteron (eds), *Science and Technology of Fructans*. CRC Press, Florida, pp. 119–139.

Jackson, K.G., Taylor, G.R.J., Clohessy, A.M. and Williams, C.M. (1999) The effect of the daily intake of inulin on fasting lipid, insulin and glucose concentrations in middle-aged men and women. *British Journal of Nutrition*, **82**, 23–30.

Kleessen, B., Sykura, B. and Zunft, H.J. (1997) Effect of inulin and lactose on fecal microflora, microbial activity and bowel habit in elderly constipated persons. *American Journal of Clinical Nutrition*, **65**, 1397–1402.

Meyer, W.J.M., Matthysen, E.W.J.M. and Borm, G.E.L. (1993) Crop characteristics and inulin production of Jerusalem artichoke and chicory. In: A. Fuchs (ed), *Inulin and Inulin-Containing Crops, Studies in Plant Science*, 3rd ed. Elsevier Science Publishers, Amsterdam, pp. 29–44.

Prosky, L. (1999) Inulin and oligofructose are part of the dietary fiber complex. *Journal of AOAC International*, **82**, 223–226.

Reddy, D.S., Hamid, R. and Rao, C.V. (1997) Effect of dietary oligofructose and inulin on colonic preneoplastic aberrant crypt foci inhibition. *Carcinogenesis*, **18**, 1371–1374.

Roberfroid, M.B. (1997) Health benefits of non-digestible oligosaccharides. In: D. Kritchevsky and C. Bonefield (eds), *Dietary Fiber in Health and Disease*. Plenum Press, New York, pp. 211–219.

Roberfroid, M.B., Cumps, J. and Devogelaer, J.P. (2002) Dietary chicory inulin increases whole-body mineral density in growing male rats. *Journal of Nutrition*, **132**, 3599–3602.

Roberfroid, M., Gibson, G.R. and Delzenne, N. (1993) Biochemistry of oligofructose, a non-digestible fructooligosaccharide: an approach to estimate its caloric value. *Nutrition Reviews*, **51**, 137–146.

Roberfroid, M., Van Loo, J. and Gibson, G. (1998) The bifidogenic nature of chicory inulin and its hydrolysis products. *Journal of Nutrition*, **128**, 11–19.

Root, H. and Baker, M. (1925) Inulin and artichokes in the treatment of diabetes. *Archives of Internal Medicine*, **36**, 126–145.

Rose, V. (1804) Über eine eigenthümliche vegetablische Substanz. *Neues allgemeines Journal der Chemie*, **3**, 217.

Rowland, I.R., Rummey, C.J., Coutts, J.T. and Lievense, L. (1998) Effects of Bifidobacterium longum and inulin on gut bacterial metabolism and carcinogen-induced aberrant crypt foci in rats. *Carcinogenesis*, **19**, 281–285.

Sachs, J. (1864) Über die Sphärokristalle des Inulins und dessen mikroskopische Nachweisung in den Zellen. *Botanische Zeitung*, **22**, 77.

Smits, G., Daenekindt, L. and Booten, K. (1994) Tiense Suikerraffinaderij, Fractionated polydisperse compositions. Patent Application WO 96 01849.

Taper, H., Delzenne, N. and Roberfroid, M.B. (1997) Growth inhibition of transplantable mouse tumours by non-digestible carbohydrates. *International Journal of Cancer*, **71**, 1109–1112.

Thomson, T. (1818) *A System of Chemistry*, 5th London edition, Abraham Small, Philadelphia, vol. 4, p. 65.

Van Loo, J., Coussement, P., De Leenheer, L., Hoebregs, H. and Smits, G. (1995) On the presence of inulin and oligofructose as natural ingredients in the Western diet. *Critical Review in Food Science and Nutrition*, **35**, 525–552.

Van Loo, J., Cummings, J., Delzenne, N., Englyst, H., Franck, A., Hopkins, M., Kok, N., Macfarlane, G., Newton, D., Quigley, M., Roberfroid, M., van Vliet, T. and van den Heuvel, E. (1999) Functional food properties of non-digestible oligosaccharides: a consensus report from the ENDO project (DGXII AIRII-CT94–1095). *British Journal of Nutrition*, **81**, 121–132.

Wiedmann, M. and Jager, M. (1997) Synergistic sweeteners. *Food Ingredients and Analysis International*, November–December, 51–56.

11 Konjac Glucomannan

Jean-Marc Parry

ABSTRACT

Konjac flour yields a high molecular weight, viscous polysaccharide: konjac glucomannan. Glucomannan is a highly soluble, neutral plant polysaccharide that has been used as a gelling agent for 2000 years in China. It gels alone or in association with other polysaccharides where it shows strong synergistic effects in viscosity, gel strength and elasticity. Konjac has a greasy mouthfeel and chewy texture that resemble fat. It is resistant to digestion and has a very low calorific value. As well as being a healthy food, it is a preferred texturing ingredient in Asia. Glucomannan has generally recognised as safe (GRAS) status in the USA and it is well established as an additive in the EU food industry. Cost-effective as a thickener, the effects on post-prandial sugar and lipid levels, low digestibility and prebiotic fermentation have placed glucomannan amongst the most exciting hydrocolloids on the 'emerging' world nutraceutical market.

11.1 INTRODUCTION

11.1.1 History

China hosts many endemic *Amorphophallus* species and konjac was discovered by the Chinese and listed as early as 206 BC in Shen Nong's Herbal Classic. Monks introduced it to Korea in 550 and then to Japan, where it was first quoted by Zuo Si in 950. It rapidly became a commodity and planting programmes spread across the country in the nineteenth century. In 1960, konjac was promoted when there was a renewal of interest in healthy traditional foods in Japan. The US Food and Drug Administration listed konjac flour GRAS (FDA, 1997) shortly followed by recognition as a food additive in the EU (1998, 2003). This coincided with rural development projects in China and renewed agronomic and taxonomic work.

Production originates mainly in China and Japan with contributions from Korea and Thailand. World production is estimated between 25 000 and 35 000 tonnes of crude powder. Half of this is purified further to meet European or US standards for use as an additive or as a botanical active in food supplements.

11.1.2 Regulatory status

In the US in 1957, GRAS status was extended to alcohol-washed grades of *Amorphophallus* flour for consumption in foods. In 1998, European regulatory approval was given to E425i 'konjac gum' and E425ii 'konjac glucomannan' for use as food additives. Under EU regulations, konjac cannot be sold above 1% concentration in an end product alone or in a blend with other thickeners. An exception exists in Belgium where regulations permit konjac as a botanical material for weight control. Standard exceptions for the use of this additive are in products with positive ingredient lists, such as honey, yoghurt, mineral water and infant and baby foods.

The FAO/WHO Codex General Standard for Food Additives (GSFA), the European Union European Food Safety Authority (EFSA), the US Food and Drugs Administration (FDA), the Mercosur National Services for Health and Quality of Agricultural and Food Products (SENASA), Canada and Switzerland have issued additive specifications for glucomannan flours. The main differences lie in the maximum permitted level of protein, ranging from 1.5% to 3% in the EU (2003) to 8% in the USA (FCC, 1996). China, Japan and other Eastern countries, with a tradition of using konjac tubers, consider it food. Australia has established konjac flour as a vegetable when used as the key ingredient (FSANZ, 2007) allowing the import of konjac noodles (see Fig. 11.1).

11.1.3 Safety

Occupational allergies occur in konjac factories due to frequent exposure to fine powder. Precautions should be taken regarding the consumption of dry powders and jellies. As with most pure dry thickeners used in instant foods, insufficient hydration may cause irritation when consumed, and if the particles swell further with water, they may develop into an obstruction and, possibly, lead to suffocation.

Fig. 11.1 Drying traditional konjac noodles. (Reproduced with permission of P. Vieille and J.M. Parry, KALYS SA.)

Fig. 11.2 Typical mini-cup jellies. (Reproduced with permission of P. Vieille and J.M. Parry, KALYS SA.)

Another risk is tract obstruction by pieces of insufficiently-chewed hard water gels. This problem led to bans on mini-cup jellies (Fig. 11.2) and candies of similar shapes, sizes, melting properties and elastic texture in the USA (FDA, 2001). The bans were implemented after warnings on packets, targeted at customers with impaired chewing ability, or children who were unaware of the potential risks, did not prove sufficient to avoid accidents.

However, konjac was only banned specifically in jelly confectionery, based on their size and shape, and not from all confectionery applications (Vieille, 2003). The emphasis later moved from prohibiting konjac to a general ban on most gelling agents used for mini-cup jellies (EFSA, 2004; EU, 2006). Mini-cup jellies are still very common in Asia as treats or as sources of fibre.

11.2 RAW MATERIALS

11.2.1 Glucomannan in the plant

Glucomannan is present in woody species and can be obtained from various tree hemicelluloses and from monocot storage organs such as leaves, tubers, bulbs, roots or seeds. Konjac from plants differs widely from yeast mannan chains (Xu *et al.*, 2005). Glucomannan is found in specific large-sized idioblast cells located in the protoplast. Raphide crystal bundles of oxalic acid are enveloped in the polysaccharide hindering any attempts to eat the tuber. During processing, focus is put on eliminating the protein membrane of these cells and removing the needle-shaped oxalic acid crystals by sieving to give residual levels of around 0.2% for crude powder and lower for refined grades.

11.2.2 Cultivation

There is no short-term alternative plant source with agricultural potential other than *Amorphophallus* species. Essentially farmed under conditions with low chemical input, nevertheless, pesticides and fungicides are used sparingly. For intensive practices, it is recommended to use ground disinfectants. 'Konjac flour' is obtained from the tubers of less than 10 species

Fig. 11.3 *Amorphophallus konjac* tubers. (Reproduced with permission of P. Vieille and J.M. Parry, KALYS SA.)

of *Amorphophallus*, a genus of around 150 species. *A. oncophyllus* and *A. variabilis* are used for iles-iles flour (Sakai, 1983). Several species of *Amorphophallus* have glucomannan concentrations in excess of 50% dry matter giving flour yields of 2–5 tonnes per hectare. Most attempts to industrialise and intensify the crop have led to difficulties in disease control and more irregular yields associated with increased seed demand. Any damage inflicted upon a seed tuber encourages pathogens and propagation of rot. Homogeneity of harvest is still an issue as the crop is collected from thousands of small farmers (Fig. 11.3) and favourable inland growing areas are quite remote during harvest in late autumn. Nevertheless, yields in Japan have improved by 50% over 30 years.

11.2.3 Crop economics

Konjac was a domestic endemic crop until the 1980s to 1990s covering over 10 000 hectares in Japan and much less in China. The Japanese konjac industry is well organised, Japan being the world's biggest consumer of konjac-based products. Japanese industry promoted a highly refined product, a pharmaceutical grade containing above 95% glucomannan, but did not access developing export markets in pet food or Western food ingredients in spite of growing demands. To compensate for a reduction from domestic production, Japan searched favourable tropical sites. Encouraged by Japanese needs and an opening of the Western food industry, it has been reported that China has been planting thousands of hectares of konjac as a cash crop for poor rural areas. Official policies in China exist since the mid-nineties encouraging rural regions, including minority-held provinces, to grow konjac for health products (6th Torch program).

After a boom, when Japan was a major client, China is now looking at both a growing internal demand for health foods and a wider demand for refined glucomannan on the domestic and neighbouring markets in Taiwan, Thailand, Korea and Singapore and in Western countries. Other Asian countries have established production facilities and they transform Chinese or local raw material, mainly for use in domestic markets.

11.3 PROCESSING

The principal component of the tuber is a high-molecular-weight, non-ionic glucomannan, present with starch, protein and minerals. Glucomannan represents 70–90% of the dry matter in commercial flours and protein and ash may be as low as 1% in total (see Table 11.1). Protein content (Kjeldahl N $\times$ 5.7) is an indication of purity and thus the EU purity levels qualify as the strictest with a maximum 3% protein. The starch content is usually much lower than 2% and ash level varies but it is lower than 5%.

Crude, dark-coloured flours are used to make Shirataki noodles for the Asian food market (see Fig. 11.1). A residual low concentration of glucoside may provide a disagreeable bitter 'bamboo shoot' taste (Suzuki, 1980) alongside undesirable amines that give a typical fish-like smell. Konjac only became a popular Japanese food after improvements were made in processing from 1750 to 1850 leading to purer and cleaner konjac through pulverising dried tubers and wind sifting.

11.3.1 Konjac flour (E425i type)

11.3.1.1 *Dry milling*

Tuber sizes and shape depend on age and cultivar and 15–20 cm diameters are frequent. Mainly 3-year-old tubers, which weigh between 300 g and 1.5 kg, are washed and peeled before slicing into chips (Fig. 11.4). Chip quality strongly influences output and flour characteristics. Sulphites are often used and appear as traces in flours. Heavy metal levels are linked with processing technology, although soil trace elements can show up in some flours.

Table 11.1 Glucomannan yields.

	Dry matter (DM) yield	**Konjac glucomannan (KGM) yield**	
	DM yield on tuber	**KGM yield on DM**	**KGM recovered from tuber**
Tuber (averages)	18%–23%	0.60	10–14%
	Chip yield on tuber	KGM yield on chips	
Chips	12.80%	0.60	7.7%
Processing of flours	Flour yield on tuber	KGM yield on flour	
Wet milling	14.12%	0.94	13.3%
Combined milling	11.40%	0.88	10.0%
High performance dry milled	8.30%	0.88	7.3%
Dry milling refined powder	7.04%	0.70	4.9%
Dry milling ultra-refined powder	5.15%	0.90	4.6%

Source: Reproduced with permission of P. Vieille and J.M. Parry, KALYS SA.

Fig. 11.4 Konjac tuber harvest. (Reproduced with permission of P. Vieille and J.M. Parry, KALYS SA.)

11.3.1.2 *Wet/dry milling*

The characteristic smell of konjac may be washed out with other alcohol-soluble ingredients by a 10–50% alcohol solution to give 'premium (cleaner) odourless grades'.

As processing steps increase in dry and semi-wet milling, the concentration of glucomannan rises as shown in Table 11.1, column 2. Extra steps do not always offer advantageous cost-effectiveness as the overall yield of glucomannan, shown in columns 2 and 3, is significantly lowered (Kalys, 2001). Wet milling is an effective but costly method of recovering glucomannan.

11.3.2 Glucomannan (E425ii type)

Glucomannan precipitates when an organic solvent is added. Shimizu and Shimahara (1973) industrialised wet milling to make a very pure konjac glucomannan (E425ii), ideal for the pharmaceutical industry but expensive in many food applications despite the higher gum content permitting lower use levels (Table 11.1).

11.3.3 Suppliers

The wide choice of commercial flours includes cost-effective grades for high-volume commodity markets, such as pet-food or growing food markets in the Asian catering industry and South American and Eastern European meat industries. Other grades include added-value qualities for food and for applications on the border between food and medicine which call for more intense selection and standardisation (Zhong *et al.*, 2006).

11.4 STRUCTURE

11.4.1 Backbone

Konjac glucomannan is mainly composed of a backbone of β-1–4-linked D-glucopyranose and β-D-mannopyranose sugars in a random order. The molecular ratio of mannose to glucose averages 1.6 to 1 and is random, though small repeats of three to five mannose blocks are frequent (Fig. 11.5).

11.4.2 Side chains

Randomly spread acetyl groups occur every 10–19 units (3–6%) on the C2, C3 or C6 carbons of mannopyranose (Fig. 11.5). Single mannose or glucose substitutions or longer side chains, reported in early publications, may be due to artefacts (Kalys, 2005). Methylation (Kishida *et al.*, 1978) and carboxylation (Matsumara *et al.*, 1991; Ohya *et al*, 1994) are documented. Although derivatisation improves chain flexibility, these products are not permitted for use in food.

11.4.3 Molecular weight

The average molecular weight is 1 million daltons and commercial samples have a molecular weight range between 200 000 and 2 million daltons. Polydispersity is usually low for the native material, except in low-quality tubers, with a ratio of weight average molecular weight to number average molecular weight (M_w/M_n) between 1.1 and 1.8. Polydispersity also increases with additional processing.

11.4.4 Characterisation of flours

Molecular chain composition and acetylation rates have been described using gas chromatography (Ohya *et al.*, 1994), HPLC (Kohyama *et al.*, 1993) FT-IR and NMR. Gels have been

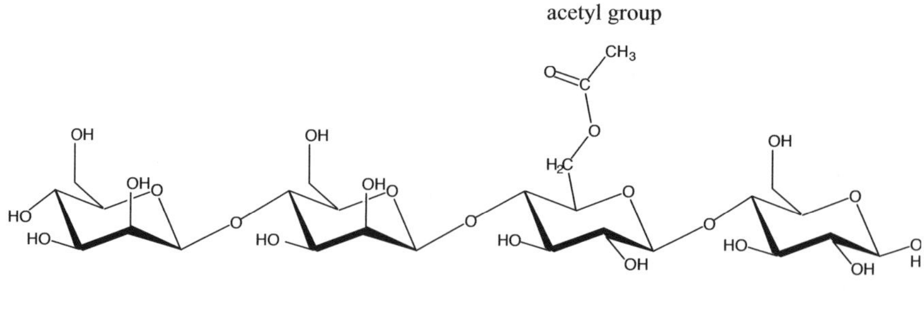

Fig. 11.5 Glucomannan structure. (Reproduced with permission of P. Vieille and J.M. Parry, KALYS SA.)

studied using X-ray diffraction, differential scanning calorimetry (Williams *et al.*, 1992), fluorescence intensity and light scattering (Case *et al.*, 1992; Li *et al.*, 2006a) as well as using traditional rheological equipment. In food, AOAC enzymatic–gravimetric methods measure glucomannan fibre in flour, though these are not specific to one type of non-starch polysaccharide. Enzymatic analysis is also available. Future developments with polymerase chain reactions might lead to more discriminate assays.

11.5 FUNCTIONAL PROPERTIES

11.5.1 Solubility in water

Konjac flours solubilise even in ice water. Hydration is accelerated by raising the temperature and speed of agitation. Viscosity and solubility can be hindered by the addition of competing solutes, low-molecular-weight ingredients such as partially hydrolysed guar gum or maltodextrin (Valerie *et al.*, 2002), branched dextrins (Tomita and Morita, 2006) and salts (Hu, 1993) or by the use of alcohols, such as ethanol or isopropyl alcohol, that will eventually precipitate the glucomannan chains from solution.

11.5.2 Water absorption

Konjac flours show remarkable water absorption. Following the standard AACC method, 1 g of konjac flour absorbs 15–20 g of water, although it is claimed in commercial documentation that 1 g can absorb up to 50–100 g water. Increasing the level of acetylation reduces both water absorbency and viscosity (Koroskenyi and McCarthy, 2001). Konjac flour absorbs between one and two times its weight of oil.

11.5.3 Degradation

Glucomannan starts to decompose around 250°C and decomposition is complete at 350°C (Dave and McCarthy, 1997). The effect of prolonged exposure above 80°C on the viscosity of konjac solutions is obvious, especially in acidic media.

Konjac glucomannan is sensitive to enzymes such as β-D-glucanase and β-D-mannanase and thus bacterial fermentation is able to occur in the intestines. Food preservatives should be used to prevent the fermentation of gels and stock solutions by air-borne microorganisms.

11.5.4 Viscosity

Viscosity measurement by food-testing instruments must be done with care as konjac is sensitive to shear thinning (Fig. 11.6) and also may be only partially hydrated when evaluated (Fig. 11.7). High purity glucomannan grades are difficult to solubilise. Konjac exhibits very viscous solutions in the entanglement area, at concentrations above 10 g/L, but it is much less viscous than guar or xanthan gums at higher dilutions. Extreme values reported for molecular weight and viscosity could be linked to the presence of microgel particles; clusters of condensed molecules produced during processing and/or purification.

Native konjac material, 'crude flour', typically available as 20–60 mesh coarse powder (250–850 μm), is still bound to the original protein membrane and this must first swell before

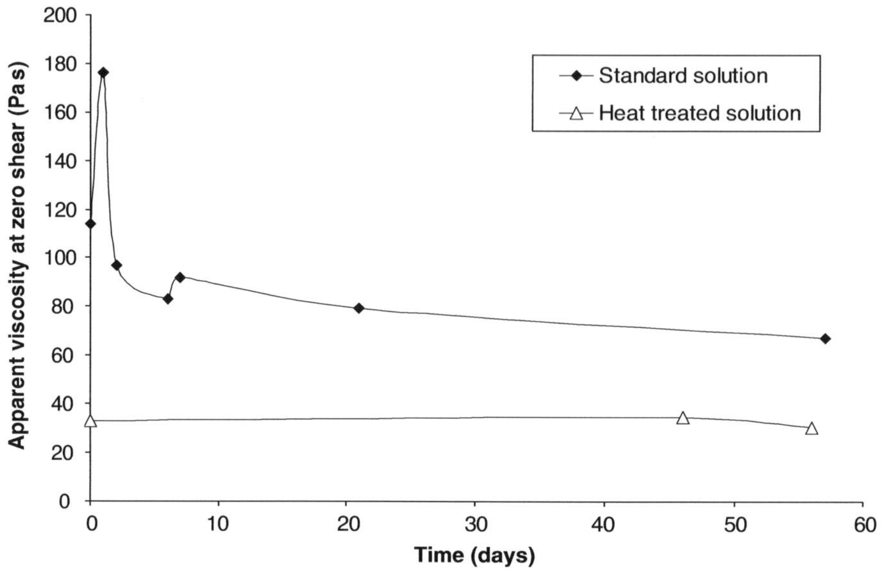

Fig. 11.6 Effect of heat treatment on the apparent viscosity at zero shear over time of a 1% konjac solution. Granules require several hours to fully disperse: viscosity first peaks then slowly diminishes. (Reproduced with permission of P. Vieille and J.M. Parry, KALYS SA.)

the konjac molecules are released into solution. Peak viscosity builds up in an average of 30–50 min and will slowly rise over time unless the flour has been washed and ground, heat treated, hydrolysed or co-dried with some form of dispersing material. Later, viscosity diminishes when the konjac is fully hydrated and aggregates are dispersed (Fig. 11.6). Viscosity is positively correlated with glucomannan content (Fig. 11.7) and degradation of the chain length reduces viscosity.

Native grades of konjac typically exhibit non-Newtonian behaviour (Fig. 11.8). The galactomannan gums (guar, locust bean and tara) have a similar rheology to konjac glucomannan and also exhibit shear-dependent viscosity.

11.5.5 Konjac gels

11.5.5.1 *Deacetylated konjac glucomannan gels*

Eliminating acetyl groups produces deacetylated konjac glucomannan which is able to build random junction zones through hydrogen bonds to form an irreversible and extremely stable gel with similar elastic recovery to polyacrylamide gels (Case, 1992). Acidic deacetylation is possible but alkali reactions are preferred to limit hydrolysis.

11.5.5.2 *Native konjac gels*

In some cases, temperature-stable gels may be obtained without deacetylation. Non-modified konjac flours show different gelling properties depending on pH, the concentration of other

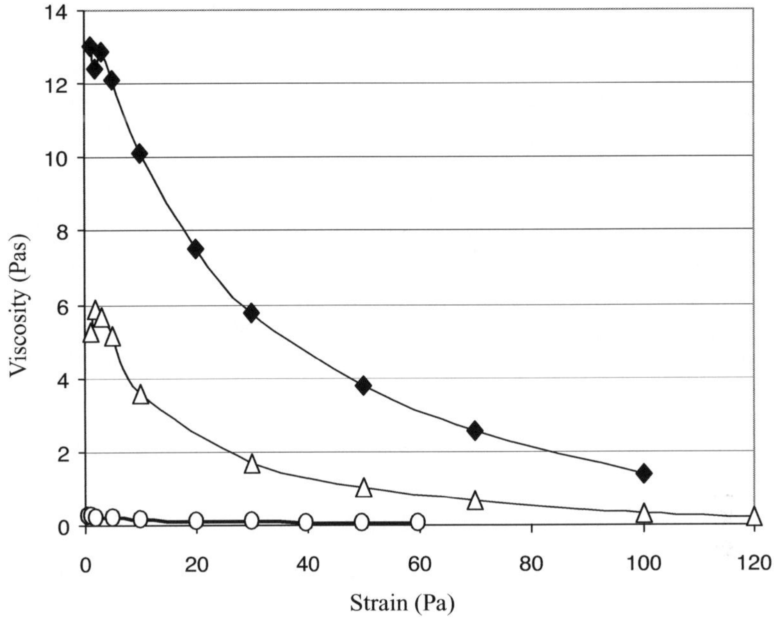

Fig. 11.7 Konjac viscosity under strain. Viscosity is correlated to purity of glucomannan material: ♦ alcohol washed grade; △ crude powder; ○ sieved-out material. (Reproduced with permission of P. Vielle and J.M. Parry, Kalys SA.).

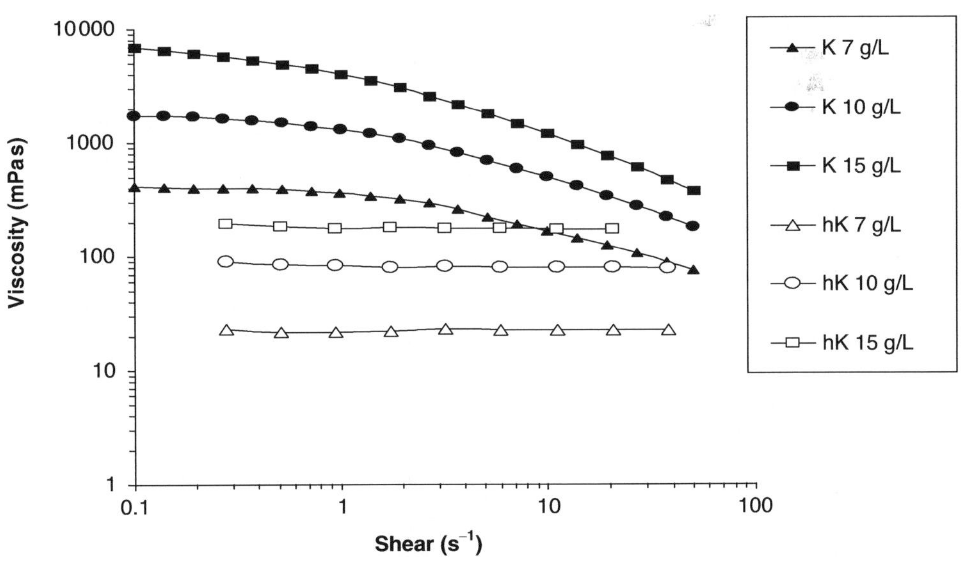

Fig. 11.8 Comparison of viscosity of konjac (K) and hydrolysed konjac (hK) under shear. (Reproduced with permission of P. Vieille and J.M. Parry, KALYS SA.)

ingredients and the presence of synergistic polysaccharides, such as semi-refined carrageenan (PES, E407a), kappa or iota carrageenan (E407) or xanthan gum. The gels differ in heat sensitivity and melt and are characterised by elasticity and hardness. For example, gels of kappa carrageenan with potassium chloride are brittle and xanthan gum does not gel alone but both give elastic, cohesive gels with konjac. Ions also interact with blends containing konjac.

High concentrations of 2–5% glucomannan may gel without any alkali (Cairns *et al.*, 1988; Dave and McCarthy, 1997). These gels do not exhibit a specific fusion temperature.

High clarity gels may be obtained by using enhanced-refining processes (Ohashi *et al.*, 2000).

Cross-linking agents have been studied in demanding applications (Richards *et al.*, 1997). Copolymer blends of rarer gelling agents or modified gums (Zhang *et al.*, 2005) are innovative areas of research (Chen *et al.*, 2005b). A better understanding of glucomannan interactions in gelling systems may lead to new forms of encapsulation of aroma or active ingredients (Li *et al.*, 2006b).

11.5.6 Interactions with other materials

It is important to properly characterise glucomannan grades. Impurities, such as starch, protein and minerals, affect the quality of products but are often less important to rheology than intrinsic characteristics. The choice of flour will depend very much on the priorities established for viscosity, smell, gel strength, price, colour, transparency, taste and so on (Fig. 11.9).

The affinity of glucomannan with iron, zinc, calcium and magnesium salts has been studied by Doi (1995). High proportions of glucomannan cause little active mineral stripping, though they may reduce the intake of minerals or even pharmaceutical actives through physical entrapment.

Lyotropic salts alter gel properties (Case and Hamann, 1994). Sodium sulphate increases the rate of gelation but lowers gel strength, and sodium nitrate increases setting time, gel strength and transparency. High sodium chloride levels affect the gel properties of konjac–carrageenan blends.

11.5.7 Konjac–carrageenan blends

Maximum peak viscosity is attained from synergy with a blend of 75–90% konjac flour with 25–10% carrageenan. Different grades of konjac flour develop different synergies with

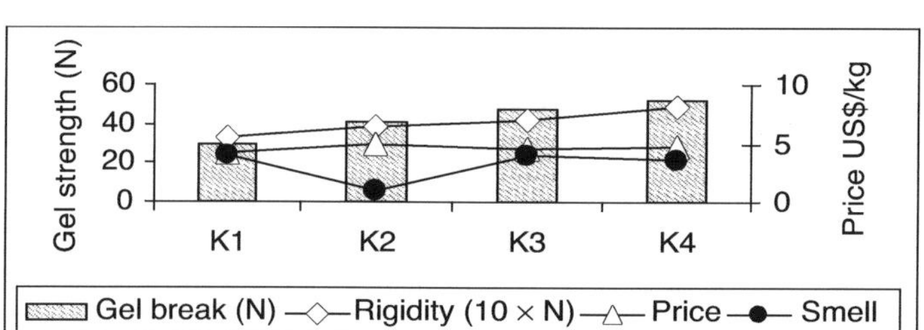

Fig. 11.9 Characteristics of konjac-kappa carrageenan blends. (Reproduced with permission of P. Vieille and J.M. Parry, KALYS SA.)

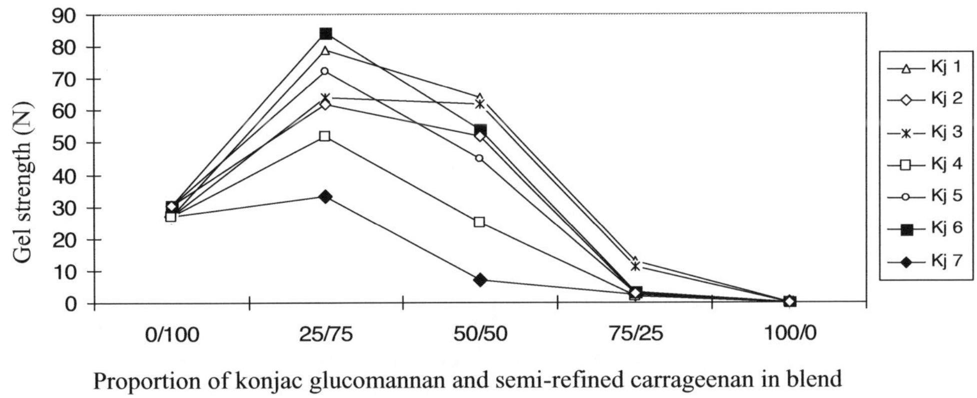

Fig. 11.10 Gel strength of blends of semi-refined carrageenan with different konjac grades. Glucomannan content of Kj 1 to Kj 4 (crude powders) is between 65% and 75%, Kj 7 is 70–75% and Kj 5 and Kj 6 is 80%. (Reproduced with permission of P. Vieille and J.M. Parry, KALYS SA.)

a standard carrageenan (Fig. 11.10). Konjac grades show different affinities depending on carrageenan type, such as kappa, iota or semi-refined.

For milk gels with iota carrageenan, the maximum increase in gel strength is obtained with a blend of 20:80 to 40:60 konjac flour and carrageenan (Fig. 11.11). More purified glucomannan is necessary to obtain maximum synergy with carrageenan. When added to blends incorporating carrageenan and high-purity, refined glucomannan gum (E425ii), sugar enhances gel strength in high concentrations of glucomannan and reduces gel strength in low concentrations.

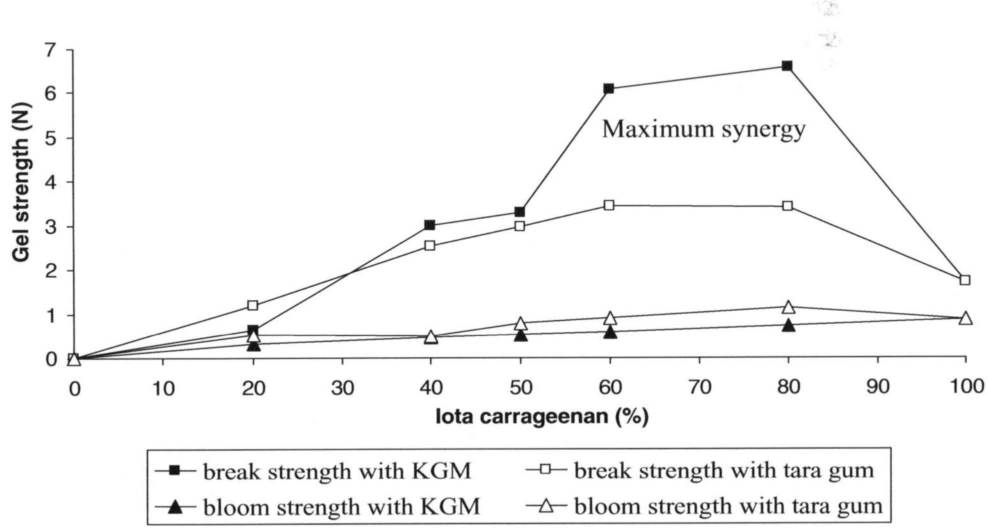

Fig. 11.11 Gel strength of blends of iota carrageenan with konjac glucomannan (KGM) or tara gum. Bloom strength is the force (g) to depress a 12.5-mm-diameter plunger 4 mm into gel; break strength is the force to break the gel using a 12.5-mm-diameter plunger. (Reproduced with permission of P. Vieille and J.M. Parry, KALYS SA.)

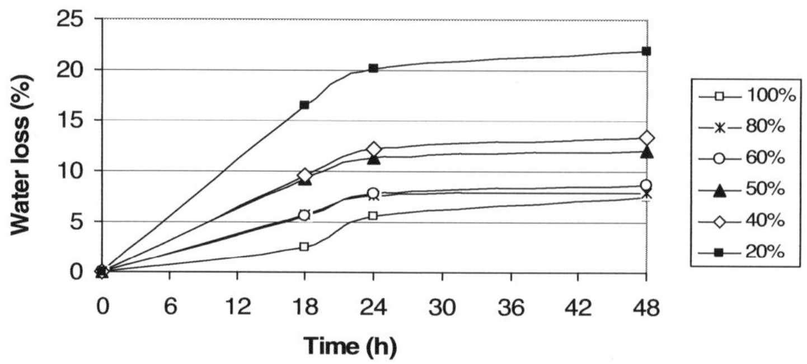

Fig. 11.12 Syneresis in an iota carrageenan–konjac gel. Different curves indicate effect of carrageenan content in gel blend. (Reproduced with permission of P. Vieille and J.M. Parry, KALYS SA.)

In brines for injected meat products, due to its cold hydration, konjac is used only at low levels to boost carrageenan gel strength, otherwise it imparts too much immediate viscosity. When possible, a compromise may allow the use of coarse mesh konjac that swells slowly. Another option is to choose a fine-mesh, low-viscosity konjac compatible with needle injection. Syneresis varies with glucomannan ratio but little increase in syneresis is observed at up to 30% replacement of iota by konjac (Fig. 11.12; Kalys, 2003). In price-sensitive applications such as injected meats, syneresis should be evaluated against the cost of a konjac–iota-carrageenan blend and inclusion rates as well as product yield.

11.5.8 Konjac–xanthan blends

Konjac processes cause variations in chain length, chain rigidity and powder purity and these determine the strength of the interaction with xanthan gum. Konjac–xanthan blends produce very elastic, thermally-reversible gels, with maximum synergy obtained for a blend of 50% konjac flour and 50% xanthan gum (Fig. 11.13) at total gum concentrations as low as 0.02%.

In Fig. 11.14, solid lines show the range of melting and gelling temperatures for 1% gels made of 50% refined konjac and 50% xanthan and can show a 15°C shift depending on

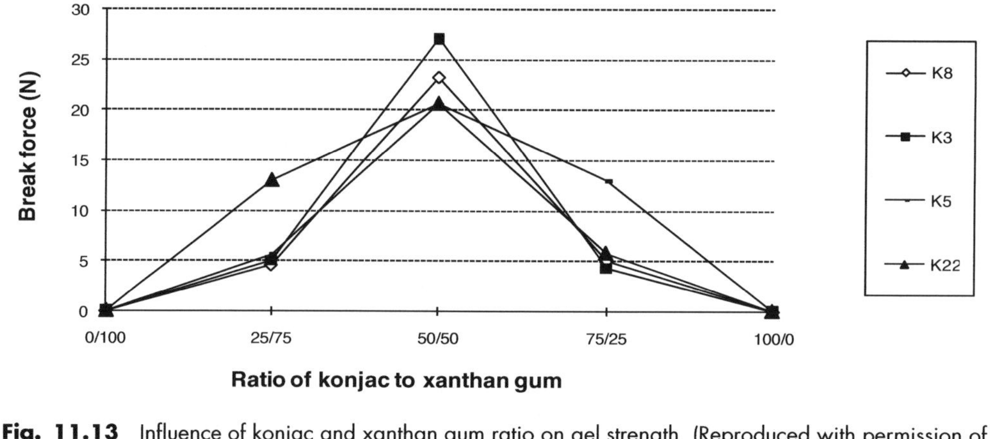

Fig. 11.13 Influence of konjac and xanthan gum ratio on gel strength. (Reproduced with permission of P. Vieille and J.M. Parry, KALYS SA.)

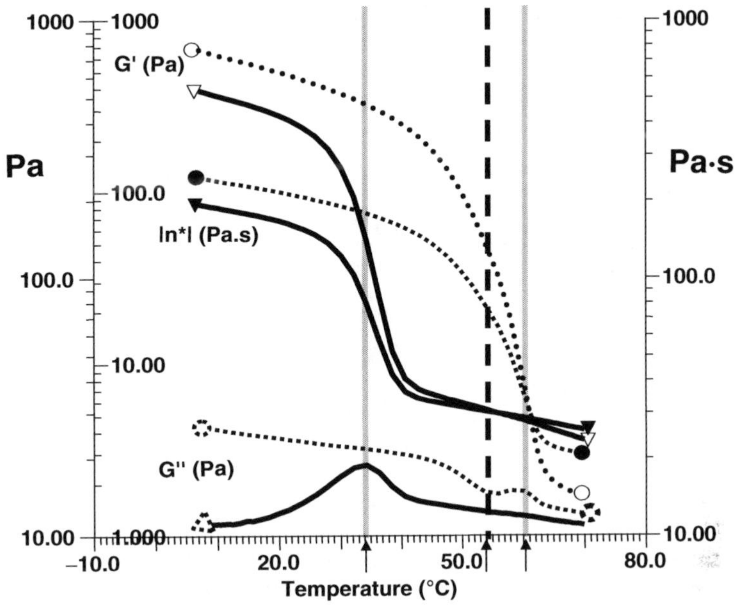

Fig. 11.14 Influence of konjac and xanthan gum ratio on gel melting characteristics. (Reproduced with permission of P. Vieille and J.M. Parry, KALYS SA.)

origins of materials, including raw materials, bacterial strain and commercial source (dotted line from Goycoolea *et al.*, 1995; Kalys, 2006).

Xanthan shows more affinity for glucomannan than the galactomannans, locust bean, tara and cassia gums. A konjac–xanthan blend is more influenced by salts than a locust bean gum–xanthan blend. In presence of 0.04 mol/L sodium chloride, xanthan polymeric chains are more available to glucomannan (Annable *et al.*, 1994) and gel at a lower temperature of 42°C.

11.5.9 Konjac–starch blends

Glucomannan–starch blends show enhanced viscosity, improved freeze–thaw stability and reduced syneresis compared to starch alone. Adding small amounts of glucomannan to starch enhances the immediate short-term retrogradation of amylopectin but reduces long-term effects with an overall positive effect (Yoshimura *et al.*, 1998).

Glucomannan promotes amylose gelation and the effect depends on concentration, water and starch type. Glucomannan (1%) and starch (9%) form thermo-irreversible gels at 10% that have the same temperature, acid and alkali stability as deacetylated glucomannan and are much stronger than gelled starch alone at 10% concentration (Tye, 1991).

11.5.10 Konjac–agar blends

A ratio of 10 to 90 of glucomannan to agar shows maximum effect in gel strength and texture; adding more konjac dilutes the agar and reduces gel strength proportionally.

11.5.11 Konjac and other gums

High-molecular-weight glucomannan strongly affects the self-association of gellan gum promoted by salts. Suggested ratios are 0.3 to 0.5 parts gellan for 1 part konjac glucomannan (Nishinari *et al.*, 1996). Konjac–acetan blends have been studied by Ridout *et al.* (1998).

11.6 FOOD APPLICATIONS

11.6.1 Thickening and binding

Konjac influences retrogradation and moisture release in soft breads, pastries and long shelf life bakery products when used at 0.1–0.5%. Konjac provides adhesion for coatings and binding in complex matrixes such as restructured meat or vegetables.

11.6.2 Gels

Commercial konjac-based blends include konjac with carrageenan, konjac with starch, konjac with carrageenan and one or several of the galactomannans and so on. These have been used in pasta, restructured foods, meat and desserts for many years and their use is now growing in precooked foods. The gel composition can be adjusted to give the required strength and melting temperatures, offering a versatile range of mouthfeel properties (Penroj *et al.*, 2006). Typical food gels use konjac at 0.1–1.0% with other gelling agents, or alone at levels up to 5%, to give strong shear resistance with an improved cohesiveness and bite or, when used as a highly sheared gel, more creaminess.

For *Konnyaku* and *Shirataki* noodles, konjac flours are mixed with lime and heated to produce non-melting gels. Strong food bases are used to increase the pH above 9 for a rapid effect, and though solutions may reach pH 10 or above, the resulting pH in food products may be lowered by washing, by the addition of acid further in the process or simply by buffering from other ingredients (Mandava *et al.*, 1997). Processors may use any of the following alkalis in a ratio of 10% weight to konjac flour: calcium hydroxide, calcium carbonate (for an opaque gel), sodium bicarbonate, potassium hydroxide, potassium carbonate (for a translucent gel), sodium carbonate, potassium orthophosphate and sodium tripolyphosphate or by dosing sodium hydroxide to obtain the required pH. The final pH may even be neutral to slightly acidic (Cheney *et al.*, 1982). Drying films prior to the neutralising acid wash enhances gel elasticity and strength.

A substantial level of solids and other ingredients can be incorporated and formed into chunks of meat, vegetable or fruit to give products that resist sterilisation and deep frying. These gels are highly stable to salt. The minimum konjac flour concentration for such gels is 0.5–2% for flours containing 65–99% glucomannan.

More recent developments using deacetylated konjac glucomannan include deep-fried food (Harada and Ikeda, 1996), neutral low-calorie fibre (Suratto, 1999), freeze–thaw stable gels (Shiota and Yoshida, 2005) and freeze-dried food (Shimizu *et al.*, 2006). Partial deacetylation leads to weaker gels (Zhang *et al.*, 2001).

Different grades of konjac and alkali yield different types of noodles: they may be a translucent white or a darker colour, sold dehydrated or ready-to-use sealed in liquid.

A recommended recipe for deacetylated konjac glucomannan for noodles might use sodium or phosphate salts which give less off-flavour than calcium hydroxide. In the recipe, gelling agents (alkali, konjac) represent 1–5% and preservatives, flavours and sweeteners increase the dry matter to 10–20%. This gelled product may be formed or blended with other ingredients, such as starch and protein. The pH is raised above 10 before washing to eliminate excess alkali and replace entrapped water.

Improved non-melting firm gels, for retort-stable meat products, have been obtained with konjac and other gums without alkali given sufficient heat and cooking time above pH 6.5.

Adding konjac to carrageenan improves elasticity and gel strength in jellies, surimi, films, fillings and other products. Gel strength is affected by retorting which may degrade the gel, especially in acidic solutions, if low concentrations of gelling agents are used for cost-effectiveness. For pH values below 4, as in fruit jellies or gums, in presence of carrageenan it is better to use a combination of citric, adipic and malic acids to limit the reduction in gel strength.

11.6.3 Suspension

Some grades show some suspension activity and have been applied to tea (Gong *et al.*, 2006), fruit juices (Westphal *et al.*, 2006) and smoothies in place of cellulose or dilute gelling agents.

When konjac is used with xanthan or carrageenan to form gels and then sheared, the resulting mouthfeel is oily or similar to cream. Once mixed with liquid or oil, these gels provide low-cost suspension to pulp-rich drinks, improved whipping and pouring properties in thixotropic solutions and can be whisked or foamed. If deacetylated konjac gum is used, the resulting sheared gel might be used as a shortening or as filler in ice cream.

11.7 NUTRITIONAL APPLICATIONS

11.7.1 Obesity and cholesterol management

Many articles refer to glucommanan for treating diabetes (Walsh *et al.*, 1984; Huang *et al.*, 1990; Li *et al.*, 2002), type 2 diabetes and insulin-resistance syndrome (Renard *et al.*, 1991; Vuksan *et al.*, 1999). Konjac lowers cholesterol without any adverse effects and can be three times as effective as other polymers (Vuksan *et al.*, 2000). Also it is synergistic with stanols and sterols for low-density lipoprotein (LDL) reduction (Yashida *et al.*, 2006).

11.7.2 Prebiotic properties

Substantial published evidence supported the decision by the Chinese Academy of Science to put mannan oligosaccharides (konjac) on the 6th Torch Program in the 9th Five Year Plan from 1997 to 2002. Konjac oligosaccharides improve intestinal balance, intestinal integrity and immunity but require advanced technology to decrease the molecular weight by several orders of magnitude. Konjac oligosaccharides are very expensive compared to fructo-oligosaccharides but could be complementary in some applications.

Although konnyaku gel acts like an insoluble fibre, long-chain konjac is a potential colonic food and prebiotic because it shows very little degradation in the digestive tract and is fermented by beneficial human bacterial strains (Takahashi *et al.*, 1984; Choi and Park, 2004). Current work suggests the use of medium-length polysaccharides and oligosaccharides for the best prebiotic effect (Chen *et al.*, 2005a).

11.7.3 Low-calorie soluble fibre

Konjac has a mechanical function in slowing food intake and reducing appetite (Chen *et al.*, 2006). As well as having a very low effective caloric value, konjac can replace high-calorie thickeners at very low use levels.

11.7.4 Improved chewing abilities and quality of life

Firm chewy konjac-based products, as described in Section 11.1.3, can help in the recuperation of the ability to masticate (Kawamura and Horio, 1989). Konjac is a natural choice to thicken drinks for patients with impaired handling or swallowing of fluids (dysphagia) (Sliwinski, 2006).

11.8 FUTURE DEVELOPMENTS

Konjac is a soluble vegetable fibre that is very effective as a fat mimetic and for regulating blood sugar and low-density lipoprotein. It has nutritional properties and is accepted widely amongst ethnic and religious communities as a gelling and thickening agent, it can be Halal and Kosher, including Badatz. It has a substantial record as a common healthy ingredient in Asia and will, no doubt, find wider application as it becomes more available and familiar to food technologists worldwide.

References

Annable, P., Williams, P.A. and Nishinari, K. (1994) Interaction in xanthan–glucomannan mixtures and the influence of electrolyte. *Macromolecules*, **27**, 4204.

Cairns, P., Miles, M.J. and Morris, V.J. (1988) X-ray fiber diffraction studies on konjac mannan-kappa carrageenan mixed gels. *Carbohydrate Polymer*, **8**, 99–104.

Case, S.E. (1992) *Rheology and Chemistry of Konjac Mannan*. PhD thesis, North Carolina State University, Raleigh.

Case, S.E. and Hamann, D.D. (1994) Fracture properties of konjac mannan gel. *Food Hydrocolloids*, **8**(2), 147–154.

Case, S.E., Knopp, A., Hamann, D.D. and Schwartz, S. (1992) Characterisation of gelation of konjac mannan using lyotropic salts and rheological measurements. In: G.O. Phillips, P.A. Williams and D.J. Wedlock (eds), *Gums and Stabilisers for the Food Industry 6*. IRL Press at Oxford University Press, Oxford, pp. 489–500.

Chen, H.-L., Cheng, H.-C., Liu, Y.-J., Liu, S.-Y. and Wu, W.-T. (2006) Konjac acts as a natural laxative by increasing stool bulk and improving colonic ecology in healthy adults. *Nutrition*, **22**(11/12), 1112–1119.

Chen, H.-L., Fan, Y.-H., Chen, M.-E. and Chan, Y. (2005a) Unhydrolyzed and hydrolyzed konjac glucomannans modulated cecal and faecal microflora in Balb/c mice. *Nutrition*, **21**(10), 1059–1064.

Chen, L.-G., Liu, Z.-L. and Zhuo, R.-X. (2005b) Synthesis and properties of degradable hydrogels of konjac glucomannan grafted acrylic acid for colon-specific drug delivery. *Polymer*, **46**(16), 6274–6281.

Cheney, P.A., Stares, J. and Vernon, A.J. of Mars Limited (1982) Food product thickened or gelled with carrageenan and glucomannan, US Patent 4427704.

Choi, J.-Y. and Park, G.-G. (2004) Metabolism activity of *Bifidobacterium* spp. by degree of polymerization of konjac glucomannan hydrolysates, *Han'guk Sikp'um Yongyang Kwahak Hoechi. Korean Society of Food Science and Nutrition*, **33**(7), 1186–1191.

Dave, V. and McCarthy, S. (1997) Review of konjac glucomannan. *Journal of Environmental Polymer Degradation*, **5**, 237–241.

Doi, K. (1995) Effect of konjac fibre (glucomannan) on glucose and lipids. *European Journal of Clinical Nutrition*, **49**(3), 190–197.

EU. (1998) Directive 98/72/EC published 4th Nov 1998 modifying European Parliament and Council Directive 95/2/EC of 20 February 1995 on food additives other than colours and sweeteners.

EU. (2003) CONSLEG: 1996L0077 – 20/11/2003.

EFSA (2004) Opinion of the Scientific Panel on food additives, flavourings, processing aids and materials in contact with food (AFC) on a Request from the Commission related to the use of certain food additives in Jelly mini cups. Question number EFSA-Q-2004-054. Available at http://www.efsa.europa.eu/EFSA/efsa_locale-1178620753812_1178620761905.htm (accessed 25 May 2009).

FCC. (1996) Food and Nutrition Board Committee on Food Chemicals Codex, June 13, 1996.

FDA. (1997) GRAS title 21 CFR 170.30.

FDA. (2001) 4th October Import Alert IA3315. FIARS.

FSANZ. (2007) Record of views formed by the FSANZ Novel Foods Reference Group, May 2007.

Gong, J.-S., Liu, P.-Y. and Liu, Q.-J. (2006) Study on stabilizing mechanism of konjac glucomannan in tea infusions. *Journal of Food Science*, **71**(9), E358–E364.

Goycoolea, F.M., Richardson, R.K., Morris, E.R. and Gidley, M.J. (1995) Stoichiometry and conformation of xanthan in synergistic gelation with locust bean gum or konjac glucomannan: evidence for heterotypic binding. *Macromolecules*, **28**(24), 8308–8320.

Harada, S. and Ikeda, M. of Uni Colloid Kabushiki Kaisha (1996) Coated fried food product, US Patent 5552166.

Hu, H. (1993) Method for inhibition of water absorption and swelling of glucomannan, CN 1068832 published 02.10.1993.

Huang, C.Y., Zhang, M.Y., Peng, S.S., Hong, J.R., Wang, X., Jiang, H.J., Zhang, F.L., Bai, Y.X., Liang, J.Z. and Yu, Y.R. (1990) Effect of konjac food on blood glucose level in patients with diabetes. *Biomedical and Environmental Science*, **3**(2), 123–131.

Kalys (2001) European FAIR Program 1998–2001, KALYS SA.

Kalys (2003) Presentation on Tara gum and konjac gum as LBG substitutes, KALYS SA.

Kalys (2005) personal communication by Dr S. Dorthe, KALYS SA.

Kalys (2006) personal communication by R. Nervo, KALYS SA.

Kawamura, Y. and Horio, T. (1989) Effects of chewing exercise on the maximum biting force and chewing performance. *Shika Kiso Igakkai Zasshi*, **31**(3), 281–290.

Kishida, N., Okimasu, S. and Kamata, T. (1978) Molecular weight and intrinsic viscosity of konjac glucomannan. *Agricultural and biological Chemistry*, **42**, 1945.

Kohyama, K., Ida, H. and Nishinari, K. (1993) A mixed system composed of different molecular weights konjac glucomannan and kappa carrageenan: large-deformation and dynamic viscoelastic study. *Food Hydrocolloids*, **7**, 213.

Koroskenyi, B. and McCarthy, S.P. (2001) Synthesis of acetylated konjac glucomannan and effect of degree of acetylation on water absorbency. *Biomacromolecules*, **2**(3), 824–826.

Li, B., Xie, B. and Kennedy, J.F. (2006a) Studies on the molecular chain morphology of konjac glucomannan. *Carbohydrate Polymer*, **64**(4), 510–515.

Li, B., Xu, Z. and Xie, B.J. (2006b) Preparation and temperature effect on the swelling behavior of konjac glucomannan–methylcellulose blend film. *European Food Research and Technology*, **223**(1), 132–138.

Li, M., Zhang, M.-Y., Hong, J.-R., Xiao, Q.-W., Wang, X.-H., Ye, L., Zhu, Y.-L. and Gao, Z.-F. (2002) The influence of a konjac food on human lipid and glucose metabolism. *West China Medical Journal*, **1**, 36–38.

Mandava, R., Fernandez, I. and Juillerat, M.A. of Nestec (1997) Process for preparing gelled meat chunks, US patent 5693356.

Matsumura, S., Nishioka, M. and Yoshikawa, S. (1991) Enzymatically degradable poly(carboxylic acid) derived from polysaccharide. *Makromoleculare Chemie, Rapid Communication*, **12**, 89–94.

Nishinari, K., Miyoshi, E., Takaya, T. and Williams, P.A. (1996) Rheological and DSC studies on the interaction between gellan gum and konjac glucomannan. *Carbohydrate Polymer*, **30**(2), 193–207.

Ohashi, S., Shelso, G.J., Moirano, A.L. and Drinkwater, W.L. (2000) FMC Corporation, Clarified konjac glucomannan, US Patent 6162906.

Ohya, Y., Ihara, K., Murata, J.-I., Sugitou, S. and Ouchi, T. (1994) Preparation and biological properties of dicarboxy-glucomannan: enzymatic degradation and stimulating activity against cultured macrophages. *Carbohydrate Polymer*, **25**(2), 123–130.

Penroj, P., Hill, S.E., Mitchell, J.R. and Garnjanagoonchorn, W. (2006) Effect of kappa carrageenan on konjac glucomannan gelation. In: P.A. Williams and G.O. Phillips (eds), *Gums and Stabilisers for the Food Industry 13*. RSC, Cambridge, pp. 211–226.

Renard, E., Bringer, J., Jaffiol, C., Orsetti, A. and Mirouze, J. (1991) Non-insulin diabetes and glucose intolerance: effect of glucomannan fibres on blood sugar and insulin. *Semaine Des Hôpitaux de Paris*, **67**(6), 153–157.

Richards, J.C., Hills, W.A. and Snow, W.C. (1997) FMC Corporation, Stable peracid sols, gels and solids, US patent 5597791.

Ridout, M.J., Cairns, P., Brownsey, G.L. and Morris, V.J. (1998) Evidence for intermolecular binding between deacetylated acetan and the glucomannan konjac mannan. *Carbohydrate Research*, **309**(4), 375–379.

Sakai, W.S. (1983) Aroid root crops; *Alocasia, Cyrtosperma* and *Amorphophallus*. In: H.-T. Chan (ed), *Handbook of Tropical Foods*. Marcel Dekker, New York, pp. 29–83.

Shimizu, M. and Shimahara, H. of Kabushiki Kaisha Shimizu Manzo Shoten (1973) Method of selective separation of konjac flour from the tubers of *Amorphophallus* konjac, US Patent 3767424.

Shimizu, K., Oka, Y., Kaneko, T., Shono, F. and Saki, M. (2006) Production method of thermal non-reversible dehydrated foods. 6 pp. CODEN: JKXXAF JP 2006149340 A2 20060615.

Shiota, T. and Yoshida, H. (2005) Food production from Konnyaku powder or glucomannan gelation product (Japan). PCT Int. Appl., 13 pp. WO 2005065466 A1 20050721.

Sliwinski, E.L. (2006) Thickener composition for dysphagia patients (N.V. Nutricia, Netherlands). PCT Int. Appl., 21 pp. WO 2006054886 A1 20060526.

Suratto, P. (1999) Food material having high vegetable fiber content by utilizing konjac mannan, Japanese Patent 11169104, published 29 June 1999.

Suzuki, M. (1980) 3–4-Dihydroxybenzaldehyde-D-glucoside, the irritant substance of konnyaku. *Journal of Food Science*, **45**(4), 1075.

Takahashi, R., Kusakabe, I., Kusama, S., Sakurai, Y., Murakami, K., Maekawa, A. and Suzuki, T. (1984) Studies on the mannanase of Streptomyces. Part IV. Structures of glucomanno-oligosaccharides from the hydrolytic products of konjac glucomannan produced by a β-mannanase from Streptomyces sp. *Agricultural and biological Chemistry*, **48**(12), 2943–2950.

Tomita, K. and Morita, Y. (2006) Method for dissolution of glucomannan and/or galactomannan, and gelling/thickening agents containing them with improved solubility. 11 pp. CODEN: JKXXAF JP 2006061070 A2 20060309.

Tye, R. (1991) Konjac flour: properties and applications. *Food Technology*, **45**, 11–16.

Valerie, J. Ryan, R.Y., Chienkuo and Crosby, G.A. of Opta Food Ingredients, Inc. (2002) Methods for lowering viscosity of glucomannan compositions, CN 1352530, published 2002.06.05.

Vieille, P.H. (2003) Europe bans konjac in jelly sweets, *Confectionery News* 12/02/2003. http://www.confectionerynews.com/Formulation/Europe-bans-konjac-in-jelly-sweets (accessed 5 August 2009).

Vuksan, V., Jenkins, D.J., Spadafora, P., Sievenpiper, J.L., Owen, R., Vidgen, E., Brighenti, F., Josse, R.G., Leiter, L.A. and Bruce-Thompson, C. (1999) Konjac-mannan (glucomannan) improves glycemia and other associated risk factors for coronary heart disease in type II diabetes: a randomised controlled metabolic trial. *Diabetes Care*, **22**, 913–919.

Vuksan, V., Sievenpiper, J.L., Owen, R., Swilley, J.A., Spadafora, P., Jenkins, D.J., Vidgen, E., Brighenti, F., Josse, R.G., Leiter, L.A., Xu, Z. and Novokmet, R. (2000) Beneficial effect of viscous dietary fibre from konjac mannan in subjects with the insulin resistance syndrome: results of a controlled metabolic trial. *Diabetes Care*, **23**(1), 9–14.

Walsh, D.E., Yaghoubian, V. and Behforooz, A. (1984) Effect of GM in obese patients: a clinical study. *International Journal of Obesity*, **8**(4), 289–293.

Westphal, K.A., Mattson, P.H. and Casteel, R.M. of California Giant, Inc. (2006) Nutritious strawberry fruit juice drink and method of making the same, US Patent 7037539.

Williams, P.A., Clegg, S.M., Langdon, M.J., Nishinari, K. and Phillips, G.O. (1992) Studies on the synergistic interaction of konjac mannan and locust bean gum with kappa carrageenan. In: G.O. Phillips, P.A. Williams and D.J. Wedlock (eds), *Gums and Stabilisers for the Food Industry 6*. IRL, Oxford, pp. 209–221.

Xu, G., Wu, Y. and Jiao, C. (2005) Preparation of glucomannan oligosaccharide by culturing *Trichodema viride*. *Faming Zhuanli Shenqing Gongkai Shuomingshu*, 4 pp.

Yashida, M., Vanstone, C.A., Parsons, W.D., Zawitowski, J. and Jones, P.J. (2006) Effect of sterols and glucomannan on lipids in individuals with and without type II diabetes. *European Journal of Clinical Nutrition*, **60**(4), 529–537.

Yoshimura, M., Takaya, T. and Nishinari, K. (1998) Rheological studies on mixtures of corn starch and konjac–glucomannan. *Carbohydrate Polymer*, **35**, 71–79.

Zhang, H., Yoshimura, M., Nishinari, K., Williams, M.A.K., Foster, T.J. and Norton, I.T. (2001) Gelation behaviour of glucomannan with different molecular weights. *Biopolymers*, **59**(1), 38–50.

Zhang, Y.-Q., Xie, B.-J., Gan, X. (2005) Advance in the applications of konjac glucomannan and its derivatives. *Carbohydrate Polymer*, **60**(1), 27–31.

Zhong, G., Zhang, S., Zhang, W. and Li, H. (2006) Production of konjac oligo-glucomannan with high purity. *Faming Zhuanli Shenqing Gongkai Shuomingshu*, 5 pp.

12 Microcrystalline Cellulose

Domingo C. Tuason, Gregory R. Krawczyk and Greg Buliga

ABSTRACT

Microcrystalline cellulose is purified cellulose, produced by converting fibrous cellulose to a redispersible gel or aggregate of crystalline cellulose using acid hydrolysis. This material is dried to a pure, fine-particle form for powdered grades or coprocessed with a water-soluble polymer, such as cellulose gum, which acts as a barrier dispersant, so that the copolymer can be readily dispersed in water using high shear. This dispersion will reconstitute to deliver a colloidal form of microcrystalline cellulose. These colloidal dispersions are unique when compared to other soluble food hydrocolloids. They exhibit a variety of desirable characteristics including suspension of solids, heat stability, ice crystal control, emulsion stabilization, foam stability, texture modification and fat replacement. Food applications include frozen desserts and ice cream, whipped toppings, low-fat mayonnaise and salad dressings, ambient-stable dairy and non-dairy beverages, nutritional drinks, ready-to-use bake-stable bakery fillings and fruit fillings and dairy and non-dairy creams.

12.1 INTRODUCTION

The early development work on microcrystalline cellulose (MCC) began in the 1960s with Dr O.A. Battista of the American Viscose Corporation, which later became part of the FMC Corporation. MCC, also known as cellulose gel, is purified, naturally occurring cellulose, produced by converting fibrous cellulose to a redispersible gel or aggregate of crystalline cellulose. This is accomplished by a simple acid hydrolysis to a level-off degree of polymerization (LODP). This material may be dried to a pure, fine-particle form for powdered grades or it may be coprocessed with a barrier dispersant, followed by drying, to provide a grade that will reconstitute to deliver a colloidal form of MCC.

The physical properties of colloidal MCC dispersions are unique when compared to typical soluble hydrocolloids commonly used in the food industry. The usefulness of these properties has led to a highly developed application technology of colloidal MCC and an extensive study of the nature of its functional properties. Several of these characteristics include ice crystal control, texture modification, emulsion stabilization, heat stability, foam stability, suspension of solids, bulking agent and fat replacement.

All-natural line extensions of colloidal MCC-based products with enhanced physical properties have recently been developed to fit specific needs in a variety of food systems. Today, a range of MCC products is used extensively for food, pharmaceutical and industrial

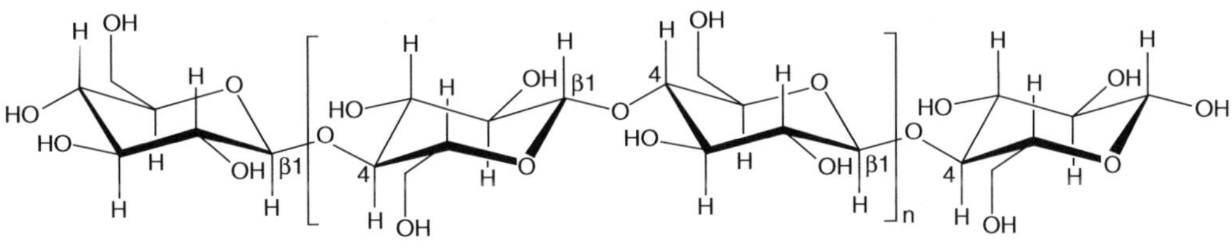

Fig. 12.1 Structure of cellulose. (Reproduced with kind permission from FMC Corporation.)

applications. The properties and uses of this ingredient for the food industry depend on both the composition and the manufacturing process.

12.2 MCC PRODUCT TECHNOLOGIES

12.2.1 Chemical composition

Cellulose, in the chemical sense, is a polysaccharide of sufficient chain length to be insoluble in water or dilute acids and alkalis at ordinary temperatures (Ward, 1954). It consists of anhydroglucose units linked together through the 1 and 4 carbon atoms with a beta-glycosidic linkage. Thus, it has the repeating unit shown in Fig. 12.1, with *n* having values ranging from about 50 to 5000 or more (Ott and Tennent, 1954).

The raw material for the production of MCC is a selected refined alpha (α-) cellulose derived from specialty grades of wood pulp. The cellulose fibers are composed of millions of microfibrils. The individual microfibril is composed of two regions, the paracrystalline region, which is an amorphous and flexible mass of cellulose chains, and the crystalline region, which is composed of tight bundles of cellulose chains in a rigid linear arrangement.

12.3 MANUFACTURING PROCESS

12.3.1 MCC process

The first step requires cooking the purified pulp with a dilute mineral acid in water. The acid preferentially attacks the less-ordered or amorphous regions of the cellulose polymer chain, thereby exposing and freeing the crystalline sites, which form cellulose crystallite aggregates. The process is carried to a point at which a LODP is attained. After completion of the hydrolysis, the cake obtained is neutralized and thoroughly washed to remove impurities. The resulting wetcake is freed of water and the dried crystalline cellulose aggregate is recovered and ground to give powdered grades of MCC.

12.3.2 Colloidal MCC process

For the production of colloidal MCC products, the washed cake is then subjected to mechanical disintegration to break up the aggregates and release additional microcrystals. The attrition should be sufficient to produce a mass in which not more than about 1% by weight of the particles has an average length greater than 1 μm as determined by electron microscopy.

When this is achieved, the material is then capable of forming a stable dispersion in aqueous and other media.

After the mechanical disintegration step, the attrited microcrystals are coprocessed with a hydrophilic dispersant, such as sodium carboxymethyl cellulose (CMC). The CMC prevents the microcrystals reaggregating due to hydrogen bond formation during drying (Durand *et al.*, 1970). CMC also functions as a dispersant when the dry product is added to water.

After coprocessing with CMC, the material can be dried in a number of ways. Bulk-dried colloidal MCC–CMC products need homogenization to activate the microcrystals and, therefore, these are appropriate for use in frozen desserts and aerated food systems, such as whipped toppings. The spray-dried versions need shear from a high-speed mixer for effective dispersion.

Another colloidal MCC product has been designed for use in dry mix, reconstituted food systems where insufficient shear is available to obtain complete dispersion of standard MCC–CMC grades. This readily dispersible stabilizing agent is produced by adding sweet whey powder to the colloidal MCC–CMC product before drying (McGinley, 1982).

12.4 COLLOIDAL MCC PRODUCT LINE EXTENSIONS

Other line extensions of colloidal MCC have been designed to provide unique functionalities for specific uses.

12.4.1 MCC–guar gum

The coprocessing of colloidal MCC with guar gum results in powdered particles which are essentially water insoluble, shear resistant and spherical. The processing is carried out in an aqueous medium by forming an intimate mixture of the homogeneously dispersed MCC and guar gum under controlled high shear conditions to obtain flocculated MCC–guar gum particles of the desired size. The cellulose–guar aggregates enhance the quality of reduced-fat and nonfat food products by simulating the rheological, textural and mouthfeel properties of fat (McGinley and Tuason, 1993; Tuason *et al.*, 1995).

12.4.2 MCC–calcium alginate

A colloidal MCC grade coprocessed with an alginate salt complex was developed to provide colloidal MCC that could be dispersed in dry mixes and in milk systems. The alginate–calcium complexing reaction is exploited to provide a method for preparing the coprocessed MCC–alginate blend in a form useful for producing water-dispersible MCC particles (Tuason and McGinley, 1994).

12.4.3 Calcium carbonate–MCC–CMC

Coprocessing calcium carbonate with MCC is effective in grinding the MCC aggregates during the coattrition process to prepare ultra-fine MCC particles and, thus, a very effective colloidal ingredient. The product provides a low-viscosity-suspending network particularly suited for use in calcium-fortified, milk-based beverages (Venables and Buliga, 2000).

12.4.4 MCC–CMC for retort applications

Special coprocessing of MCC with CMC was shown to be effective in providing a change in the rheology under the severe heat treatment of rotary and static retort processing. This modified colloidal MCC–CMC stabilizer allows the processor to meet and exceed the thermal F_0 death time requirements of retort sterilization while minimizing total process time. This is especially important in static retort processing where standard MCC–CMC grades may rapidly form a network and thicken the product, the net result of which may reduce heat penetration and, thereby, increase the process time required. This modified MCC–CMC product provides a low-viscosity-suspending network particularly useful in retorted/canned and UHT/retort-processed beverages and foods (Tuason *et al.*, 2002).

12.4.5 MCC–high methyl-esterified pectin

A colloidal grade of MCC coprocessed with high methyl-esterified (HM) pectin was developed to provide colloidal MCC stabilization in low-pH, UHT-processed protein-based beverage systems. The coprocessed powder consists primarily of MCC particles intimately associated with HM pectin and an inorganic salt. The inorganic salt facilitates the formation of colloidal MCC during co-extrusion with HM pectin by enabling it to be subjected to a high work profile which breaks up the aggregates and releases additional colloidal MCC particles.

The HM pectin functions as a barrier dispersant by allowing dry, ultra-fine attrited MCC particles to be formed without reaggregation. New and/or improved colloidal cellulose properties such as low pH stability, fruit pulp suspension, protein stability, emulsion stability, bake stability and colloidal MCC that may be dispersed in milk systems suggest new product applications. Potential areas of application include specialty low pH protein-based beverages, drinkable yogurt and cultured products, fruit sherbets, low-pH sauces, baked goods, such as fruit fillings, pie and pastry fillings and products labeled 'all natural'.

A patent pertaining to a composition and method to produce a co-dried MCC and HM pectin as a stabilizing agent in aqueous food systems is currently pending (Krawczyk *et al.*, 2005).

12.5 PHYSICAL MODIFICATION – THE ALLOYING CONCEPT

Several potentially useful 'alloys' were identified by coprocessing MCC with other food-approved ingredients, to provide novel functionalities for specific food uses. The wet-end processing produced the physical interaction of the components and this was reinforced by the subsequent drying step.

12.5.1 MCC–iota carrageenan

The use of iota carrageenan produces a coprocessed MCC–iota carrageenan powder that can be readily dispersed. It can be dry blended with other ingredients and can be dispersed in water, in the presence of these ingredients, with minimal agitation, such as spoon stirring, stirring with a wire whisk or shaking. Other significant new properties of the coprocessed

MCC–iota carrageenan product are acid stability and salt stability. The powder will disperse and remain stable regardless of the acidity of the system (Tuason *et al.*, 2002).

12.5.2 MCC–maltodextrin

Coprocessed MCC–maltodextrin involves using maltodextrin alone as a barrier for the MCC particles during drying. This cost-effective MCC-based product potentially can be used where colloidal grades of MCC are used, including frozen desserts, salad dressings, beverages and bakery products. Other potential uses are as a bulking agent in nonaqueous food systems, for example in peanut butter and in low-moisture food systems, as an inactive carrier or excipient for chewable tablets, and for masking the taste of drug actives such as aspirin and acetaminophen (APAP) in acetaminophen in pharmaceutical applications (Buliga *et al.*, 2002).

12.5.3 MCC–surfactant

Colloidal MCC–surfactant compositions, composed of MCC and one or more surfactants in which the surfactant is adsorbed onto the surface of the cellulose, were developed to provide improved dispersibility in low-moisture or in nonaqueous food applications. These composites can be used as bulking agents in low-moisture foods or in the oil phase of food products, and are especially useful in reduced-calorie foods (McGinley *et al.*, 1995).

12.6 PHYSICAL AND FUNCTIONAL PROPERTIES

12.6.1 Powdered MCC grades

In 1957 Dr O.A. Battista at the American Viscose Corporation first studied the functional properties of MCC as a potential stabilizer and fat replacer. The powdered grades of MCC were originally used as sources of fiber in low-calorie foods but later found wide acceptance within the pharmaceutical industry for use in direct compression tablets.

The powdered grades are white, odorless, tasteless, relatively free-flowing powders that are virtually free from organic and inorganic contaminants. They are insoluble and chemically inert, crystalline in nature and very porous.

12.6.2 Specialist powdered MCC

More recently, MCC with small particles, sized less than about 30 μm, has been used in foods as bulking agents and as fat substitutes. The pure spheronized, attrited MCC has found application in very low-moisture cookie fillings and coatings. The product is distinctive in that particles are produced which are substantially smooth, having a high absolute density, a high loose bulk density (>0.40 g/cc), a low degree of oil absorptivity (<1.0) and a substantially spherical shape, as seen at 150 times magnification. This finely powdered grade of MCC is particularly suited for use as a bulking agent in oil-based foods or food components, which include nut butters, chocolates, ice creams, mayonnaise, lard and cream fillings.

12.6.3 Colloidal MCC–CMC/alginate/HM pectin products

The physical properties of colloidal MCC dispersions are quite different from the properties of gum solutions, starch gels and other water-soluble materials commonly used in food technology. When colloidal grades of MCC are properly dispersed, the cellulose crystallites set up a three-dimensional network, with the majority of the particles less than 0.2 μm. The swelling capacity of the soluble hydrocolloid provides the dispersant function by aiding in the dispersibility of the MCC particles during reconstitution as well as in the stabilization of the resulting colloidal dispersion. It is the formation of this insoluble cellulose structural network that provides the colloidal MCC grades with their functional properties of low viscosity, suspending network, etc. The gel that is formed possesses the elastic properties of a solid that exhibits relatively high yield stress and a time-dependent type of flow behavior (thixotropy). Thixotropic properties impart a variety of desirable characteristics suitable for products such as salad dressings and mayonnaise.

12.6.3.1 *Fat replacement*

The rheological properties of colloidal MCC products dispersed in water have been used to simulate fat in various food applications. By adding colloidal MCC into food systems including ice cream, salad dressings, sauces and gravies, the level of oil or fat can be effectively reduced, while preserving their physical and rheological properties (McGinley *et al.*, 1984). Basic studies have shown that a simple emulsion containing 60% soybean oil has similar rheological properties and stability characteristics to a 20% soybean oil emulsion containing 1.0–1.5% colloidal MCC. Because of the special orientation of the colloidal MCC particulates at the oil-in-water interface, the emulsions acquire a yield value, which make them stable against 'oiling-off' or creaming.

12.6.3.2 *Emulsion stability*

The strong affinity of colloidal MCC for both oil and water results in the preferential orientation of the solid particulates at the oil-in-water interface. Several basic studies, involving the use of brightfield and polarized light microscopy, freeze-etch electron microscopy and rheological measurements, have established that this colloidal network can provide a mechanical barrier of considerable strength and durability at the interface (Oza, 1988). In addition, colloidal MCC thickens the water phase between the oil globules preventing their close approach and ensuing coalescence. These attributes clearly demonstrate the potential of using colloidal MCC in stabilizing emulsion-based food systems that require good shelf stability.

12.6.3.3 *Heat stability*

The excellent heat stability of colloidal MCC particles ensures minimal or no product breakdown at elevated temperatures. This heat stability permits great flexibility in a variety of heat processes utilized in food manufacturing such as batch pasteurization, UHT treatment, hot filling. Figure 12.2 illustrates the heat stability of a dispersion of MCC–CMC across a wide range of temperatures from freezing point through chill and ambient to boiling point. This property is central to the utilization of MCC–CMC for ambient-stored beverages, dairy and nondairy creams, sauces, ready-to-use bake-stable bakery fillings and fruit fillings.

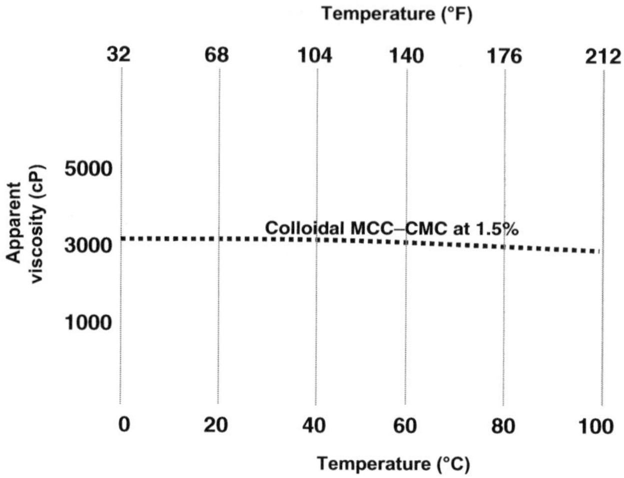

Fig. 12.2 Heat stability of a 1.5% dispersion of MCC–CMC in water. (Reproduced with kind permission from FMC Corporation.)

The addition of colloidal grades of MCC to both pectin-based and starch-based bakery fillings produces a modified gel structure with improved texture, spreadability and heat stability characteristics. It is believed that the development of a distinct colloidal MCC network within the pectin gel system sustains the fibrous pectin network as a heat-stable structure (McCormick, 1974). Several studies have also demonstrated the value of using colloidal MCC in retortable oil-based salad dressings. Colloidal MCC has the ability to preserve textural consistency, viscosity and emulsion stability of oil-based dressings prepared under sterilization processing conditions of 240°F (116°C) for 1 h.

12.6.3.4 *Foam stability*

In aerated food systems, the insoluble colloidal MCC particulates will stabilize the water phase between air cells and provide a supplementary structure to the protein film surrounding the air cells. Several application studies have demonstrated the use of colloidal MCC in improving the stand-up ability, stiffness and foam stability of both nondairy and dairy whipped toppings. In addition, it has been found that colloidal MCC can effectively stabilize marshmallow toppings and confectionery products and control overrun in frozen desserts.

12.6.3.5 *Freeze–thaw stability*

Colloidal MCC, with its tremendous surface area and wicking action, has the unique ability to manage the free water produced in frozen desserts during freeze–thaw cycles. Its unique water-adsorption capability compensates for the inability or slowness of the other solids in the mix to reabsorb free water upon partial melting of ice crystals during temperature fluctuations (Keeney, 1979). The use of colloidal MCC produces frozen dessert products with improved body and texture qualities, improved extrusion qualities and good 'heat shock' resistance, the industry term for the cyclical temperature conditions that occur during storage and transport.

12.6.4 MCC–guar gum

The properties of MCC–guar gum aggregates are very different from the other colloidal grades. Spherical MCC–guar aggregates, produced by controlled interaction between these two polymers, are shown to have certain characteristics similar to those obtained when fats and/or oils are emulsified into an aqueous phase. It is the formation of a guar layer adsorbed onto the cellulose that gives the MCC–guar aggregates the ability to function as a fat-like substance in aqueous-based food systems. In this situation, the aggregates have the smoothness and lubricity properties most closely approximating the physical and organoleptic properties associated with fat dispersed in water. As well as a smooth, bland mouthfeel, the MCC–guar aggregates have no detectable aftertaste or residual effect.

12.7 LEGISLATION AND NUTRITION

In the US, MCC has 'generally recognized as safe' (GRAS) status and has been used safely in foods for over 30 years. In Europe, MCC is listed in Annex 1 of the European Parliament and Council Directive 95/2/EC of 18 March 1995 on Food Additives other than Colours and Sweeteners. It is approved as E460i in the list of permitted emulsifiers, stabilizers, thickening and gelling agents for use 'quantum satis', the level required to achieve a given technological benefit. Purity standards of food grade materials are listed in the EEC Directive 82/504/EEC. MCC has been evaluated by both the EC Scientific Committee for Food (SCF) and the Joint FAO/WHO Expert Committee for Food Additives (JECFA). Both committees determined the maximum acceptable daily intake (ADI) to be 'not specified'.

The copolymers used with MCC, CMC, calcium alginate, pectin and guar gum, are also listed in Annex I of the directive 95/2/EC and are approved for use under numbers E466, E404, E440 and E412, respectively. Recently, cellulose gum was authorized as a synonym for CMC which aligns EU and US ingredient declarations for this material.

The MCC component of powdered and colloidal grades is composed of insoluble fiber only and does not contribute any other material of nutritional significance. It contributes zero calories to foods.

12.8 FOOD APPLICATIONS

As the product type, use level and processing method vary with the application, a brief description of a number of food applications is necessary. An introduction is needed to describe how the MCC is incorporated into foods.

12.8.1 Preparing MCC dispersions in foods

The key to developing the functional properties from the insoluble microcrystals is the creation of a three-dimensional gel network through an adequate dispersion of the colloidal MCC. This dispersion is formed by shear. The amount of shear required to create an effective, functional dispersion depends on the product type. Products have been developed to suit any process from high-pressure homogenization to simple dry blends.

A number of key factors may interfere with the dispersion and effective use of colloidal MCC:

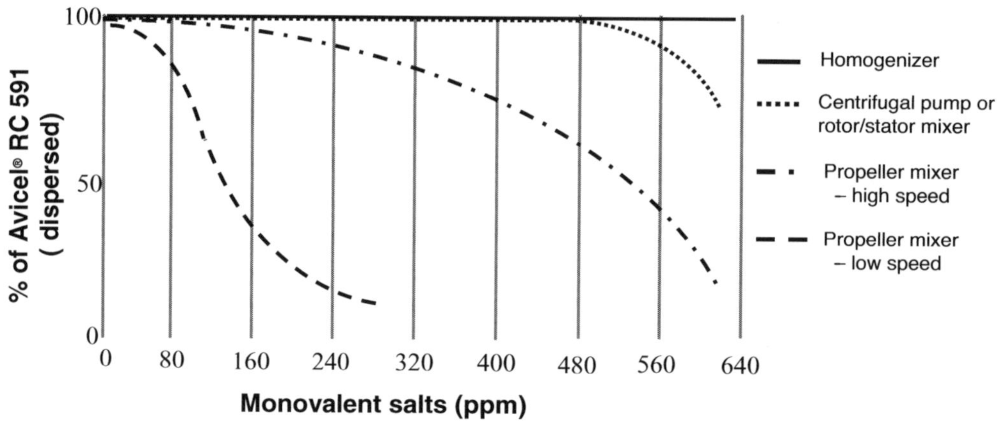

Fig. 12.3 Dispersion of MCC–CMC at varying levels of monovalent salts, such as acids and sodium and potassium salts. (Reproduced with kind permission from FMC Corporation.)

1. Adequate shear must be used, that is, the shear applied during the process must be sufficient to separate and disperse the particles of the grade of MCC selected.
2. The preferred order of addition is to add the colloidal MCC to water before the other ingredients, especially other thickeners and gelling agents, which can bind water and prevent the distribution of the insoluble particles throughout the aqueous phase. In fact, as MCC does not strongly bind water, it will not prevent the hydration of other ingredients: starches and gums can readily hydrate in a dispersion of MCC in water. If gums or starches are added at the same time as MCC, the time to achieve complete dispersion is extended.
3. Hard water and dissolved electrolytes can inhibit the dispersion of colloidal MCC, so dispersing in soft water will give complete dispersion in the minimum time. Figure 12.3 shows that if clean water is not used to prepare solutions, or if mixing tanks are not completely washed between batches, rotor–stator mixers or homogenizers are needed to fully activate the MCC–CMC dispersion.

   The effect on activation is more pronounced for divalent salts, as shown in Fig. 12.4. Products containing high levels of divalent cations, such as ice cream mixes, milk or reconstituted milk beverages, require high-pressure homogenization to ensure that the MCC product is efficiently utilized.

   New grades have been developed that activate more easily than older products. However, shear is important to activate MCC–CMC products and the dispersion cannot be over-sheared. Prolonged mixing times or very high shear pressures, such as 4000 psi (250 bar), do not affect MCC–CMC viscosity and suspending properties:
4. If a dispersion of MCC is acidified below pH about 4.5, a protective colloid is necessary to prevent flocculation. Most soluble gums have a protective effect but the most effective is xanthan gum at a level of approximately 10% by weight of the MCC. However, many other gums, already included in the food formulation for additional thickening, gelling and mouthfeel effects, can also act as protective colloids.

MCC–guar aggregates must be well dispersed to achieve full functionality in the aqueous phase. The shear required depends on the aqueous phase viscosity and composition, the concentration and type of aggregates and the method of addition. Generally, however, it is usually independent of the aqueous phase temperature and pH.

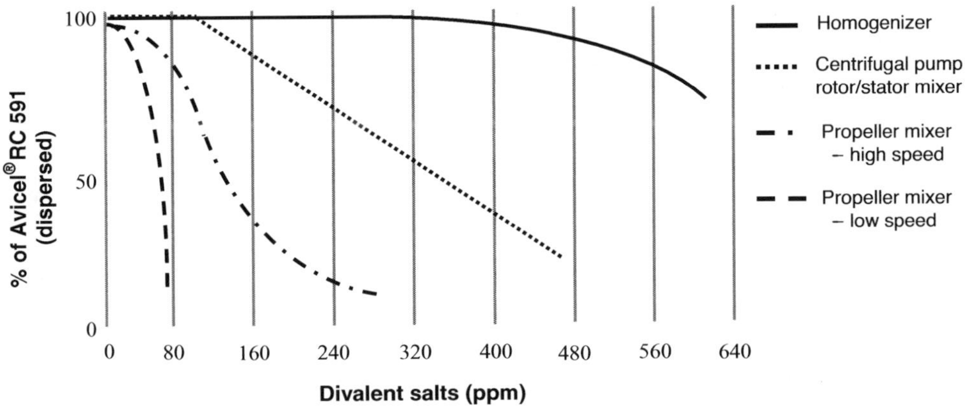

Fig. 12.4 Dispersion of MCC–CMC at varying levels of divalent salts, such as calcium and magnesium salts. (Reproduced with kind permission from FMC Corporation.)

As a general guideline, MCC–guar aggregates, added dry or as part of a dry blend with other food ingredients, will readily disperse in water by applying relatively low shear, such as using a propeller mixer at approximately 1600 rpm for 5 min. Even with minimal shear, such as spoon stirring for dry mix applications, much of the functionality of the aggregates is obtained. Very high shear is able to disrupt the spheres and give fibrous rods but these conditions would only occur in solutions subjected to multiple passes in a homogenizer.

12.8.2 Dressings, sauces and spreads

One of the most common applications of MCC is to prepare low-fat mayonnaise and salad dressings. Formulas are shown below for a low-fat, pourable dressing made with low-viscosity colloidal MCC, which provides a smooth structure and creamy mouthfeel (Formulation 12.1), and a low fat spoonable dressing using medium viscosity colloidal MCC to give more structure and a softly gelled consistency (Formulation 12.2).

Formulation 12.1 Low-fat pourable dressing

Ingredients	Quantity (%)
Water	q.s.
Corn syrup 42 DE	12.00
Soybean oil	5.00
Vinegar 120 grain	5.00
Sugar	4.00
Cultured buttermilk powder	4.00
MCC–CMC	1.65
Salt	2.00

(continued)

Formulation 12.1 Low-fat pourable dressing (continued)

Maltodextrin 10 DE	1.00
Powdered egg yolk	0.50
Xanthan gum	0.40
Mono sodium glutamate	0.30
Garlic powder	0.18
Onion powder	0.18
Potassium sorbate	0.10
Mustard powder	0.05
Ground black pepper	0.03
Parsley	0.02
Calcium disodium EDTA	0.01
Ribotide	0.01
Flavors	as required
Total	100

Formulation 12.2 Low-fat spoonable dressing

Ingredients	Quantity (%)
Water	q.s.
Sugar	6.00
Soybean oil	5.00
Vinegar 120 grain	4.00
Starch	3.25
Corn syrup solids (28 DE)	2.50
Salt	2.50
MCC–CMC	2.00
Powdered egg yolk	1.50
Lemon juice concentrate	1.00
Xanthan gum	0.45
Mustard flour	0.25
Iota carrageenan	0.20
Potassium sorbate	0.10
Garlic powder	0.01
Onion powder	0.01
Calcium disodium EDTA	0.01
Color (1.4% β-carotene solution)	0.001
Total	100

Developed to match the organoleptic profile of an 80% fat mayonnaise, MCC contributes a variety of properties. The colloidal material functions as an effective emulsion stabilizer with thixotropic flow properties and it imparts a full-fat texture, appearance, mouthfeel and body in low-fat dressings.

Typically, a rotor/stator mill will be used in industry to manufacture these types of products. Clean water is placed in the milling chamber under a vacuum of 200–400 mbar. The colloidal

MCC is introduced and milled at 3500 rpm for 3–4 min. A blend of xanthan gum and sugar is added next and incorporated by milling for a further 4 min. To increase plant throughput, these initial dispersion steps often may be carried out in a premix tank using a high shear mixer. Adequate shear is essential: mixing for 5–10 min at a minimum of 1800 rpm is usually sufficient but complete dispersion may be confirmed by microscopic inspection. Starch is then added, either as a starch paste or in dry form if a pregelatinized starch is used. The egg yolk, salt, preservatives, mustard and spices can then be added, followed by the oil, while continuously milling. The vinegar and other acids are added last and the dressing may be pasteurized and hot or cold filled.

Similar MCC products can be used to give creamy, low-fat sauces which do not separate during cooking and maintain viscosity and cling at serving temperatures. For solid-reduced fat products, such as low-fat spreads, MCC can stabilize the aqueous phase in oil-continuous spreads down to 20% fat. MCC–guar gum aggregates can be added to give further body and mouthfeel.

12.8.3 Processed cheese

MCC technology is also commonly utilized in the manufacture of reduced-fat processed cheese slices and blocks. A recipe for a reduced-fat processed cheddar cheese with MCC–guar aggregates is given in Formulation 12.3.

Formulation 12.3 Reduced-fat processed cheese

Ingredients	Quantity (%)
MCC–guar gum aggregates	1.0–2.0
Skim milk cheese	52.00
Sweet whey powder	6.00
Trisodium citrate	1.20
Disodium phosphate dihydrate	0.70
Salt	0.85
Carrageenan	0.60
Enzyme-modified cheddar cheese	0.50
Potassium sorbate	0.15
Color	0.04
Buttermilk, 0.5% fat	qs
Total	100.0

Containing less than 1% fat compared to over 30% fat in many processed cheeses, the cellulose-based aggregates contribute a full-fat mouthfeel, body and opacity. In addition, the aggregates interrupt the elastic protein structure which tends to form in the absence of fat. By interrupting and lubricating the protein structure, the aggregates improve both the melting and eating properties of low-fat processed cheese. MCC can also be used to greatly improve the eating quality of low-fat soft cheese.

For large-scale manufacture of processed cheese, a dispersion would be made in a premix tank followed by processing in a jacketed cheese cooker. The dry ingredients are first dry blended and added to the buttermilk in the premix tank. The ingredients are incorporated with a planetary (or higher shear) mixer at approximately 1800 rpm for 4–8 min until smooth. In the jacketed cheese cooker, the cheese is shredded and the premix added while heating. The product is then heated to approximately 170°F (75°C) under agitation and vacuum-packed hot.

12.8.4 Frozen desserts and ice cream

The original application for MCC was in frozen desserts and ice cream. Colloidal MCC has been used for many years as a stabilizer to control ice crystal growth in standard and low-fat ice cream. A recipe for a low-fat frozen dessert is given in Formulation 12.4.

Formulation 12.4 Low-fat frozen dessert

Ingredients	Quantity (%)
Sugar	12.0
Milk solids nonfat	9.0
Corn syrup solids 36 DE	8.0
Butterfat	4.0
Whey solids	3.0
Stabilizer: MCC-CMC and carrageenan	0.35–0.40
Emulsifer: Mono- and diglycerides of fatty acids and polyoxyethylene (20) sorbitan monooleate (Polysorbate 80)	0.15–0.20
Total solids	36.55
Total	100.0

Colloidal MCC is able to prevent separation in the premix, to stabilize the emulsion and foam during freezing and aeration, to improve extrusion and hold the product shape and to control ice crystal growth and reduce shrinkage during storage. More recently, it has been used to improve the resistance to thermal shock in standard ice cream, stored and distributed in countries with high ambient temperatures and to avoid shrinkage caused by pressure differences during transport at high altitude. The structuring properties of MCC–CMC also maintain shape, texture and mouthfeel in recipes with reduced dairy protein levels. In reduced-fat frozen desserts, these contributions become even more critical. Additionally, colloidal MCC imparts a full-fat body, mouthfeel and appearance and can be supplemented further with MCC–guar aggregates to give the creaminess and body of a standard ice cream.

12.8.5 Dairy products

Creams can be stabilized using low-viscosity colloidal MCC. This avoids emulsion separation during storage, particularly under high ambient storage, and contributes to the mouthfeel and body of the product. The structure imparted to whipped creams and vegetable fat-based toppings prevents foam drainage and improves stability. In low-fat toppings (see Formulation 12.5), the cellulose microcrystals strengthen and mimic the structure of fat crystals and, hence, improve the stability of the whipped toppings, especially at elevated temperatures. This technology has been extended to aerated dairy desserts, such as mousse, shown in

Formulation 12.6, where the inclusion of MCC is able to stabilize the foam and give a 'crisp' bubble structure.

Formulation 12.5 Low-fat whipped topping

Ingredients	Quantity (%)
Water	59.85
Hydrogenated vegetable fat	20.00
Sugar	8.00
Corn syrup solids 62 DE	6.00
Milk solids nonfat	5.00
MCC–CMC	0.60
Polyoxyethylene (20) sorbitan monostearate (Polysorbate 60)	0.30
Distilled monoglycerides of fatty acids	0.15
High-viscosity CMC	0.10
Total	100.0

Formulation 12.6 Cold instant gelatin-free mousse

Ingredients	Quantity (%)
Sugar	11.0–13.0
Cocoa	5.0–6.0
Whipping base (vegetable fat, emulsifier and milk protein)	3.6–4.4
MCC–Alginate blend with tetrasodium pyrophosphate and calcium sulfate	1.4–4.8
Instant starch	0.8–1.8
Milk	up to 100%
Total	100

12.8.6 Beverages

Low-viscosity chocolate milks are traditionally stabilized with very low levels (0.015–0.025%) of κ-carrageenan. If heat-treated milks are not cooled below about 60°F (15°C) before filling, or if the ambient storage temperature is high, the stabilizing network of κ-carrageenan–κ-casein cannot maintain a stable suspension of cocoa particles. MCC at 0.25–0.40% can be used to improve the stability of these milks so that higher temperatures during filling and storage can be tolerated.

A variety of colloidal MCC products are used to provide effective stability with low-viscosity impact in beverages. As a general rule, the colloidal MCC–CMC products are used to provide effective suspension of insoluble components. In Formulation 12.7, the stabilizer provides uniform color and flavor delivery of cocoa solids for chocolate-flavored beverages. In the second example, Formulation 12.8, this property provides uniform nutrient delivery in beverages fortified with insoluble calcium components.

Formulation 12.7 Chocolate milk

Ingredients	Quantity (%)
Sugar	6.0–9.0
Cocoa powder	1.5–2.0
MCC–CMC + carrageenan	0.4–0.5
Flavor	as required
Milk	up to 100
Total	100

Formulation 12.8 Calcium-fortified dairy beverage

Ingredients	Quantity (%)
Skim milk	92.47
Sugar	6.00
Cocoa powder	1.25
MCC–CMC	0.22
Calcium carbonate	0.13
Kappa (κ-) carrageenan	0.03
Total	100

Calcium-fortified milk stabilized with MCC–CMC demonstrates a very stable, low viscosity over time (see Fig. 12.5) with no gelation and only a trace of sediment after 6 months storage at ambient and chill temperatures (72 and 39°F or 22 and 4°C). Carrageenan is widely used to control protein aggregation and prevent serum separation in heat-treated beverages. Combining a low level of carrageenan with MCC–CMC produces a beverage with a creamy mouthfeel, rich texture and excellent stability, although the viscosity increases slightly during 6 months' storage.

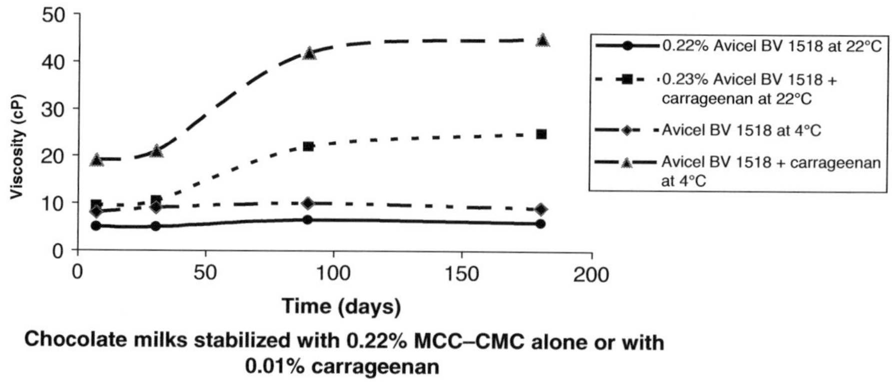

Fig. 12.5 Viscosity changes in chocolate milks at ambient and chill temperatures. (Reproduced with kind permission from FMC Corporation.)

12.8.6.1 *Retorted beverages*

In this more specialized case, a particular coprocessed MCC has been developed to provide rheological stability under retort processing while providing a low fluid viscosity to maximize thermal penetration during heat treatment. This property allows manufacturers to meet the thermal F_0 death time requirements of retort sterilization while minimizing total process time. This is especially important in static retort processing; other colloidal MCC grades may form a network and thicken before processing is complete, thereby reducing heat penetration which increases the process time or temperature required for sterilization, which may lead to the development of undesirable color and flavor notes. After retorting, this specialized MCC–CMC product provides a low-viscosity-suspending network particularly useful in retorted/canned, and UHT/retort processed beverages and creams, such as the retorted adult nutritional beverage shown in Formulation 12.9.

Formulation 12.9 Retorted adult nutritional beverage

Ingredients	Quantity (%)
Corn syrup solids 24 DE	8.60
Sucrose, dry granular	6.60
Skim milk powder	3.30
Corn oil	2.50
Soy protein isolate	0.71
Cocoa, red dutched	0.60
Cocoa, natural	0.60
MCC–CMC (for retorted products)	0.5–0.8
Potassium citrate, monohydrate	0.30
Soy lecithin	0.28
Natural and artificial vanilla flavor	0.25
Potassium chloride	0.23
Dipotassium phosphate	0.20
Vitamin and mineral premix	0.103
Calcium carbonate	0.10
Sodium chloride	0.07
Lambda carrageenan	0.01–0.02
Water	up to 100
Total	100.00

12.8.6.2 *Low-pH protein beverages*

For low-pH protein beverages, the stability requirements for suspension are more daunting. This stabilizer consists of MCC coprocessed with HM pectin, which are both considered 'natural' stabilizers. It provides superior stability and viscosity control when compared to other stabilizers, such as pectin, xanthan gum and propylene glycol alginate, in many of today's high growth products, including soy-based drinks, drinkable yogurts, functional drinks, energy drinks and acid milks.

MCC–HM pectin can be activated with low shear in the protein phase, the makeup water or in the low-pH phase including juice and juice concentrates. The order of addition and the process conditions must be chosen carefully in order to provide protein protection and full shelf life stability.

At protein levels ranging from 3 to 7 g/236 mL (1.25–3.00%) serving in milk-juice, acidified soy-juice or whey-juice beverages, low stabilizer use levels achieve suspension and provide long-term stability (see Formulation 12.10). Pectin by itself may appear stable as it gives minimal serum separation. However, higher levels may be needed to prevent sedimentation over time, especially at higher protein levels, which may result in higher viscosity beverages. Pectin as the sole stabilizer may not provide complete stability in low-pH, high-protein beverages and can result in sedimentation.

Formulation 12.10 Acidified milk – juice beverage

Ingredients	Quantity (%)
Skim milk	20.00
Sugar	8.00
Orange juice concentrate	4.21
Non-fat milk powder	1.73–5.03
MCC–HM pectin	0.40
High methyl-esterified pectin	0.35
Citric acid	0.25–0.40
Water	up to 100
Total	100.0

12.8.7 Meat products

MCC has recently been studied as a stabilizer and fat replacer in cooked meats and emulsified products, such as sausages and pate. The addition of cellulose is able to raise product yields by retaining moisture during cooking, even at the high temperatures required for microbial safety in poultry products. The insoluble cellulose particles interrupt the protein structure in the cooked meats and avoid cold shortening, the rubbery consistency produced by excessive protein extraction by seasonings during fresh meat storage. In pate, MCC gives a smooth consistency to the product which spreads easily giving a good flavor release and full-fat texture.

12.8.8 Powdered cellulose applications

The powdered grades of MCC are commonly used in foods as a high-quality, inert fiber source and a zero-calorie bulking agent. Very finely powdered MCC can be readily suspended in ready-to-drink dietetic products to provide fiber. It can also impart much of the body and opacity usually contributed by fat. It has a clean flavor release with no significant flavor contribution or flavor-masking properties.

Additionally, the porous and free-flowing nature of powdered grades of MCC makes them ideal as carriers of liquid materials such as essential oils. This characteristic has been widely used to prevent caking in shredded cheese.

12.8.9 Bakery and confectionery fillings, toppings and coatings

Fine particle grades of MCC have been specifically developed for reduced fat/calorie and low-moisture products, such as confectionery products and biscuit fillings. Because of the water activity and texture limitations, structured water cannot be used as a direct replacement for fat, as is the practice with emulsion systems. These applications, in contrast, require high levels of noncaloric or low-calorie bulking agents to achieve a sufficient calorie and fat reduction. New grades of cellulose microcrystals have been manufactured to reduce the porosity and surface area of the particle so that absorption properties are minimized. These grades can be used up to 15%, often in conjunction with sugar syrups, to function as a high-quality replacement for fat in low- and very low-moisture foods with water activities restricted to the range 0.45–0.6, as shown in Formulation 12.11 for nougat confectionery.

Formulation 12.11 Nougat-style confectionery

Ingredients	Quantity (%)
Part I	
Corn syrup 42 DE	10.0
Water	8.7
Powdered sugar 10×	5.0
Egg white solids	2.0
Fine-mesh MCC (for low-moisture products)	1.0
Part II	
Corn syrup 42 DE	32.5
Sugar	31.6
Water	8.4
Cocoa powder 10–12% fat	0.8
Total	100

12.9 FUTURE DEVELOPMENTS

The widespread use of MCC as a stabilizer in low-calorie dressings and mayonnaise underlines the preeminence of this product for providing structure and a creamy mouthfeel in low-fat systems. The rapidly growing use of MCC as a stabilizer in dairy, soy and nutritional beverages, ice cream mixes and long-life creams demonstrates its value for giving effective suspension, improved tolerance to heat processing and long-term ambient stability in addition to improved mouthfeel properties. New uses have been identified in low-moisture systems for retaining texture and mouthfeel in bakery fillings and coatings and in sugar confectionery. Improved moisture retention and texture stability in meat products and superior stability and mouthfeel in aerated and nonaerated dairy products, such as ready-to-use bakery creams and mousse, are also being exploited. New alloys and gum combinations, such as MCC–konjac aggregates (McGinley and Tuason, 1995), are being developed to extend the range of uses in new food applications as commercial demand increases for this unique stabilizer.

References

Buliga, G.S., Tuason, D.C. and Venables, A.C. (2002) Texture and stabilizer composition, US Patent 6,440,474.

Durand, H.W., Fleck, E.G. and Raynor, G.E. (1970) Dispersing and stabilizing agent comprising beta-1,4 glucan and CMC and method for its preparation, US Patent 3,539,365.

Keeney, P.G. (1979) Confusion over heat shock, *Food Engineering Magazine*, **52**(6), 116–118.

Krawczyk, G.R., Tuason, D.C. and Fischer, G.A. (2005) MCC/hydrocolloid stabilizers and edible compositions comprising the same, US Patent Application 20050233046.

McCormick, R.D. (1974) Controlling Heat Stability for Bakery Fillings, Toppings, *Food Product Development Magazine*, **8**(10), 24, 27.

McGinley, E.J. (1982) Stabilizing agent for dry mix food products, US Patent 4,311,717.

McGinley, E.J., Krawczyk, G.R. and Selinger, E. (1995) Coprocessed particulate bulking and formulating aids: their composition, production, and use, US Patent 5,441,753.

McGinley, E.J., Thomas, W.R., Champion, J., Philipps, G.O. and Williams, P.A. (1984) The use of microcrystalline cellulose in oil in water emulsions. In: G.O. Phillips, D.J. Wedlock and P.A. Williams (eds), *Gums and Stabilisers for the Food Industry 2*. Pergamon Press, Oxford, pp. 241–249.

McGinley, E.J. and Tuason, D.C. (1993) Fat-like bulking agent for aqueous foods comprising microcrystalline cellulose and a galactomannan gum, US Patent 5,192,569.

McGinley, E.J. and Tuason, D.C. (1995) Microcrystalline cellulose and glucomannan aggregates, US Patent 5,462,761

Ott, E. and Tennent, H.G. (1954) Introduction. In: E. Ott, H.M. Spurlin and M.W. Grafflin (eds), *Cellulose and Cellulose Derivatives, Part I*. Interscience Publishers, Inc., New York, 1–8.

Oza, K.P. (1988) Colloidal microcrystalline cellulose-stabilized emulsions. Ohio State University, Columbus, *Diss. Abstr. Int. B*, 1988, **49**(2), 372–373.

Tuason, D.C. and McGinley, E.J. (1994) Colloidal microcrystalline cellulose and barrier dispersant composition and method for manufacture, US Patent 5,366,742.

Tuason, D.C., McGinley, E.J. and Selinger, E. (1995) The development of Novagel™ RCN. *Food Technology International Europe*, Sterling Publications Ltd., London, England.

Tuason, D.C., Selinger, E., Krawczyk, G.R., Sewall, C. and Hogan, D.T. (2002) Rapidly peptizable microcrystalline cellulose-based stabilizing agents, US Patent 6,391,368.

Venables, A.C. and Buliga, G.S. (2000) Ultra-fine microcrystalline cellulose compositions and process, US Patent 6,037,380.

Ward, K. (1954) Occurrence of cellulose. In: E. Ott, H.M. Spurlin and M.W. Grafflin (eds), *Cellulose and Cellulose Derivatives, Part I*. Interscience Publishers, Inc., New York, 9–27.

13 Pectin

Sarah M. Brejnholt

ABSTRACT

Pectin is a polysaccharide that is naturally present in most land plants, although commercial pectin is primarily extracted from citrus peel and apple pomace. Two forms of commercial pectin are available: high methyl- and low methyl-esterified pectin; and two versions of the latter exist: a conventional and an amidated form. High methyl-esterified pectin forms gels in high soluble solids and acidic systems, whereas low methyl-esterified pectin forms gels in a much broader pH and soluble-solids range, but requires the presence of divalent cations for gelling. As a consequence, each type has its own particular function. Nevertheless, general attractive features include excellent flavour release, good processing characteristics and stability at low pH. Its traditional and major function is to act as a gelling agent in foods but, nowadays, it also serves as a thickening and stabilising agent. The application of pectin is diverse and covers fruit-based products, dairy products, acidified milk drinks and other beverages, confectionery, bakery products, various fine foods and spreads. Additionally, pectin finds use in the pharmaceutical industry. Finally, increasing consumer awareness of healthy life-style habits and the emerging trend to produce functional foods increases the significance of the status of pectin as a water-soluble dietary fibre.

13.1 INTRODUCTION

13.1.1 Historical perspective

Pectin is a natural constituent of all land plants where, together with cellulose, it plays a key role in the cell wall structure. It comprises a group of polysaccharides rich in galacturonic acid units and, to a lesser extent, various neutral sugars. Pectin extracted from plants has been used as a gelling agent in food for many years. In fact, the invention of using pectin as a gelling agent dates from the 1820s when the Frenchman Henri Braconnot prepared a synthetic jelly with alkali-extracted pectin. However, the first recorded commercial production of pectin extract was in Germany in 1908, after which the process spread to the US, where Robert Douglas obtained a patent in 1913 (Douglas, 1913). The centre of production is currently located in Europe and citrus-producing countries such as Mexico and Brazil. Historically, apple pomace was the major pectin source, but in recent years, an increasing use of citrus peel has taken place. An additional but much less important source of pectin is sugar beet pulp. In recent years, new application opportunities have emerged and pectin is no longer just a gelling agent but also used as a stabiliser and thickener.

13.1.2 Commercial pectin

According to the EU directive of 11 November 1998, the *Food Chemicals Codex*, 5th Edition, published in 2003 and FAO Food and Nutrition Paper 52 in 2001, at least 65% of the pectic substance must be galacturonic acid in order for material to be classified as commercial pectin. The galacturonic acid units may or may not be esterified with a methyl group, and depending on the degree of methyl esterification, pectin is divided into two groups comprising high methyl-esterified (HM) and low methyl-esterified (LM) pectin, with the percentage of methyl ester galacturonic acid units being higher or lower than 50%, respectively. Additionally, amidated LM pectin may be prepared. In some pectin sources, including sugar beet pulp, the galacturonic acid units may be *O*-acetylated. This type of pectin behaves quite differently from the traditional pectin extracted from apple pomace and citrus peel as it cannot gel. Consequently, it is used as an emulsifier or stabiliser. Commercial pectin is, in general, standardised with sugar prior to marketing. Furthermore, it may be buffered with suitable food-grade salts including ammonium, sodium, potassium and calcium salts, which may be required for pH control or desirable setting characteristics.

13.1.3 Current pectin market

With a growth rate of approximately 3.5% per year, the world pectin market has doubled during the last 20 years. Figure 13.1 illustrates the steady development of the pectin market from 1982, where the annual sales volume of pectin was 16 000 metric tonnes, to 2005, where the annual volume is estimated at 34 000 metric tonnes (CP Kelco, 2005). Covering this market, there is a range of industrial producers with the major players comprising CP Kelco, Cargill, Danisco, Herbstreith & Fox and Obipektin.

The traditional application of pectin is as a gelling agent in jams and jellies. However, over the past years, new opportunities have emerged. Pectin applications are diverse and plentiful and comprise a variety of fruit-based products, dairy products, confectionery products, different beverages, including fermented and acidified milk and soya drinks, bakery products, fine foods and various spreads. Finally, pectin is finding more and more use in the pharmaceutical industry. Figure 13.2 depicts an estimated picture of the world market of pectin applications in 2005. At present, the greatest developmental trend, relative to market

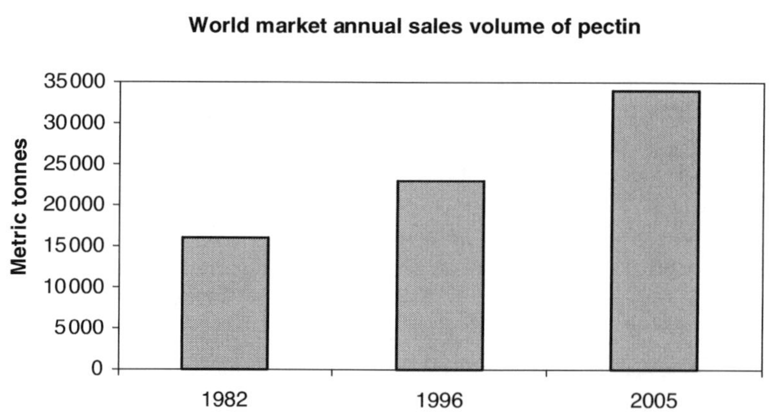

Fig. 13.1 Development of the annual world pectin market from 1982 to 2005. (Reproduced with permission from CP Kelco.)

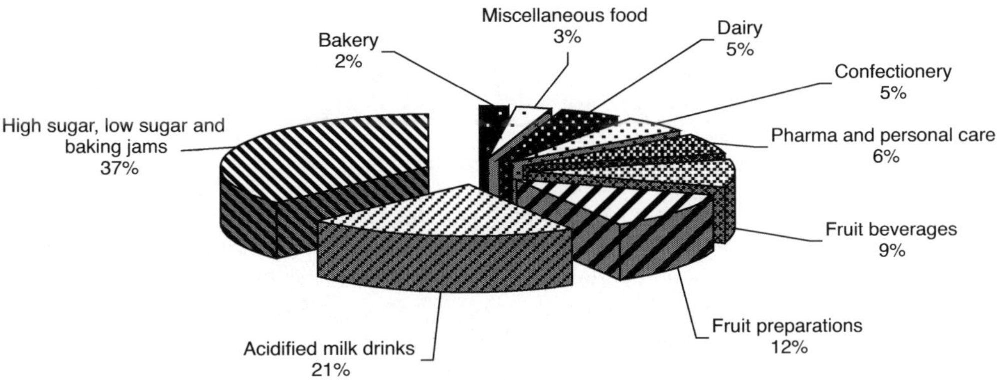

Fig. 13.2 Estimated world market of pectin applications 2005. Bakery includes bread, cake and various prepared mixtures; Miscellaneous food includes mayonnaise, sauces, low-fat spreads, meat preparations, prepared foods and water desserts; Dairy includes yoghurts and desserts; and Confectionery includes sugar jellies and wine gums. Pharmaceutical and personal care encompasses all applications in pharmaceutical and personal care products. Fruit beverages include juices, carbonated drinks, non-dairy drinks, liqueurs and neutral milk drinks. Fruit preparations mainly comprise applications in fruit yoghurt. Acidified milk drinks cover milk and soya drinks. The last group, high sugar, low sugar and baking jams, includes the traditional application in jams and jellies. (Reproduced with permission from CP Kelco.)

share, is found in acidified and fermented milk and soya drinks. Nevertheless, the largest application remains the traditional uses in jams and jellies.

13.1.4 Regulatory status

At present, pectin is regarded as a safe food additive and has been given an acceptable daily intake (ADI) of 'not specified' by the Joint FAO/WHO Expert Committee on Food Additives (JECFA) as well as the European Union Scientific Committee for Food. Furthermore, the US Food and Drug Administration (FDA) accords pectin 'generally recognised as safe' (GRAS) status. It should be noted that the Codex Alimentarius and US specifications include both amidated and non-amidated pectins in one class. EU legislation distinguishes between non-amidated and amidated pectins. Nevertheless, both pectin types are labelled E440 and permitted uses are identical except for organic foods, where only non-amidated pectin is permitted.

13.2 RAW MATERIALS

13.2.1 Pectic materials

Pectin raw materials presently used in industrial processes include apple pomace, citrus peel – comprising lime, lemon and orange – and, to a lesser extent, sugar beet pulp. Citrus peel and apple pomace are by-products of the fruit juice industry and sugar beet pulp of the refined sugar industry. Both wet and dry citrus peels are used for pectin production, whereas mainly dry apple pomace is used. Apple pomace must be dried immediately in order to prevent disintegration into a thick mash from which it is uneconomic to extract pectin. Additionally, wet apple pomace is very susceptible to degradation by yeast which produces pectolytic enzymes (Voragen *et al.*, 1995). Orange peel is particularly rich in the enzyme pectin methyl esterase (May, 1990) and thus is more susceptible to methyl de-esterification than lime and lemon.

13.2.2 Pectic enzymes

All raw materials used for commercial pectin contain various enzymes which modify the pectin in vivo in the plant, as well as post harvest, thus changing the pectin chemistry during ripening. Enzyme modification in vivo may benefit the extraction of pectin as it encourages the release of pectin from the cell wall matrix. However, at the same time, the degree of methyl esterification is reduced progressively during ripening which, in some cases, is unfavourable as slight methyl de-esterification gives rise to calcium-sensitive pectin. Furthermore, de-polymerisation takes place during ripening and quickens as the degree of methyl esterification is reduced (Pilnik and Voragen, 1991; Tucker and Seymour, 2002). Thus, failure to inactivate endogenous enzymes can have a significant influence on the extracted pectin.

Pectin methyl esterase is present in plants as well as microorganisms. It cleaves methyl ester groups from the pectin galacturonic acid methyl ester backbone, transforming it gradually into low-ester pectin and eventually pectic acid. Plant methyl esterase attacks pectin next to a free carboxyl group and then proceeds along the molecule by a single chain mechanism creating consecutive blocks of de-esterified carboxylic acid units. In contrast, most microbial enzymes de-esterify in a random manner, thus creating random segments of de-esterified carboxylic acid groups (Pilnik and Rombouts, 1981; Pilnik, 1990).

In addition to the esterases, there are various polymerases which cleave galacturonic acid linkages. Polygalacturonase hydrolyses glycosidic linkages next to free carboxylic acid groups and is primarily a microbial enzyme, although it can also occur in some plants. Pectate lyases and pectin lyases are both microbial enzymes that split glycosidic linkages by β-elimination, but pectate lyases split linkages next to free carboxylic acid groups, whereas pectin lyases split linkages next to esterified carboxylic acid groups (Pilnik and Voragen, 1991; Voragen *et al.*, 2001). Obviously, these enzymes may degrade the molecular weight drastically. Additionally, there are also rhamnogalacturonan-modifying enzymes which are important in relation to their role in fragmenting the so-called *hairy regions* of pectin, which comprise a range of glycosidically linked neutral sugars in the pectin molecule. In the juice industry, the activity of these enzymes is exploited as they facilitate fruit juice extraction and fruit juice ultra-filtration (Voragen *et al.*, 2001).

Methyl de-esterification and de-polymerisation of the pectin molecule is also achieved by chemical means during processing and is described in Section 13.5.

13.2.3 Standardisation

Depending on geographical origin, as well as seasonal variations, the ripeness and quality of the raw material varies. Consequently, processing of raw material to a consistent end product may be cumbersome and manufactured pectin often varies greatly from batch to batch. Thus, in order to ensure end product consistency for food use, commercial pectin is usually blended from different batches and generally standardised with sucrose. To control pH, so that pectin can be used in specific applications, it may be blended with buffer salts such as citrates, tartrates or phosphates.

Different pectin types are standardised according to various specifications. Commercial HM pectin is standardised to a uniform jelly grade which expresses how many kilograms of sugar can be gelled by 1 kg of pectin to give a standard gel with a specific composition and gel strength. Various methods are used to measure gel strength but the most common one is the SAG method: the gel is prepared and left to harden in special glasses, after which it is inverted and the deformation is measured. Most HM pectin is standardised to 150 grade USA–SAG, which means that 1 kg of standardised pectin will gel 150 kg sugar provided that pH is between 2.2 and 2.4 and the soluble solids level is 65%. Commercial LM pectin is normally standardised in a test jelly system with various calcium concentrations to reflect the variation in calcium content between different fruit types. The gel strength and break strength are determined by compression tests on a texture analyser. Commercial pectin which is to be used as a stabiliser in drinkable yoghurts or other acidified milk beverages may be graded according to sediment formation: a model beverage system is prepared and the amount of sediment formed upon centrifugation is used to grade the pectin.

13.2.4 Storage and stability of powdered pectin

Pectin is optimally stored in a dry and well-ventilated area in a vapour tight package. At 70% relative humidity, HM pectin generally reaches moisture equilibrium at around 12%, and as most pectin is manufactured to less than 10% moisture, it is likely to pick up moisture if not protected by vapour-tight packaging. During storage, powdered HM pectin loses approximately 5% of its jelly grade per year at 20°C and it is also slowly de-esterified. When the storage temperature is increased from 20°C to 30°C the degradation and methyl de-esterification rates increase significantly. The stability of LM pectin is considerably better as it is scarcely possible to detect any loss of functional properties over a year when the product is kept at 20°C.

13.3 PROCESSING

When producing commercial pectin, the aim is to obtain water-soluble pectins of a specified degree of methyl esterification with as high a molecular weight and yield as possible. Most pectin is produced by extraction with hot aqueous acid followed by precipitation in an organic solvent. Extraction time and temperature varies with raw material and desired end product properties. Figure 13.3 shows a simplified scheme of the production process.

After pre-treatment of the raw material, the peel or pomace is treated with hot aqueous acid in which the cell wall-bound pectin is released by chemical action caused by low pH and high temperature. In some processes, soaking peel in acidic solution takes place prior to the actual acid extraction, promoting pectin release from the cell wall matrix. Pectin yield increases with high temperatures, long treatment time and acidity. However, de-polymerisation and methyl de-esterification of the pectin are also favoured under these conditions. Thus, in order to obtain a satisfactory yield of high-molecular-weight pectin, a compromise must be found when determining extraction conditions. Typical extraction conditions are combinations in the range of 50–90°C for 3–12 h at pH 1–3. When producing LM pectin, a combination of low pH and low temperature is chosen as these conditions favour hydrolysis of ester linkages over hydrolysis of glycosidic linkages. As efficient pectin extraction and separation of solid waste material are more easily accomplished with a large amount of liquid, a compromise must also be drawn between the volume of extraction liquid and processing cost, which

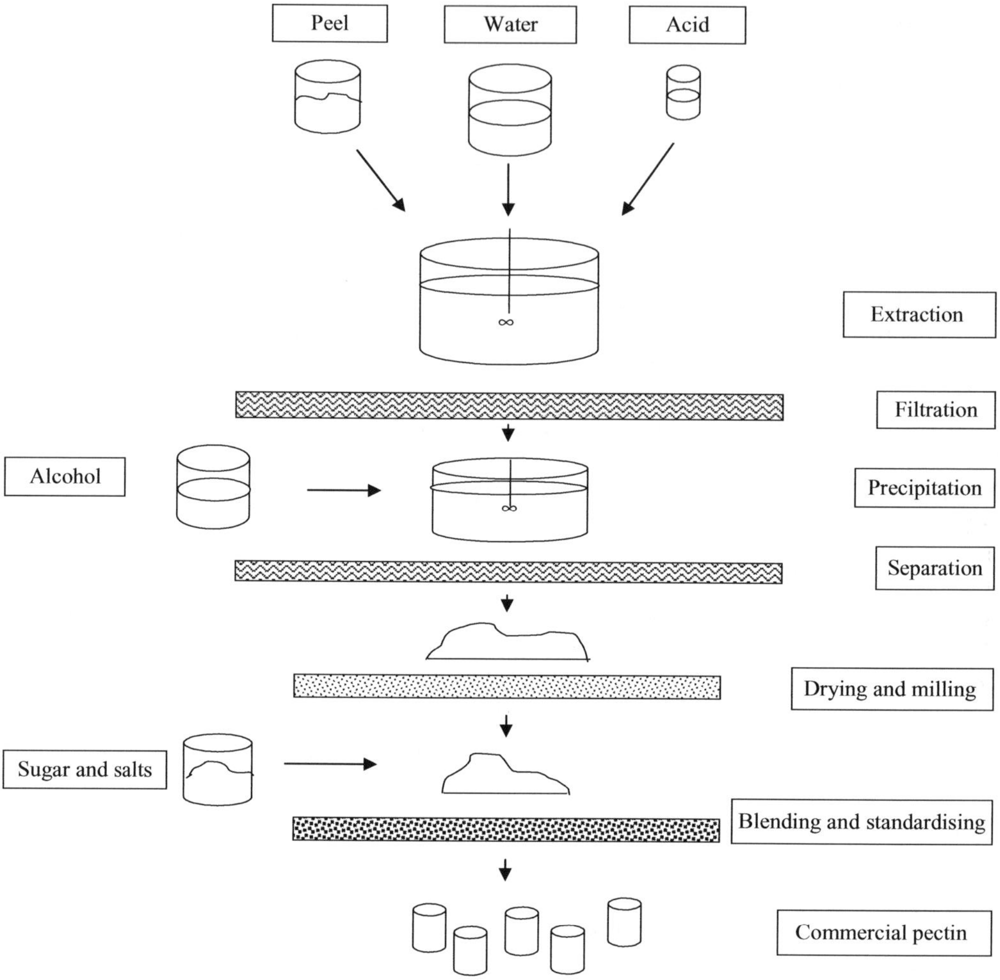

Fig. 13.3 The pectin production process. Extraction of pectin is generally carried out with hot aqueous acid followed by filtration and precipitation in alcohol. The isolated pectin is dried and then milled before it is standardised. (Reproduced with permission from CP Kelco.)

is reduced by producing a more concentrated extract. Pectin may also be extracted with alkali. However, in neutral and alkali conditions, even at ambient temperatures, pectin is very susceptible to β-elimination, whereby the glycosidic bonds of the pectin backbone are cleaved, consequently reducing the molecular weight. Thus, extraction in alkali is not generally used for the production of commercial pectin. The β-elimination mechanism is illustrated in Fig. 13.7.

The extraction process is followed by filtration to separate the acidic aqueous pectin extract from the remaining insoluble plant tissue. Efficient filtration requires relatively low viscosity, so depending on pectin type, the extract cannot contain more than 0.6–1.0% pectin. Thus, it is apparent that a lot of water must be removed in the downstream process which, consequently, makes the process rather energy consuming. Filtration may be aided by using filter aids such as wood cellulose or perlite. The waste raw material finds use as cattle feed.

The clarified pectin may be evaporated to a higher concentration before precipitation, so that the amount of solvent required for precipitation is reduced. Pectin precipitation is usually carried out with an organic solvent in which pectin is insoluble but in which many of the impurities remaining in the extract are soluble. International food standards permit the use of methanol, ethanol or isopropanol as organic solvents. An alternative to alcohol precipitation is aluminium precipitation, first carried out by Joseph and Havighorst in 1952 (May, 1990). The advantages include elimination of the concentration step and certain impurities being more readily removed. However, due to environmental problems posed by effluents rich in aluminium salts and poor precipitation of highly esterified pectin, it is not a commercially favourable extraction method and it is rarely, if at all, used in industry any more.

Precipitation in organic solvent is followed by washing in a dilute acid solution which removes all remaining alcohol-soluble contaminants including acids, sugars and polyphenols. In order to adjust pH, alkali may be added. Once the pectin is separated from as much organic solvent as possible, it is dried and ground to a fine powder.

In addition to HM and LM pectin, amidated LM pectin may be produced. It is typically produced by amidating conventional HM pectin with ammonia in an alcoholic suspension. The process requires careful control in order to obtain the desired relative rates of methyl de-esterification and amidation, whilst at the same time, minimising the rate of polymer chain degradation, which may occur in alkali conditions.

A more recent technology includes methyl de-esterification with biocatalysts in which pectin may be significantly de-esterified with marginal loss in molecular weight, in contrast to conventional methyl de-esterification with acid. This leads to low-ester pectins providing higher gel strengths than conventional low-ester pectins (Ishii *et al.*, 1980; Christensen *et al.*, 2004).

13.4 COMPOSITION

13.4.1 Commercial pectins

When discussing commercial pectin, one refers to a linear polysaccharide comprised mainly of $\alpha(1\rightarrow4)$ linked anhydrogalacturonic acid with partial methyl esterification of the carboxyl groups. However, the chemical structure of native pectin is much more complex and includes many other components including the neutral sugars arabinose, galactose, rhamnose and xylose (McNeil *et al.*, 1979; Voragen *et al.*, 1995). The chemical structure and composition of pectin varies from plant to plant but as citrus, apple and, to a minor extent, sugar beet are the main commercial pectin sources, these pectins are described in this section. Table 13.1 depicts the main constituents of pectin in lemon and apple.

13.4.2 Pectin in the cell wall

Pectin is a natural constituent of all land plants. It is present in the middle lamella and primary walls of plant cells and, depending on the plant type, around one-third of the dry matter in the cell wall may be pectin (Jarvis *et al.*, 1988). Pectin plays an important role in plant growth and development which, together with cellulose and hemicellulose, determine the plant structural properties. In addition to the polysaccharides, the plant cell wall contains structural glycoproteins (extensins), phenolic esters (ferulic and coumaric acids), ionically and covalently linked minerals including boron and calcium and, finally, expansins, oxidative

Table 13.1 Composition of typical commercial pectins expressed as percentage weight of dry matter (Kravatchenko *et al.*, 1992).

Component	Lemon	Apple
Galacturonic acid[a]	76.4	60.8
Methyl ester groups (% DE)	4.4 (71.5)	3.6 (74.3)
Acetyl groups (% DAc)	0.26 (1.4)	0.72 (5.0)
Total neutral sugars	8.5	27
Proteins (N × 6.25)	3	1.6
Total phenol	0.18	0.59
Ash	2.38	1.89
Total	95.1	95.9

Values in parentheses refer to degree of methyl esterification (DE) and degree of acetylation (DAc).
[a] Values are recorded as anhydro residues.

and hydrolytic enzymes. Pectin is understood to interact with other pectin molecules and cell components through covalent and non-covalent bonding including ionic bonding, hydrogen bonding, hydrophobic interactions and Van der Waals forces forming various cross-linkages. Covalent bonds include linkages between pectin and hemicellulose and other polymers or cell wall proteins (Mort, 2002; Schols, 2005). Various models of the structure and interactions in the cell wall have been presented. However, due to the diverse properties of cell walls, to date it is still unknown exactly how the components of the cell wall interact with each other (Schols, 2005).

13.4.2.1 *The composition of native pectin*

Explaining the structural properties of native pectin is a rather complex task and a considerable amount of work remains to be carried out in elucidating the exact chemical structure. Not only does the structure of pectin vary among different plant species, but it is also influenced by differing environmental conditions and plant maturity (Van Buren, 1991; Voragen *et al.*, 1995). However, the following section will offer a simplified explanation of the proposed composition and structural properties of pectin in general.

The native pectic substance in the plant wall is largely composed of three polysaccharide structures, the homogalacturonan (HGA), rhamnogalacturonan I (RG-I) and rhamnogalacturonan II (RG-II) domains. Figure 13.4 shows simplified diagrams of these domains.

Homogalacturonan is a linear homopolymer of (1→4)-α-linked-D-galacturonic acid and is estimated to contain around 100–200 galacturonic acid units. The polymer appears to be synthesised in the Golgi apparatus in the cell after which it is transferred to the plant cell middle lamella and then to the primary cell wall. When initially synthesised, it is proposed that each pectin molecule is highly esterified. However, once transferred to the cell wall, pectin methyl esterases de-esterify the units, resulting in a degree of methyl esterification of the galacturonic acid units of 60–90% depending on the origin of the pectin, the state of the cell wall and the ripeness of the plant (Van Buren, 1991). The more ruptured the cell wall and the more mature the plant, the lower the degree of methyl esterification. Figure 13.5 illustrates a partly methyl-esterified galacturonic acid unit.

In addition to methylation at C-6, the galacturonic acid units can be *O*-acetylated at C-3 and occasionally at C-2. In citrus plants, very few acetyl groups are present, whereas

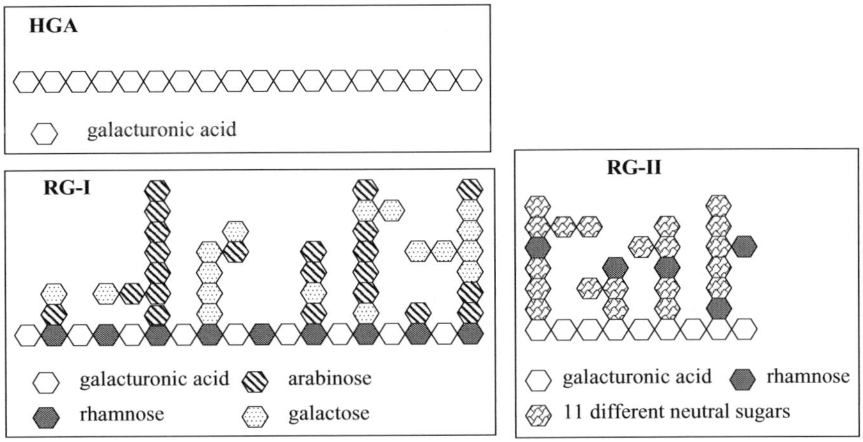

Fig. 13.4 A simplified illustration of the three major pectic polysaccharide domains, the homogalacturonan (HGA), rhamnogalacturonan I (RG-I) and rhamnogalacturonan II (RG-II) domains, which are present in the cell wall. (Reproduced with permission from CP Kelco.)

in specific plants, including potato tubers (Pauly and Scheller, 2000) and sugar beet roots (Mohnen, 1999), acetylation occurs frequently.

The RG-I domain is very diverse and highly branched, thus, often referred to as the *hairy region*. The backbone consists of up to 100 or more repeating units of the disaccharide (1→2)-α-L-rhamnose-(1→4)-α-D-galacturonic acid. The ends are glycosidically linked to the homogalacturonan domains. It is proposed that 20–80% of the rhamnose residues are substituted at C-4 or C-3 with side chains of neutral sugars which vary in length from 1 to more than 50 residues. The composition of neutral sugars varies among plant sources with the most common sugars being galactose and arabinose (Albersheim *et al.*, 1996). An interesting aspect related to the RG-I domain is the ability to cross-link to other RG-I domains by oxidative coupling of ferulic acid units given that ferulic acid is found in considerable quantities (Oosterveld *et al.*, 1997).

The structure of the RG-II domain is highly compact and is composed of a homogalacturonan backbone with around nine galacturonan units, to which four structurally different polymeric side chains are linked. Apart from rhamnose, the side chains contain 11 rare sugars including apiose, 3-*O*-methyl-l-fucose, 2-*O*-methyl-d-xylose, 3-*C*-carboxy-5-deoxy-l-xylose (aceric acid), 3-deoxy-D-manno-octulosonic acid (KDO) and 3-deoxy-D-lyxoheptulosaric acid (Voragen *et al.*, 2001; O'Neill *et al.*, 2004). An interesting feature of RG-II is that it is involved in cross-linking two pectin molecules within the cell wall by borate ester links. The RG-II domain is also included in the *hairy region*.

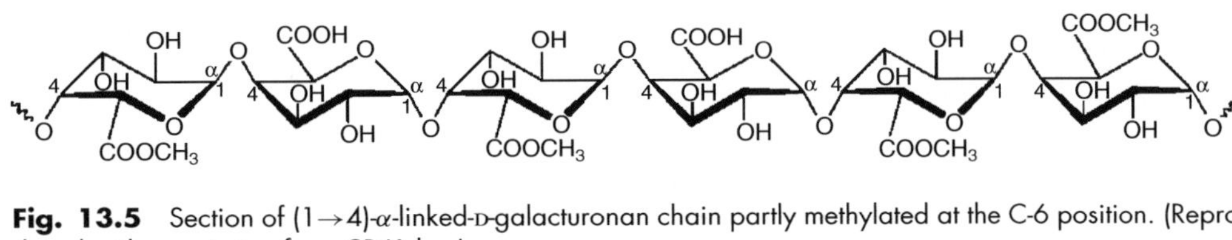

Fig. 13.5 Section of (1→4)-α-linked-D-galacturonan chain partly methylated at the C-6 position. (Reproduced with permission from CP Kelco.)

In addition to the three major polysaccharide domains, arabinogalactans, arabinans and xylogalacturonans are also found in native pectin (Albersheim *et al.*, 1996; Voragen *et al.*, 2001).

13.5 CHEMICAL PROPERTIES

13.5.1 Esterification

When extracting pectin for commercial use, most of the neutral sugar side chains, comprising the RG-I and RG-II domains present in native pectin, are removed. Thus, commercial pectin is often referred to as the homogalacturonic backbone. By weight, typically more than 70% of commercial pectin is galacturonic acid, and depending on the pectin quality and origin, up to 75% of the galacturonan groups are methyl esterified (Voragen *et al.*, 1995). The ratio of methyl esterified galacturonic acid groups to total galacturonic acid groups is termed the degree of methyl esterification (DE). Commercial pectin types are divided into HM and LM pectins and typically have a DE of 55–75% and 20–45%, respectively. Additionally, amidated LM pectin can be prepared by treating pectin with ammonia during processing to convert some of the C-6 methyl ester groups to amide groups. The degree of amidation (DA) is defined as the ratio of amidated galacturonic acid groups to total galacturonan units. Typical DE and DA values for amidated LM pectin are 30% and 20%, respectively. Finally, the degree of acetylated pectin (DAc) is defined as the ratio of acetylated galacturonic acid groups to total galacturonan units, although this material is largely limited to sugar beet pulp. The four building blocks which make up the homogalacturonic backbone are shown in Fig. 13.6.

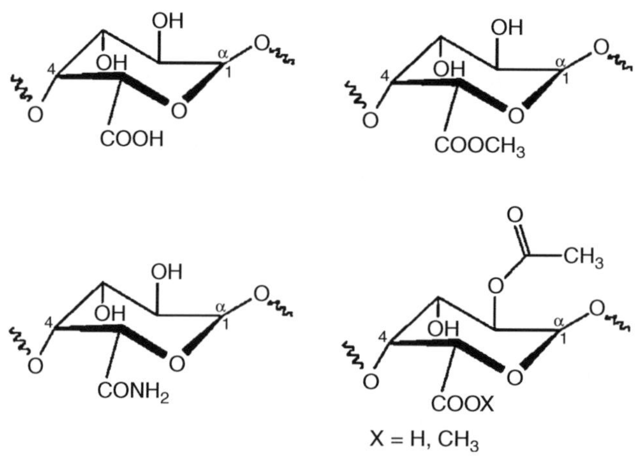

Fig. 13.6 The four galacturonic acid units which comprise the building blocks in the homogalacturonic backbone in a pectin molecule. From left to right: de-esterified, methyl-esterified, amidated and acetylated galacturonic acid units. (Reproduced with permission from CP Kelco.)

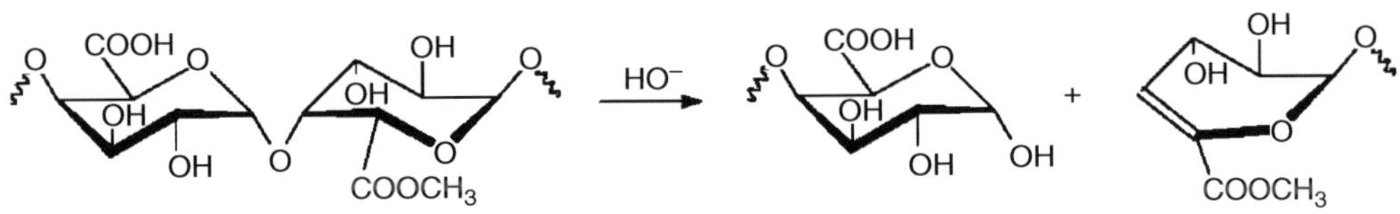

Fig. 13.7 De-polymerisation of the homogalacturonic backbone by β-elimination in which the backbone glycosidic bonds at C-4 on methylated galacturonic acid units are cleaved, reducing the molecular weight. The process is catalysed by alkali. (Reproduced with permission from CP Kelco.)

13.5.2 Acid properties

Pectin is a weak acid with polyelectrolyte behaviour. At neutral pH, pectin is negatively charged and as the pH decreases zero charge is approached. Because of the polyelectrolyte behaviour, the apparent pK_a changes with degree of dissociation of the carboxylic groups on the homogalacturonic backbone. Consequently, it is not possible to define an exact pK_a for pectin. Instead an intrinsic pK_a is obtained and reported to be in the range of 2.9–3.3, close to the pK_a for galacturonic acid which is 3.5 (Rolin, 1993; Voragen *et al.*, 1995). Because of the negative charge of pectins, they react with positively charged polymers including proteins. This feature of pectin is utilised in acidified milk drinks and discussed in further detail in Section 13.6.

13.5.3 Stability

If exposed to unsuitable conditions of pH and temperature, pectin may degrade rapidly by de-esterification and de-polymerisation. Furthermore, pectin is vulnerable to enzymatic attack as described previously in Section 13.2. Pectin has optimal stability at pH 3.5–4.0 and slowly degrades outside this range. It is especially vulnerable to degradation at high temperatures. In neutral and alkaline conditions, the homogalacturonic backbone depolymerises by β-elimination, a process in which the backbone glycosidic bonds at the C-4 position of methylated galacturonic acid units are cleaved. The reaction is illustrated in Fig. 13.7.

As a consequence, HM pectin is more vulnerable to β-elimination than LM pectin. At elevated temperatures the de-polymerisation of the pectin backbone starts at pH 5, although pectin is more robust at room temperature. Additionally, methyl ester and acetyl groups may be removed by saponification. In acidic conditions ($pH < 3$), methyl ester and acetyl groups are cleaved and the neutral sugars hydrolysed even at low temperatures. With elevated temperatures, the reactions accelerate and hydrolytic cleavage of the glycosidic bonds in the pectin backbone proceeds more quickly. At 20°C, pectin solutions are completely stable for months, at 60°C pectin solutions may, dependent on pH, be stable for days, whereas at higher temperatures long holding times should be avoided as a significant loss in jelly grade may occur. Figure 13.8 illustrates the stability of a HM pectin solution at different pH and temperature combinations. Evidently, high pH and high temperature are unfavourable parameters for the pectin stability as shown by the decline in molecular weight with time.

13.5.4 Solubility

The solubility of pectin depends on various parameters including soluble solids, type of counterions, ionic strength, pH and temperature. In general, pectin is soluble in water and

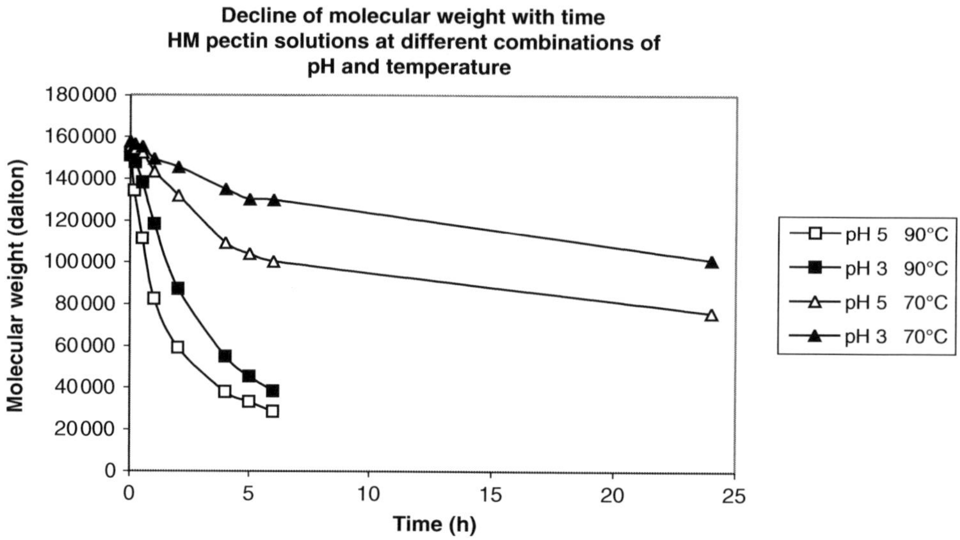

Fig. 13.8 Stability of high methyl-esterified pectin solution measured by the decline in molecular weight with time. (Data from CP Kelco.)

insoluble in alcohol and most organic solvents. However, when powdered pectin is added to water, lump formation easily occurs rendering the pectin difficult to hydrate. Consequently, certain procedures should be followed when preparing a pectin solution. As concentrated pectin solutions exhibit non-Newtonian behaviour, that is viscosity decreases with increasing rates of shear, a high-speed mixer is useful. For instance, when using a high-speed mixer solutions of up to 10% can be prepared, but without it, it is difficult to obtain smooth pectin solutions with concentrations above 3–4%, due to the higher viscosity hindering dispersion. Alternatively, to improve dispersion in water without lumping, one may dry blend the pectin powder with five parts of sugar or the pectin powder may be dispersed in a liquid in which it is insoluble (Rolin, 1993).

Pectins form both soluble and insoluble salts; thus sodium pectinates are more soluble than pectic acids, which, in turn, are more soluble than calcium pectinates. LM pectins are soluble only as the sodium or potassium salt. Sodium, potassium and other monovalent ions are bound electrostatically to the pectin backbone, whereas divalent cations are involved in a process named the 'egg-box' model, the mechanism for gelling LM pectin. Further details regarding the process are found in Section 13.5.6.2. As a general rule, high levels of soluble solids and salts decrease pectin solubility. Improved hydration under such adverse conditions may be achieved by adding sequestering agents such as pyro- or orthophosphate to the pectin. Alternatively, keeping temperature and pH outside the gelling range facilitates solubility.

13.5.5 Rheology

Like other water-soluble polymers, pectin creates viscous solutions. The viscosity depends upon pectin concentration and becomes very pronounced above the concentration at which the molecules entangle as shown in Fig. 13.9.

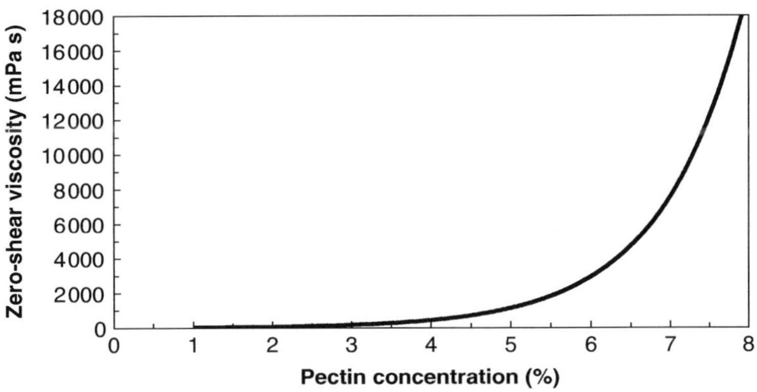

Fig. 13.9 The zero-shear viscosity of amidated low methyl-esterified pectin dissolved in deionised water (pH 4.0, 25°C/77°F). Increased concentrations result in an exponential increase in zero-shear viscosity. (Data from CP Kelco.)

A weak pectin solution (<0.5%) exerts almost Newtonian behaviour, whereas concentrated pectin solutions are shear-thinning and non-Newtonian. However, once the shear is reduced or stopped, the pectin solution regains its viscosity immediately. The viscosity of a pectin solution is dependent not merely on pectin concentration but also on pectin type, solvent, pH, temperature and the presence of salts. High-molecular-weight pectin molecules tend to increase viscosity and rigid pectin molecules exhibit higher viscosity than compact pectin molecules. The tertiary structure is influenced by ionic strength as increased ionic strength may result in a lower viscosity because of charge shielding of the polymer chain. However, polyvalent ions like calcium generally increase viscosity because of cross-linking of the polymer chains (Voragen *et al.*, 1995; Rolin *et al.*, 1998).

13.5.6 Gelation

The prime physicochemical property of pectin is its ability to gel and gel strength generally increases with increased molecular weight. HM pectin gels in acidic conditions and in the presence of sugar, whereas LM pectin may gel over a broader pH range and lower concentration of sugar but requires the presence of cations which, when referring to food, is generally calcium. Accordingly, one refers to sugar and calcium gelling, respectively. The two gelling mechanisms are distinguished from each other by their degree of methyl esterification.

13.5.6.1 *Gelling mechanism of HM pectin*

Gelling HM pectin requires low pH and low water activity. Generally, pH must be between 2.5 and 3.8 and the soluble solids content should be between 55% and 85%. However, the soluble solids content can only be as low as 55% when the degree of methyl esterification is high and pH must be below 3.8 when the soluble solids or the degree of methyl esterification is high. These requirements are explained as follows. A high content of soluble solids creates low water activity which promotes pectin–pectin interactions, rather than pectin–solvent interactions, and low pH reduces the dissociation of the carboxyl groups, thus diminishing

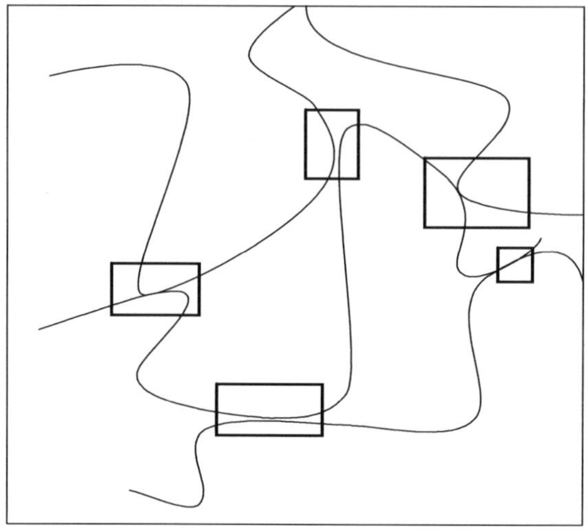

Fig. 13.10 The gelling mechanism of high methyl-esterified pectin. The mechanism relies on junction zones between pectin polymers, highlighted in the boxes. Junction zones are formed as a result of hydrogen bonding between non-dissociated carboxyl groups and secondary alcohol groups as well as hydrophobic interactions between methyl ester groups. (Reproduced with permission from CP Kelco.)

electrostatic repulsion. The gelling mechanism is thought to rely on hydrogen bonding between non-dissociated carboxyl groups and secondary alcohol groups (Morris *et al.*, 1980) together with hydrophobic interactions between methyl ester groups (Oakenfull and Scott, 1984). Accordingly, an increased degree of methyl esterification and low pH enhance the ability to gel. The interactions between the pectin polymers create so-called 'junction zones', which generate the basis of the three-dimensional pectin gel, illustrated in Fig. 13.10.

Because of the importance of the degree of methyl esterification for gelling HM pectin, further groupings of commercial HM pectin, based on setting time and temperature, have been established. The sub-groups include ultra rapid set, rapid set, medium rapid set, slow set and extra slow set pectin, with the degree of methyl esterification ranging from 74–77% for an ultra rapid set type to 58–60% for an extra slow set pectin. For a given pH, setting time increases and setting temperature decreases with decreasing degree of methyl esterification. Because of the high degree of methyl esterification, ultra rapid set pectin gels at a higher pH than slow set pectin.

13.5.6.2 *Gelling mechanism of LM pectin*

The gelling mechanism of LM pectin is referred to as calcium gelling as it relies on the presence of divalent cations, usually calcium. The mechanism is illustrated by the 'egg-box' model (Grant *et al.*, 1973) shown in Fig. 13.11.

There are contradictory theories describing this calcium-gelling mechanism, but the egg-box model is the most widely accepted. It is based on each pectin chain having twofold symmetry thus forming a series of electronegative gaps into which divalent cations associate with different affinities. Accordingly, the cations form dimers of polygalacturonic chains by ionic interaction with the free carboxylic groups on the pectin backbone. Magnesium ions show a very small affinity and thus do not create gels, whereas calcium ions exert a

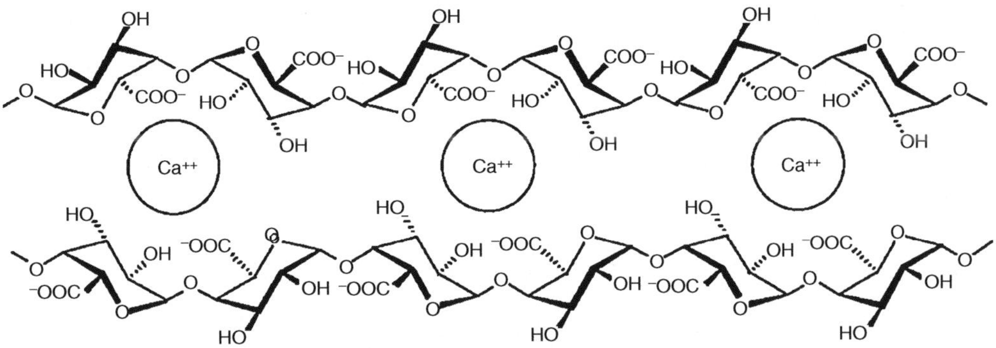

Fig. 13.11 The 'egg-box' model illustrating the gelling mechanism for low methyl-esterified pectin in which calcium ions bind pectin homogalacturonic regions together. (Reproduced with permission from CP Kelco.)

much stronger affinity and form gels. Since the cations react with de-esterified galacturonic acid units, a low degree of methyl esterification enhances the ability to gel and the more calcium sensitive the pectin becomes (Thibault and Rinaudo, 1986). It is worth noting that the relationship between gelling and degree of methyl esterification is the opposite in the case of HM pectin. Also, in contrast to the HM pectin gelling mechanism, low pH and high soluble solids content are not strictly necessary for gelling to take place. Although LM pectin gelation is favoured at low pH, gelling may take place up to pH 6 or even slightly higher.

In addition to the degree of methyl esterification, the distribution of non-esterified carboxylic groups is of significance for the gelling mechanism. This is because a sequence of non-esterified carboxylic acid units is more calcium sensitive than a few consecutive units. Hence, two pectin molecules having the same degree of methyl esterification but differing distributions of non-esterified carboxylic acid groups may have different gelling properties and setting temperatures (Thibault and Rinaudo, 1986). Accordingly, in recent years, a substantial amount of attention has been devoted to establishing the distribution of methyl ester groups along the homogalacturonic backbone.

Amidated LM pectin gels at lower pectin concentrations and requires less calcium to gel than non-amidated LM pectin and is less likely to precipitate at high calcium levels (Racapé *et al.*, 1989; Voragen *et al.*, 1995). The introduction of amide groups into the pectin molecule decreases its hydrophilic character and, consequently, enhances hydrophobic interactions. Additionally, the blocks of amide groups on the pectin backbone associate through hydrogen bonding (Racapé *et al.*, 1989).

13.5.6.3 *Oxidative cross-linking*

An alternative gelling mechanism involves oxidative cross-linking (Thibault and Rombouts, 1986). The mechanism relies on cross-linking ferulic acid groups and, thus, is limited to pectin with high ferulic acid content originating from sources such as sugar beet. So far there is no significant commercial use for this gelling mechanism. Nevertheless, the mechanism is interesting as it opens up the possibility for gelling sugar beet pectin, which will not usually gel due to its high degree of acetylation.

13.5.7 Emulsifying properties

The presence of acetyl groups may prevent pectin gelling, but at the same time, they make pectin a useful emulsifier and stabiliser. This is due to the acetyl groups enhancing the hydrophobic nature of the molecule and providing it with a surface-active character. This means that it may act as an interfacial agent in oil/water and air/water systems, while maintaining viscosifying characteristics (Dea and Madden, 1986).

13.5.8 Interactions with other polymers and proteins

Generally, pectin does not show any synergy with other hydrocolloids and combining pectin with other hydrocolloids may adversely influence the pectin gel strength. Accordingly, there are few commercially interesting interactions between pectin and other polymers or proteins. Among these, it is worth mentioning that pectin–alginate mixtures form cohesive gel networks favoured by pectin with a high degree of methyl esterification and alginate high in polyguluronic acid content. However, the pH must be less than 4 as a higher pH hinders gel formation. Interestingly, the system works well under cold-setting conditions (Morris and Chilvers, 1984). Under very acidic conditions (pH < 2.8), the gelation of LM pectin and alginate is also possible, indicating that the chains must be sufficiently uncharged before the interaction can occur and that methyl esterification assists gelation by reducing electrostatic repulsion between chains (Voragen *et al.*, 1995). Attention has also been given to interactions between pectin and positively charged molecules, including chitosan and poly-L-lysine, where the interactions rely on the positively charged molecules ability to cross-link pectin molecules (Marudova *et al.*, 2004). Such interactions may find use in films.

Furthermore, like other polysaccharides, pectin may be used to purify and concentrate protein solutions. The explanation for this phenomenon is that when polysaccharide and protein solutions are mixed, there is segregation into two phases: a protein-dominated phase and a polysaccharide-dominated phase (Tolstoguzov, 1988). In pectin–milk systems, where the pH is above the isoelectric point of the protein, pectin and protein repel each other and the milk protein will precipitate because the pectin-enriched phase has stronger affinity for water than the protein-dominated phase. Conversely, if the pH is below the isoelectric point of the protein, pectin and protein are attracted to each other and interact. This relationship is utilised in low-pH milk drinks, that is fermented milk drinks and mixtures of fruit juice and milk, where pectin is used for stabilising denatured casein (Rolin *et al.*, 1998).

13.6 APPLICATIONS

Pectin is used as a gelling, thickening and stabilising agent in foods and, to a lesser extent, in pharmaceuticals. Basically, pectin is used to control water in products and help to create the desired texture. The traditional and major application is as a gelling agent in jams and jellies which utilises the ability of HM pectin to form gels at low pH and high sugar levels and the ability of LM pectin to form gels at low sugar levels in the presence of calcium. One of the attractive features is that the pH at which pectin has optimal stability matches the natural pH of fruit preserves. Compared to other hydrocolloids, this feature is unique to pectin. Another advantage is related to texture which is physically and also organoleptically optimal and, finally, pectin gives an excellent flavour release due to its relatively small molecular weight when compared to other hydrocolloids. The thickening effect of pectin is

Table 13.2 Key features of HM, non-amidated and amidated LM pectin gels.

	HM	Non-amidated LM	Amidated LM
pH range	2.5–3.8	2.5–6.0	2.5–6.0
Soluble solids range	55–85%	25–55%	5–65% (75%)
Gel setting temperature	25–90°C	40–100°C	30–70°C
Thermal reversibility	No	Yes/No*	Yes
Gel texture below pH 3.5	Firm	Spreadable with increase in firmness as pH is lowered	Semi firm, similar to HM but more rubbery
Gel texture above pH 3.5	No gel is formed but viscosity is provided	Spreadable, slightly soft	Spreadable, similar to non-amidated LM

* The thermal reversibility depends on the calcium level in the gel.
HM, high methyl-esterified; LM, low methyl-esterified.

used in recombined fruit beverages where HM pectin restores the mouthfeel to that of, for example, freshly squeezed orange. The stabilising effect is seen in multiphase systems where permanent stabilisation of emulsions, suspensions and foams can be obtained. Regarding LM pectin specifically, benefits are seen in relation to the broader pH working range, which means it is suitable in neutral flavour products, and the ability to form heat-reversible gels, which is used in baking jams and other applications. Table 13.2 summarises the key features and working ranges of HM, non-amidated LM and amidated LM pectin.

13.6.1 Choosing a pectin

When choosing the most suitable type of pectin for a specific application, a number of factors must be considered. The first consideration is to specify which functional property is desired: is it gelling, thickening, emulsifying or stabilising or is it to function as a protein protector? Once this is established, one must look upon the extrinsic factors which play a role in the pectin performance. These include soluble solids content, salt content, ionic strength, pH and, when referring to fruit applications, the natural pectin content in the system. Additionally, the process must be taken into consideration, for example does the product need to be pumpable or firm for handling? These considerations are illustrated in Fig. 13.12.

The soluble solids content in the product is crucial as sugar and similar solutes diminish the amount of solvent available and dehydrate the pectin molecules: the higher the soluble solids content, the greater the tendency to gel. Due to the strong dehydration effect, it is nearly impossible to control the gelation of any commercial pectin above 85% soluble solids content. In general, HM pectin is the choice for applications with soluble solids above 60% and generally cannot be used for applications with soluble solids contents below 55%. LM pectin is used in applications with soluble solids content below 60%, although if an especially soft, spreadable or thixotropic texture is desired LM pectin may be preferred with soluble solids above 60%.

A too-low filling temperature may result in pre-gelling and, consequently, produce a non-homogeneous gel structure because the molecular interactions are ruptured and will not reform. With LM pectin, pre-gelling is caused not only by a too-low filling temperature but also by an uncontrolled reaction between calcium ions and pectin.

Increasing or decreasing the amount of added pectin leads to products with a firmer or softer consistency. Alternatively, a change from a very reactive pectin type to a less reactive

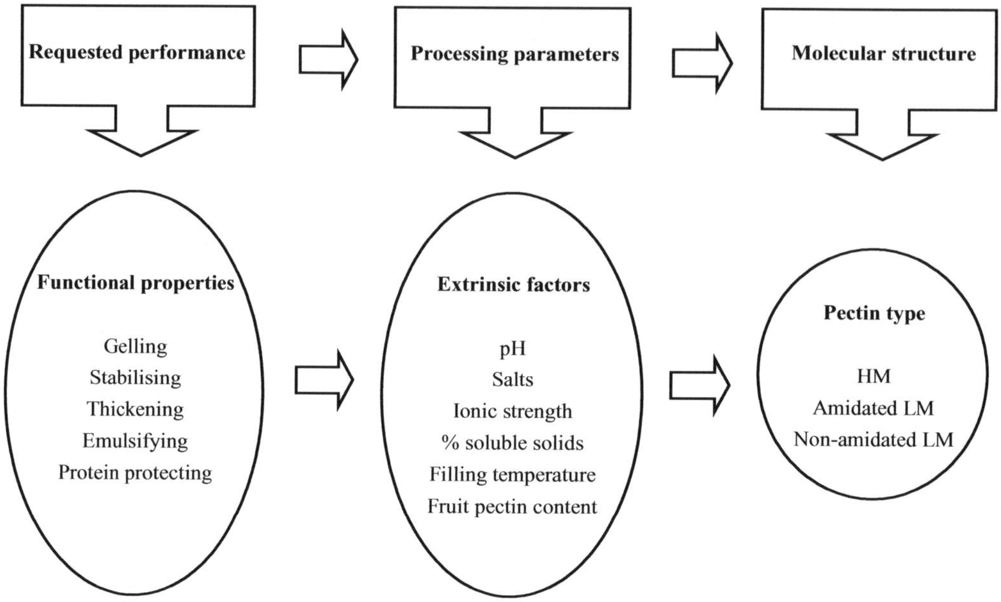

Fig. 13.12 Factors affecting the choice of pectin. HM, high methyl-esterified pectin; LM, low methyl-esterified pectin. (Reproduced with permission from CP Kelco.)

pectin type may produce a softer gel and vice versa: changing from a less reactive pectin to a very reactive pectin will result in a harder gel. Adding salts to pectin solutions generally encourages gelation. With non-amidated LM pectin, the presence of divalent cations, usually calcium, is necessary for gelling. Adding extra calcium to the product may give a firmer texture. This effect is utilised with LM pectin but one must be aware that adding extra calcium to the system may increase syneresis and pre-gelling. This effect is especially difficult to control when using readily soluble calcium salts such as calcium chloride and may partly be overcome by using less soluble calcium salts including calcium phosphates. Adding calcium to the system may be necessary in products where the fruit has very low calcium content. Alternatively, when using fruit with a very high content of calcium, it may be necessary to add calcium-binding salts in order to obtain the desired consistency and prevent syneresis and pre-gelation from taking place. Frequently used sequestering agents include sodium citrate, sodium pyrophosphate and sodium orthophosphate.

In general, non-amidated and amidated LM pectins can be used interchangeably; thus, in the following section, the designation LM pectin includes amidated as well as non-amidated pectin. However, as amidated pectin generally gives stronger gels and has a broader calcium working range than non-amidated pectin, there may be a preference to use one rather than the other LM type in specific applications.

13.6.2 High- and low-sugar jams and jellies

The role of pectin in jams and jellies is to impart a gel texture. In jams, the pectin must set quickly after filling in order to ensure a uniform distribution of fruit pieces, whereas in jellies the pectin must set slowly to allow for any air bubbles to escape. In Table 13.3, the pectin types recommended for different applications are listed, though it is important to keep in mind that the suggested pectin type is not definitive, but merely a general recommendation.

Table 13.3 Pectins recommended for fruit applications including high-sugar and low-sugar jams and jellies.

Product	Function		Recommended pectin type
High sugar (%SS > 60)	Jams in small jars	Gelling at around 85–95°C	HM rapid set
	Jams in larger containers (5–10 kg)	Gelling at around 70–75°C	HM medium rapid set
	Jams in large containers (>10 kg)	Gelling at around 60°C	HM slow set
	Jellies	Gelling and avoiding entrapment of air bubbles before gelling	HM slow set
Low sugar (%SS < 60)	Jams	Gelling without addition of calcium	Amidated LM
	Low sugar jams	Gelling in low sugar jams and jellies	Amidated and non-amidated LM
	Sauces (ripples and toppings)	Thickening	Amidated and non-amidated LM

HM, high methyl-esterified; LM, low methyl-esterified; SS, soluble solids.

In high-sugar jams and jellies, where a firm texture is typically desired, HM pectin is generally the choice. As the degree of methyl esterification influences the gelling properties significantly, HM pectin is further categorised as rapid set, medium rapid set, slow set and extra slow set pectin. Excluding extra slow set pectin, which has the lowest degree of methyl esterification, the categories are presented in Table 13.4.

HM pectin with a high degree of methyl esterification gels at a higher temperature than HM pectin with a lower degree of methyl esterification. This is because the ester group is less hydrophilic than the carboxylic group and, consequently, hydrophobic interactions are enhanced in HM pectin with a high degree of methyl esterification.

The desired filling temperature of the jam or jelly determines which pectin type is recommended for a certain product. Consequently, a rapid set pectin is recommended when filling above 85°C, at 75°C medium rapid set, at 65°C slow set and at 60°C extra slow set is recommended. If the filling temperature is too high there is a risk of fruit separation, whereas if it is too low and near the gel point pre-gelation is likely to occur. Pre-gelation leads to

Table 13.4 Time for gelation for HM pectins with different degrees of methyl esterification at pH 3.0, soluble solids 65% and pectin concentration = 0.43% after boiling and subsequently holding at the specified temperature.

	Degree of methyl esterification (%)	95°C	85°C	75°C	65°C
Rapid set	73.5	60 min	10 min	Pre-gelling	Pre-gelling
Medium rapid set	69.5	No gelling	40 min	5 min	Pre-gelling
Slow set	64.5	No gelling	No gelling	No gelling	30 min

Source: Data are obtained from CP Kelco.

either a soft-set jam or a broken-set jam with syneresis. Notably, the final strength of a jam or jelly based on slow set pectin is not reached until after approximately 1 week and longer for extra slow set pectin.

To ensure an optimal stability of the finished product and to limit the tendency of fruit separation, it is important to add sugar to the fruit at an early stage so that sugar can equilibrate between the fruit and surrounding jelly.

High pH decreases the tendency of HM pectin to gel, because the increased ratio of dissociated to non-dissociated acid groups enhances the hydrophilic character of the pectin molecule. Accordingly, gelation is enhanced at lower pH. At higher pH, pectin gives a softer gel than at lower pH. However, by adding more pectin to the gel at high pH, similarly strong gels may be obtained. HM gels are non-reversible and will not melt when heated, in contrast to LM pectin.

In low-sugar jams and jellies, where the soluble solids content is below 55%, HM pectin does not normally gel. Thus, LM pectin is recommended at soluble solids below 60%. LM pectin is very dependent on calcium or other cations for gelation. Soluble solids, pH and degree of methyl esterification are also very important. Pectin becomes more reactive as the degree of methyl esterification is lowered. For applications above 50% soluble solids, LM pectin with relatively low reactivity may be used. However, at soluble solids below 50%, or at higher filling temperatures, medium reactive LM pectin should be used unless there are special requirements for a very soft and spreadable texture for which a low reactivity LM pectin is suitable.

When the soluble solids level is below 25%, reactive LM pectin may initially yield perfect gels but, when the gel is cut, syneresis may occur due to insufficient water binding. To overcome this problem, combinations of LM pectin and other water binders, including locust bean gum and xanthan gum, are often used and gels with soluble solids down to 5% may be obtained. Pure pectin-based systems for gelling at low-soluble solids can be obtained but this requires fairly high pectin use levels.

Ten to 20 per cent less amidated LM pectin can be used in place of non-amidated LM pectin to obtain the same gel strength. This is the result of introducing amide groups into the LM pectin molecule decreasing the hydrophilic character and enhancing hydrophobic interactions. Additionally, the amide groups associate through hydrogen bonding. Generally, amidated pectin is more calcium reactive and may be used over a wider range of calcium levels compared to non-amidated LM pectin. Increasing the degree of amidation increases gelling temperatures and the gels are completely heat reversible. Thus, some advantages are seen when using amidated pectin as opposed to non-amidated pectin. However, a more soft and spreadable texture may be achieved when using non-amidated LM pectin.

13.6.3 Baking jams

In baking jams, bake stability and pumpability are the most common requirements. Table 13.5 depicts the recommended pectin types for different baking products. Traditionally, HM medium rapid set pectin has been used due to its good thermal stability. However, as such gels tend to synerise after being pumped or sheared at soluble solids below 60%, LM pectin could be a superior option. At soluble solids between 50% and 60%, LM pectin will produce a pumpable, shear and heat-stable baking filling. If a heat-reversible jam is required, LM pectin must be used. Depending on pectin type and calcium content, the pectin gelling and re-melting profile may be manipulated. When using LM pectin, gelation at higher

Table 13.5 Pectins recommended for baking jams.

Product	Function	Recommended pectin type
Cold setting jelly	Gelling	HM rapid set
Oven-resistant baking jams at high soluble solids	Gelling maintaining stability at high temperatures	HM slow set/HM medium rapid set
Glazing jelly with soluble solids range 45–65%	Gelling	Amidated LM
Oven-resistant baking jams at low-soluble solids (down to 30%)	Gelling maintaining stability at high temperatures	Amidated LM/non-amidated LM
Pumpable jams and fillings for use after baking	Gelling giving spreadable texture and counteracting the diffusion of water from a fruit filling into a baking article	LM with low calcium reactivity/ HM slow set

HM, high methyl-esterified; LM, low methyl-esterified.

temperatures is favoured by a low degree of methyl esterification and/or high degree of amidation and is enhanced further by higher calcium levels. A cold-setting jelly is produced with HM rapid set pectin by exploiting the ability to gel at low pH. A pectin–sugar syrup can be made with 61% soluble solids and pH 4.0, which is above the gelling range. The pH will drop after a concentrated fruit acid solution is added to the pectin syrup and gelation is induced within a few minutes, depending on the exact amount of added acid.

13.6.4 Fruit toppings and gelling powder/sugar

In sauces such as fruit-based ripples and toppings, where a thick pumpable texture is desired, LM pectin is used as a thickener. Additionally, the pectin confers a thixotropic behaviour and improves mouthfeel. In gelling powder or sugar, various pectin types may be used (see Table 13.6).

13.6.5 Fruit preparations for yoghurt

In fruit preparations for yoghurt, LM pectin creates a soft, thixotropic gel which is sufficiently firm to ensure uniform fruit distribution even in large containers, but at the same time, allows the fruit preparation to be easily stirred into the yoghurt (Table 13.7). Additionally, the pectin may reduce colour migration from the fruit into the yoghurt phase. Combining the pectin with other plant gums, such as locust bean gum, may further reduce the colour migration and also counteract syneresis. In preparations for 'fruit on the bottom' type yoghurt, pectin delays diffusion rates of fruit material into the yoghurt phase. However, in this type of application, it is important to be aware of a possible migration of calcium ions from the yoghurt into the

Table 13.6 Pectins recommended for fruit toppings and gelling powder/sugar.

Product	Function	Recommended pectin type
Fruit toppings	Thickening	Amidated and non-amidated LM
Gelling powder/sugar	Thickening	HM rapid set, amidated and non-amidated LM

HM, high methyl-esterified; LM, low methyl-esterified.

Table 13.7 Pectins recommended for fruit preparations for yoghurt.

Product	Function	Recommended pectin type
Fruit preparations for yoghurt	Thickening, producing a soft and pumpable product enabling uniform fruit distribution, reducing colour migration	Amidated and non-amidated low methyl-esterified

fruit phase, which may rapidly gel producing an undesirable firm fruit layer. The problem may be overcome by using low-calcium-reactive pectin and saturating the pectin in the fruit preparation with calcium. The application of HM pectin in fruit preparations for yoghurt is scarce as it is limited to preparations with soluble solids above 60% and pH below 3.5, whereas LM pectin may be used across the entire solids and pH range encountered in fruit preparations. In preparations where the pH is in the range 3.6–4.0, LM pectin with medium calcium reactivity is recommended. However, when the soluble solids content is below 50%, LM pectin with high calcium reactivity may be necessary. Non-amidated LM pectin must be used in higher concentrations than amidated LM pectin but it gives a greater degree of thixotropy.

13.6.6 Confectionery

Pectin fruit jellies are delicate confectionery products with a low water content of around 20%. Recommended pectins are shown in Table 13.8. Normally, the jellies are prepared from combinations of sugar and glucose syrup, flavoured with synthetic flavours and acidified with citric acid. However, the jellies may also contain fruit pulp or juice. Pectin is used in the concentration range 1–3%. Pectin provides a superior texture and flavour release compared to other gelling agents but may be troublesome to handle during processing due to gel formation at high temperatures ($\sim$70°C) and because of its sensitivity to pH and soluble solids content. In order to gel HM pectin, acid must be added to reduce the pH from 4.1 to 3.5. The acid must be added as a solution (maximum 50% w/v), otherwise pre-gelation will occur around each acid crystal and make the jelly grainy. When producing non-fruit flavoured pectin jellies, for flavours such as toffee and peppermint, the pH working range must be 4.0–4.5 as the neutral flavours are more compatible with this range. For this type of application, certain amidated LM pectin types can be used. For aerated confectionery products such as fruit-flavoured marshmallows, which have a soluble solids content of approximately 80%, HM pectin confers texture and stabilises the foam in the product. Amidated LM pectin may also be used.

Table 13.8 Pectins recommended for confectionery.

Product	Function	Recommended pectin type
Pectin jellies, fruit flavour	Gelling	HM extra slow set
Pectin jellies, neutral flavour	Gelling	Amidated LM
Aerated confectionery products	Gelling, thickening, stabilising	HM extra slow set or amidated LM

HM, high methyl-esterified; LM, low methyl-esterified.

Table 13.9 Pectins recommended for acidified and fermented milk drinks.

Product	Function	Recommended pectin type
Yoghurt drinks	Stabilising milk proteins	HM
Juice-milk drinks	Stabilising milk proteins	HM
Acidified whey drinks	Stabilising milk proteins	HM
Acidified soya drinks	Stabilising milk proteins	HM

HM, high methyl-esterified.

13.6.7 Beverages

13.6.7.1 *Acidified and fermented milk drinks*

Pectin has two distinct functions in dairy products and dairy analogues. The most common application in dairy products is where HM pectin is used to stabilise protein dispersions (see Table 13.9). A minor application is where LM pectin is used as a gelling agent and is discussed in Section 13.6.8. Acidified and fermented milk drinks, including yoghurt drinks, mixtures of fruit juice and milk, acidified whey and soya drinks, form a group of products which is an increasingly important application area for HM pectin. These products are especially popular in Japan and other Asia-Pacific countries. More recently, the products are finding a growing interest in Europe and especially North America. The use of HM pectin results from its ability to stabilise protein particles at low pH. Protein-containing products at a pH around 4 will aggregate and create lumps when heat treated and, consequently, produce an unstable drink. At pH 4, HM pectin is negatively charged and most proteins, including casein particles, are positively charged. Thus, an interaction takes place between pectin and protein and protein aggregation is hindered, even during heat treatment. The optimal stabilising effect is seen in the pH range 3.8–4.2 (Glahn and Rolin, 1994). This is explained by the fact that if pH is too low, the carboxylic acid groups are not sufficiently dissociated and, consequently, do not attach sufficiently to casein particles, whereas, if pH is too high, the protein does not possess enough positively charged areas to induce interactions with pectin.

13.6.7.2 *Fruit beverages*

In recombined fruit beverages, the viscosity and mouthfeel properties of HM pectin are used to restore the mouthfeel to that of the fresh juice. The ability of pectin to confer texture to a liquid is also utilised in low-calorie drinks where a dilute pectin solution may mimic the mouthfeel of a 15% sugar solution. By adding just 0.1% HM pectin, the same texture may be obtained in a low-calorie drink where 10–15% sugar has been removed. Additional but minor application areas include powdered soft drinks, powdered protein soft drinks, liquors and carbonated soft drinks. It is interesting that sugar beet pectin is finding use in carbonated soft drinks. This progress is attributed to the unique stabilising and emulsifying properties of beet pectin. Applications in fruit beverages are included in Table 13.10.

13.6.8 Dairy

In milk products, including yoghurts and fruit and milk desserts, LM pectin is used as a gelling agent utilising the natural content of calcium (see Table 13.11). Cold-setting milk

Table 13.10 Pectins recommended for fruit beverages.

Product	Function	Recommended pectin type
Fruit drink concentrates	Stabilising pulp suspension and oil emulsion	HM rapid set
Fruit beverages	Thickening and restoring mouthfeel if the amount of natural pectin is insufficient	HM
Low calorie fruit drinks	Thickening, to mimic mouthfeel of a sugar solution	HM
Carbonated soft drinks, etc.	Stabilising and emulsifying	HM or sugar beet pectin

HM, high methyl-esterified.

desserts can be prepared with LM pectin as a gelling agent. The pectin solution is mixed with cold milk which provides the calcium ions necessary for gelling to take place at low pH. The texture can range from brittle to very soft and creamy, depending on the pectin used. It is also possible to add fruit, chocolate, vanilla or caramel flavour to this type of milk dessert. In this case, the pectin must be dissolved in a sugar syrup together with the flavour prior to addition to the milk. This type of product gives a light texture with excellent flavour release. In stirred and set yoghurt, small amounts of LM pectin increase firmness, mouthfeel and creaminess through excellent water-binding ability, calcium reactivity and interaction with milk proteins. The water binding helps minimise syneresis and the reactivity of pectin with proteins reinforces the protein network in the yoghurt. The network is further strengthened by junction zones created by pectin–calcium interactions. At larger pectin dosages, the reactions may be so strong that de-stabilisation of the product occurs. Accordingly, use levels above 0.2–0.3% are not recommended. Generally, carrageenan is more suitable for gelled milk products as it gels milk at much lower concentrations than pectin. However, for sour milk puddings or milk desserts combined with fruit, LM pectin may be preferred to carrageenan as it does not co-precipitate with casein at reduced pH.

13.6.9 Bakery

In bakery applications, including bread and dough, pectin is used to increase bread volume and improve moisture retention during storage to ensure a soft and spongy end product (Table 13.12). These assets can be ascribed to the excellent water-binding property of pectin. Among the positive effects on dough, improved processing properties must be included.

Table 13.11 Pectins recommended for dairy products.

Product	Function	Recommended pectin type
Yoghurts	Thickening, creating creaminess and mouthfeel	Amidated and non-amidated LM
Gelled milk products	Gelling	Amidated and non-amidated LM
Dairy desserts (fruit/milk desserts)	Instant gelling created by calcium-pectin reactions	Amidated and non-amidated LM
Low-fat dairy products	Emulsifying, providing fat-like mouthfeel	SLENDID®

LM, low methyl-esterified.

Table 13.12 Pectins recommended for bakery products.

Product	Function	Recommended pectin type
Breads	Improving volume and retaining moisture and softness	HM
Frozen dough	Improving volume and freeze stability	HM
Pre-proofed dough	Improving volume and softness	HM

HM, high methyl-esterified.

Additionally, pectin increases the flexibility of bread, makes it less fragile and decreases the tendency to crumb. This effect is especially pronounced in breads like tortilla and pita bread. In frozen and pre-proofed dough, pectin improves the volume and softness when baked. In frozen dough, it stabilises the volume during freezing by delaying the degradation of the natural starch network and in frozen bread it stabilises the product so that it resembles fresh bread when reheated.

13.6.10 Miscellaneous foods

In the group of miscellaneous foods, dressings, fruit and vegetable sauces, spreads, mayonnaise, emulsified meat products and water dessert gels have been combined as the market share for pectin in these applications is small. Nevertheless, an interesting area emerges in this group: the application of pectin in low-fat products. Table 13.13 summarises the pectins recommended for this group.

In low-fat products, pectin is used as a fat substitute where it binds water and consequently improves the emulsion stability. A special type of pectin named SLENDID®, of which a few different types exist, has been developed for this kind of product. In addition to binding water, it provides a fat-like mouthfeel. SLENDID® is used in a broad range of food products including spreads, mayonnaise, salad dressings, processed meats, ice cream, processed cheeses, soups, sauces, desserts and bakery products in which fat may be fully or partly replaced. In spreads, comprising butter, margarine and other oil-continuous products, pectin is used to stabilise the water-in-oil emulsion and prevent phase separation. In emulsified meat products, such as sausages, pâtés and meat spreads, pectin enables fat reduction and by adding carrageenan in addition to pectin a superior texture may be obtained. In sauces, including barbecue sauces and ketchup, LM pectin is used as a thickener, where it confers a thixotropic

Table 13.13 Pectins recommended for miscellaneous foods.

Product	Function	Recommended pectin type
Dressings	Stabilising	SLENDID®, amidated and non-amidated LM
Sauces (barbecue and ketchup)	Thickening, improving mouthfeel	Amidated and non-amidated LM
Low-fat spreads, butter, margarine	Stabilising and fat replacement	SLENDID®
Low-fat mayonnaise	Stabilising and fat replacement	SLENDID®
Processed meats	Stabilising and fat replacement	SLENDID®
Water dessert gels	Gelling	SLENDID®, amidated LM

LM, low methyl-esterified.

behaviour and improves mouthfeel. Similarly, LM pectin is recommended in fruit-based ripples and toppings where a thick, pumpable texture is desired. By using LM pectin in tomato ketchup, the rheological properties may be varied and controlled from smooth and pourable to almost spreadable. In addition, excellent flavour release is achieved. In vegetable-based sauces, such as salsa and chutney, using amidated LM pectin alone or in combination with other hydrocolloids, such as modified starch, the texture may be varied from pourable to a thick and almost sliceable texture. Due to the low level of soluble solids, which characterises water gels, hydrocolloids other than pectin are usually preferred in this application. Nevertheless, SLENDID and amidated LM pectin may be used with successful results. In particular, the use of amidated LM pectin has given comparable results with other hydrocolloids. Furthermore, pectin is used at low pH which improves the fruity taste. Finally, pectin may be used in mouth-freshener strips, where the film-forming properties of low-molecular-weight pectin can be exploited.

13.6.11 Pharmaceuticals and personal care

In addition to the wide range of food applications, the use of pectin in the pharmaceutical market and the personal care area is, although still minor, a growing area. Applications include ostomy devices, where the water-binding ability and skin-friendliness of pectin are utilised. Furthermore, pectin is used in wound-care products, where proposed bactericidal and wound-healing effects, together with the skin-friendliness of pectin, play a role. The skin-friendliness of pectin is ascribed to a buffering effect, in which pectin helps maintain the natural skin pH of 5–6. Pectin also finds use in cosmetics, skin and hair-care products. The ability of pectin to add viscosity and stabilise emulsions is used in a number of pharmaceutical preparations including controlled and sustained drug-release formulations, colonic-specific drug-release formulations, anti-diarrhoea and anti-reflux preparations. In drug encapsulation, pectin is useful as it is acid stable and, thus, ideal for drugs which need to travel through the stomach and be released in the colon.

13.7 FUTURE DEVELOPMENTS

In addition to the technological benefits of pectin in food processing and, to a smaller extent, in industrial and pharmaceutical products, pectin can, like other dietary fibres, exhibit a variety of favourable physiological and nutritional effects in man. Some of these potential benefits include attenuating blood cholesterol level, attenuating blood glucose response, reducing gastric emptying time, anti-diarrhoea effects and improving mineral absorption (Rolin *et al.*, 1998; Thibault and Ralet, 2001). Yet, contradictory findings have been reported and caution should be exercised in ascribing benefits solely to pectin. Instead, in many cases, it may be more correct to ascribe the observed health benefits as an interaction between pectin and other factors beneficial to health, including diet and life-style habits. However, with increasing consumer awareness of a healthy life-style, and an emerging trend to produce biomedical products and functional foods, the industrial manufacturer must realise the potential health-related benefits of pectin.

The growing necessity and demand for a healthy diet is continuously promoting low-fat products and the development and production of fat-reduced foods is now a major sector of the industry. In this respect, pectin is used in ‘oil-barrier technology’ where the uptake of oil in products such as fried battered products and potatoes can be reduced up to 50%.

The technique relies on pectin cross-linking with calcium salts to form a clear, transparent film barrier (Gerrish and Carosino, 2001; Gerrish *et al.*, 2001). The barrier reduces oil absorption while preserving the organoleptic properties of the food. Similarly, the demand for low-carbohydrate products has been met by developing a specially designed pectin for use in such products (Christensen and Thoegersen, 2005). Adding to the uses of pectin in low-fat and low-carbohydrate products, a recent patent describes the use of biopolymers, including pectin, for the encapsulation and stabilisation of oils in oil-in-water emulsions (McClements and Decker, 2005). The encapsulated oils have the same physicochemical and sensory properties as the original material, thus texture, taste, stability and appearance of full-fat food product, is maintained but, due to encapsulation in a non-digestible dietary fibre, the oil is not absorbed by the human digestive system.

With advances in molecular biology, which contribute greatly to elucidating the complex pectic network and interplay of various plant cell factors, including endogenous enzymes (Willats *et al.*, 2001), new possibilities for manipulating enzyme activity and pectin structure in the plant may find increasing importance. This is especially interesting with regard to sugar beet pectin which, due to its inability to gel, finds use in a very limited number of applications. By enzymatic modification of pectin in the plant, it may be possible to create sugar beet pectin that can gel. Another possibility may include suppressing esterase activity, thus acquiring highly esterified pectin. Despite significant production costs, the molecular approach opens up an interesting potential area for the future manufacture of pectin.

At present, the advanced technology and knowledge of pectin structure are encouraging the development of pectins with tailored molecular structures which exhibit specific properties. Consequently, the range of applications is expanding continuously which corresponds well with the increasing consumer desire for nutritious products with good appearance and taste.

Acknowledgements

I thank my colleagues at CP Kelco for helpful discussions on pectin and critically reviewing the manuscript.

References

Albersheim, P., Darvill, A.G., O'Neill, M.A., Schols, H. and Voragen, A. (1996) An hypothesis: the same six polysaccharides are components of the primary cell wall of all higher plants. In: J. Visser and A. Voragen (eds), *Progress in Biotechnology*. Elsevier, Amsterdam, pp. 47–55.

Christensen, S., Hansen, K. and Trudsø, J. (2004) Process for making de-esterified pectins, their compositions and uses thereof, *WO2004005352*.

Christensen, S. and Thoegersen, A. (2005) Gelling agent for low calorie gels, Patent Application 200401568.

Dea, I. and Madden, J. (1986) Acetylated pectic polysaccharides of sugar beet. *Food Hydrocolloids*, **1**, 71–88.

Douglas, R. (1913) Food Product, US Patent 1,082,682.

Gerrish, T. and Carosino, L. (2001) Low oil food composition and method, US Patent 6,290,999.

Gerrish, T., Higgins, C. and Kresl, K. (2001) Method of making battered and breaded food compositions using calcium pectins, US Patent 6,261,618.

Glahn, P.-E. and Rolin, C. (1994) Casein–pectin interaction in sour milk beverages. In: *Food Ingredients Europe: Conference Proceedings*. Process Press Europe, Maarssen.

Grant, G.T., Morrris, E.R., Rees, D.A., Smith, P.J.C. and Thom, D. (1973) Biological interactions between polysaccharides and divalent cations: the egg box model. *FEBS Letters*, **32**, 195–198.

Ishii, S., Kiho, K., Sugiyama, S. and Sugimoto, H. (1980) Novel pectin esterase, process for its production, and process for producing demethoxylated pectin by the use of said pectin esterase, US Patent 4,200,694.

Jarvis, M.C., Forsyth, W. and Duncan, H.J. (1988) A survey of pectic content of non lignified monocot cell walls. *Plant Physiology*, **88**, 309–314.

Kravatchenko, T., Voragen, A. and Pilnik, W. (1992) Analytical comparison of three industrial pectin preparations. *Carbohydrate Polymers*, **18**, 17–25.

Marudova, M., MacDougall, A. and Ring, S. (2004) Pectin–chitosan interactions and gel formation. *Carbohydrate Research*, **339**, 1933–1939.

May, C. (1990) Industrial pectins: sources, production and applications. *Carbohydrate Polymers*, **12**, 79–99.

McClements, D. and Decker, E. (2005) Biopolymer encapsulation and stabilization of lipid systems and methods for utilization thereof, *WO2005086976*.

McNeil, M., Darvill, A. and Albersheim, P. (1979) The structural polymers of the primary cell walls of dicots. *Fortschritte der Chemie Organischer Naturstoffe*, **37**, 191–249.

Mohnen, D. (1999) Biosynthesis of pectins and galactomannas. In: B.M. Pinto (ed), *Comprehensive Natural Products Chemistry*. Elsevier, Amsterdam, pp. 497–527.

Morris, E., Gidley, M., Murray, E., Powell, D. and Rees, D. (1980) Characterization of pectin gelation under conditions of low water activity, by circular dichroism, competitive inhibition and mechanical properties. *International Journal of Biological Macromolecules*, **2**, 327–330.

Morris, V. and Chilvers, G. (1984) Cold-setting alginate-pectin mixed gels. *Journal of the Science of Food and Agriculture*, **35**, 1370–1376.

Mort, A.J. (2002) Interactions between pectins and other polymers. In: G. Seymour and J. Knox (eds), *Pectins and Their Manipulation*. Blackwell Publishing, Oxford, pp. 30–47.

Oakenfull, D. and Scott, A. (1984) Hydrophobic interaction in the gelation of high methoxyl pectins. *Journal of Food Science*, **49**, 1093–1097.

O'Neill, M.A., Ishii, T., Albersheim, P. and Darvill, A. (2004) Rhamnogalacturonan II: structure and function of borate cross-linked cell wall pectic polysaccharide. *Annual Review of Plant Biology*, **55**, 109–139.

Oosterveld, A., Grabber, J., Beldman, G., Ralph, J. and Voragen, A.G.J. (1997) Formation of ferulic acid dehydro-dimers through oxidative crosslinking of sugar beet pectin. *Carbohydrate Research*, **300**, 179–181.

Pauly, M. and Scheller, H. (2000) *O*-acetylation of plant cell wall polysaccharides: identification and partial characterization of a rhamnogalacturonan *O*-acetyl transferase from potato suspension-cultured cells. *Planta*, **210**, 559–667.

Pilnik, W. (1990) Pectin – a many splendoured thing. In: G. Phillips, D.J. Wedlock and P.A. Williams (eds), *Gums and Stabilisers for the Food Industry*. IRL Press, Oxford, pp. 209–221.

Pilnik, W. and Rombouts, F. (1981) Pectic enzymes (in food processing). In: G. Birch, N. Blakebrough and K. Parter (eds), *Enzymes and Food Processing*. Applied Science, London, pp. 105–128.

Pilnik, W. and Voragen, A. (1991) The significance of endogenous and exogenous pectic enzymes in fruit and vegetable processing. In: P. Fox (ed), *Food Enzymology*. Elsevier, London, pp. 303–336.

Racapé, E., Thibault, J.-F., Reitsma, J. and Pilnik, W. (1989) Properties of amidated pectins. II: polyelectrolyte behaviour and calcium binding of amidated pectins and amidated pectic acid. *Biopolymers*, **28**, 1435–1448.

Rolin, C. (1993) Pectin. In: J. BeMiller and R. Whistler (eds), *Industrial Gums*. Academic Press, New York, pp. 257–293.

Rolin, C., Nielsen, B.U. and Glahn, P.-E. (1998) Pectin. In: S. Dumitriu (ed), *Polysaccharides, Structural Diversity and Functional Versatility*. Marcel Dekker, New York, pp. 377–419.

Schols, H.A. (2005) Plant cell wall architecture: Pectin structure, extraction and enzymatic modification. *Internal Presentation, CP Kelco.*

Thibault, J.-F. and Ralet, M.-C. (2001) Pectins, their origin, structure and functions. In: B.V. McCleary and L. Prosky (eds), *Advanced Dietary Fibre Technology*. Blackwell Science, Oxford. pp. 369–378.

Thibault, J.-F. and Rinaudo, M. (1986) Chain association of pectic molecules during calcium-induced gelation. *Biopolymers*, **25**, 455–468.

Thibault, J.-F. and Rombouts, F. (1986) Effects of some oxidizing agents, especially ammonium peroxysulfate, on sugar beet pectins. *Carbohydrate Research*, **154**, 205–215.

Tolstoguzov, V. (1988) Concentration and purification of proteins by means of two-phase systems: membraneless osmosis process. *Food Hydrocolloids*, **2**, 195–207.

Tucker, G. and Seymour, G. (2002) Modification and degradation of pectins. In: G. Seymour and J. Knox (eds), *Pectins and Their Manipulation*. Blackwell Publishing, Oxford, pp. 150–173.

Van Buren, J. (1991) Function of pectin in plant tissue structure and firmness. In: R. Walter (ed), *The Chemistry and Technology of Pectin*. Academic Press, New York, pp. 1–21.

Voragen, A., Pilnik, W., Thibault, J., Axelos, M. and Renard, C. (1995) Pectins. In: A.M. Stephen (ed), *Food Polysaccharides and Their Applications*. Marcel Dekker, New York, pp. 287–339.

Voragen, F., Beldman, G. and Schols, H.A. (2001) Chemistry and enzymology of pectins. In: B.V. McCleary and L. Prosky (eds), *Advanced Dietary Fibre Technology*. Blackwell Science, Oxford, pp. 379–398.

Willats, W., McCartney, L., Mackie, W. and Knox, J. (2001) Pectin: cell biology and prospects for functional analysis. *Plant Molecular Biology*, **47**, 9–27.

14 Pullulan

Hiroto Chaen

ABSTRACT

Pullulan, a linear homopolysaccharide of glucose, is produced from starch by fermentation. The manufacturing process was developed in 1976, and pullulan is now widely used in the food, pharmaceutical and cosmetic industries for its specific functional properties. It easily dissolves in water to form a stable, viscous solution that does not gel. Pullulan solutions have excellent adhesive properties, binding ability and biodegradability. They readily form films with low oxygen permeability and these can protect foods from oxidation. The safety of pullulan in foods is supported by its chemical composition, the purity of the final product, a series of toxicological studies and the fact that it has a long history of use of about 30 years as a food ingredient in human foods in Japan. Recently, the demand of pullulan has rapidly increased for films and hard capsules, and its use in these fields is expected to grow.

14.1 INTRODUCTION

Pullulan is an extracellular glucan elaborated by a fungus of the genus *Aureobasidium*, commonly called 'black yeast', and was first discovered in 1938 by Bauer (1938). Its production and structure were investigated by Bender *et al.* (1959), Ueda *et al.* (1963) and Wallenfels *et al.* (1961, 1965) and the structure of pullulan elucidated in 1965. Using pullulanase, an enzyme that specifically cleaves the (1→6)-α-D-glucopyranosidic bonds, it was determined that maltotriose is the predominant component. Pullulan is a linear glucan consisting of repeating units of maltotriose joined by α-D-(1→6) linkages (see Fig. 14.1). The polymer also contains some maltotetraose units (Bender and Wallenfels, 1961; Catley and Whelan, 1971).

The Hayashibara Company Ltd. began commercial production of pullulan in Okayama, Japan, in 1976. Pullulan production gradually increased and reached about 200 metric tonnes in 2000. Thereafter, the demand for pullulan rapidly increased for hard capsules and for oral care film. The capacity of pullulan production by the Hayashibara Company is now 1000 metric tonnes per year.

The danger from animal plagues such as mad cow disease and foot-and-mouth disease has raised concerns about raw materials of animal origin being used in foods. The requests for raw materials of plant origin for use in foods, cosmetics and the medical fields have been increasing. Pullulan is produced from starch and, therefore, pullulan demand for films and hard capsules is expected to grow.

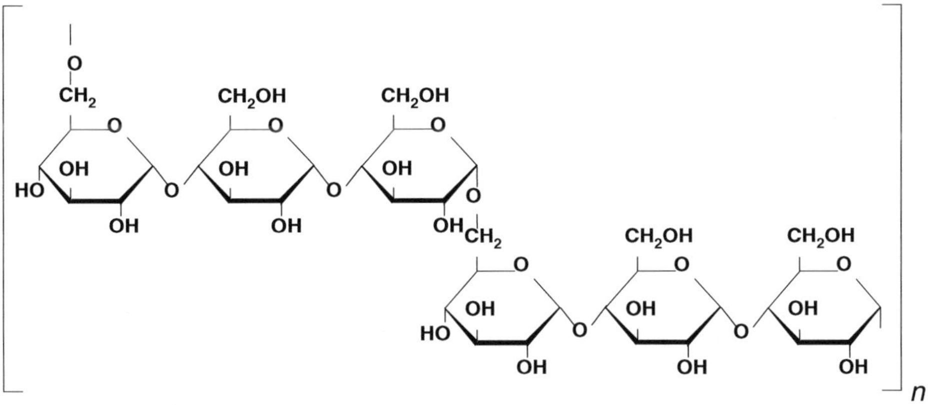

Fig. 14.1 Structure of pullulan.

It has been confirmed that pullulan is safe for use in foods, pharmaceuticals and cosmetics after observing no evidence of acute toxicity in mice, chronic toxicity in rats or in vitro mutagenicity.

In Japan, pullulan appears in the List of Existing Food Additives published on 16 April 1996 (Ministry of Health and Welfare Notification No. 120). The specifications are described in the Specifications and Standards for Foods, Food Additives, etc. (Ministry of Health and Welfare Notification No. 370 issued in 1959 revised on 30 March 2006) and listed in Japan's *Specifications and Standards of Food Additives*, 8th edition on 30 August 2006. It is also listed in the *Japanese Pharmacopoeia*, 15th edition, published on 31 March 2006 (Ministry of Health, Labour and Welfare Notification No. 285) and the Japanese Standards of Quasi-Drug Ingredients 2006 (Notification No. 0331030 issued by the Director of the Pharmaceutical and Food Safety Bureau, the Ministry of Health, Labour and Welfare).

In the USA, the FDA received the generally recognised as safe (GRAS) notification of pullulan on 1 March 2002. The Agency Response Letter issued on 1 August 2002 noted that the FDA had no question about the GRAS status of pullulan and designated it as GRAS in Notice GRN 000099.

Recently, in the EU, pullulan was approved as food additive E1204 for use in food capsules, tablets and films under directive 2006/52/EC. In addition, it is permitted in Russia, China, Korea, Taiwan, Thailand, Singapore, Vietnam and Mercosur of South America (Brazil, Argentine, Paraguay and Uruguay). In India, it has been submitted for approval as a miscellaneous food additive and is being examined by the Indian expert committee.

At the international level, the Joint FAO/WHO Expert Committee on Food Additives (JECFA) evaluated the safety of pullulan and, at the 65th meeting, announced the acceptable daily intake (ADI) as 'not specified'. The International Numbering System (INS) number 1204 was assigned to pullulan. Additionally, pullulan is approved for use in Halal and Kosher foods.

14.2 RAW MATERIALS

Pullulan is produced by the culture of *Aureobasidium pullulans* using a medium containing an appropriate sugar syrup as the carbon source. Many workers have investigated pullulan production using various sugars. Ueda *et al.* (1963) found that sucrose, maltose, xylose

Table 14.1 Pullulan yield (%) with starch syrups of different dextrose equivalent (DE).[a]

Starch syrup DE	25	30	40	50	60	70
Acid converted	45	53	68	76	75	58
Enzyme converted	47	58	65	–	–	–

[a] Yield of pullulan for different starch syrups determined by the cultivation of *Aureobasidium pullulans* in a medium containing 10% starch syrup as a carbon source for 7 days at 30°C.

and arabinose are the most suitable carbon sources. Bender *et al.* (1959) reported that glucose and sucrose are preferred carbon sources. Nowadays, starch hydrolysates, such as cornstarch syrup, are used as the carbon source for the industrial production of pullulan (Yuen, 1974; Sugimoto 1978; Tsujisaka and Mitsuhashi, 1993). No differences are observed in the pullulan yield from starch syrups produced by either acidic or enzymic conversion. The yield is dependent on the dextrose equivalent (DE) of the syrups with the highest yield obtained using a starch syrup with a DE in the range of 40–50 (see Table 14.1).

14.3 PRODUCTION

14.3.1 Production method

Microbial cultivation is conducted in a medium containing approximately 15% cornstarch syrup with a DE of about 50, nitrogen sources and various phosphates, sodium, magnesium and iron salts. The cultivation of *A. pullulans* is done with agitation and aeration at 30°C. When starch syrup is used as the carbon source, pullulan production reaches a maximum after about 100 h with a yield greater than 70%.

To purify pullulan, microbial cells are first removed by filtration through a precoated filter. After decolouration with activated carbon, the filtrate is subjected to an ion-exchange process to remove the contaminating salts and proteins and to obtain a purified solution. After concentration, the solution is drum dried and pulverised in order to produce a fine powder.

14.3.2 Product specification

Dry pullulan powders are white and readily dissolve in hot or cold water. In an atmosphere with a relative humidity of less than 70%, pullulan powder has an equilibrium moisture content of 10–15% with no hygroscopicity or tackiness. The specification of 'pullulan' manufactured by the Hayashibara Company describes a white powder that is tasteless and odourless, with a loss on drying of not more than 6.0%, and giving a viscosity of 100–180 mPa s for a 10% w/w solution at 30°C.

14.4 FUNCTIONAL PROPERTIES

14.4.1 Physicochemical properties

14.4.1.1 *Solubility*

Pullulan easily dissolves in cold or hot water to form a stable, viscous solution that does not gel. It is insoluble in organic solvents, with the exception of dimethylformamide and dimethylsulfoxide. Pullulan can be modified with ethers or esters so that, depending on the

Table 14.2 Viscosities of various polysaccharide solutions.

Polysaccharide	Viscosity (mPa s)[a]
Pullulan (PI-20)	2
Gum arabic	1–5
Methyl cellulose	200
Tamarind gum	100–200
Sodium alginate	200–700
Guar gum	2000–3000
Locust bean gum	2000–3000
Xanthan gum	2000–3000

[a] 1% aqueous solution at 30°C.

degree of substitution, it may be insoluble in water and soluble in other organic solvents such as acetone or chloroform.

14.4.1.2 *Stability*

A pullulan solution is stable over a wide range of pH and is also relatively stable to heat.

Pullulan powder is degraded by heating above 250°C and carbonises at 280°C similar to other polysaccharides, but no toxic gas or exothermic heat is generated, unlike some chemical polymers.

Pullulan is a biodegradable polymer. It is easily metabolised by many microorganisms found in nature to give carbon dioxide and water.

14.4.1.3 *Viscosity*

Pullulan solution has a relatively low viscosity, like gum arabic, when compared with other polysaccharides, as shown in Table 14.2.

A pullulan solution is a Newtonian fluid with a surface tension approximating that of water, 74 dyne/cm^2. The solution viscosity is essentially unaffected by pH over a wide range from pH 2 to 11 and is stable in the presence of most metal ions. The viscosity shows a sharp increase in the presence of certain ions, such as borate, that are capable of complexing with hydroxyl groups but, even under such conditions, pullulan does not gel (see Table 14.3).

Table 14.3 Effect of 100 ppm metal ions on the viscosity of a pullulan solution.[a]

Metal ion	Viscosity (mPa s)
None	51
Fe^{2+}	52
Cu^{2+}	55
Ca^{2+}	60
Mn^{2+}	55
Ti^{4+}	400
B^{3+}	320

[a] Solution concentration: 20% w/w, pullulan molecular weight: 7×10^4.

A significant feature of pullulan is its high stability to sodium chloride. Heating pullulan in 30% sodium chloride solution at 100°C for 6 h produces no noticeable change in viscosity.

14.4.1.4 *Adhesiveness*

Pullulan is highly adhesive when dissolved in water and it has remarkable binding properties. This function is used for binding and agglomerating powders. The solution imparts a high strength to paper or wood, and it adheres to inorganic substances such as glass, metal and concrete when applied and dried. Pullulan shows superior adhesiveness on wood compared to cornstarch or modified cellulose (Tsujisaka and Mitsuhashi, 1993).

14.4.1.5 *Film-forming characteristics*

Pullulan readily forms films and can be used as a coating for foods or processed into a strong transparent, edible film. The film is impermeable and anti-static, and therefore retains flavours and resists penetration by oils. Pullulan film protects foods from oxidation because of its low oxygen permeability (Yuen, 1974; Sugimoto, 1978; Nakamura, 1985).

14.4.1.6 *Mouldability*

Pullulan powder is directly compressible with heat and pressure in the presence of moisture. Using this property, edible shaped articles and filaments can be produced.

14.4.2 Physiological properties

Pullulan is a type of dietary fibre. Okada *et al.* (1990) reported that pullulan is partially hydrolysed by salivary amylase, pancreatic amylase and enzymes in the small intestine, and that most of the residue is fermented by the microorganisms in the large intestine to short-chain fatty acids. A large proportion of dietary pullulan reaches the large intestine and functions as a prebiotic, selectively promoting the growth of beneficial bifidobacteria (Yoneyama *et al.*, 1990). Wolf *et al.* (2003) reported that a food grade pullulan is slowly digested in humans, and that half of administered pullulan is absorbed as blood sugar. Pullulan does not affect calcium absorption or bone mineral content in young rats (Kang *et al.*, 2003).

14.5 FOOD APPLICATIONS

The amount of pullulan used in food products is about 30% of the total supply (Fig. 14.2).

The use levels and functional properties of pullulan in foods are shown in Tables 14.4 and 14.5, respectively.

14.5.1 Thickener

Pullulan is used as a thickener in products such as beverages, creams, icings, frostings, soy sauce and other sauces. It also stabilises emulsions. This property is useful in creating foods, especially dressings and seasonings, with a smooth and viscous texture. Furthermore, the stability of pullulan to sodium chloride and pH are utilised to impart viscosity and gloss to foods with a high salt content such as soy sauce, barbecue sauces, and pickled or preserved

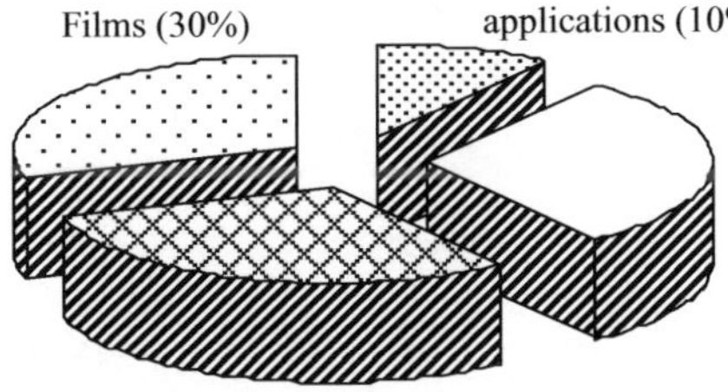

Fig. 14.2 Pullulan sales volume by application.

Table 14.4 Use levels of pullulan in various food products.

Food category	Examples	Use level
Confectionery	Candies, gummy jelly, chocolate, sugar coated snacks	2.0–3.0%
Japanese confectionery	Rice sweets, bean sweets	0.4–5.0%
Western confectionery	Tablet candies, candies, chewing gums	0.2–5.0%
Desserts	Ice cream, syrups for crushed ice	0.4–0.6%
Processed marine products	Kelp products, dainties, seaweeds, dried seafood, boiled fish paste	0.3–3.0%
Processed meat products	Ham and sausages, grilled chicken products, egg products	0.2–0.4%
Extruded and formed foods	Noodles, soy bean cake (tofu), rice cakes	0.4–3.0%
Sauces and dressings	Gravies, sauces, soy sauce, soups, miso, mayonnaise, dressings, pickles	0.3–5.0%
Instant beverages	Instant tea, coffee and soups	0.3–0.4%

Table 14.5 Functional properties of pullulan in foods.

Function	Typical applications
Thickener	Beverages, soups, seasonings, dressings, sauces, pickles, bakery mixes (e.g. cookies, doughnuts), fillings, frostings and icings, confectionery
Moisture binder	Processed and frozen meats, egg products, bakery products
Binding agent, texture improver	Ham and sausage, meat and fish snacks, granulated powders, precooked rice and noodles, tofu, baked ground fish and meat
Glazing agent	Frozen fish and shellfish, processed seaweed, snacks
Coating agent	Confectionery coatings, tablet coating
Film former, oxygen barrier	Fruits, egg products, instant chow mein, fish, meat, nuts, snacks
Tablet binder	Vitamin tablets, food supplements, coated confectionery
Dietary fibre, low calorie filler	Nutritional foods, artificial rice, noodles, confectionery, bakery products

vegetables and fruits. Pullulan may be used on its own or may be combined with other thickeners or gelling agents. The stringiness of pullulan may be a disadvantage for some applications, but this can be prevented by adding a small amount of another polysaccharide such as carrageenan or xanthan gum.

14.5.2 Moisture binder

The moisture binding properties of pullulan give an improvement of production yield in some processed food products.

Pullulan has the function of repressing water loss or drip from frozen foods. When a frozen omelette is thawed, water may be lost which significantly decreases its commercial value. The addition of 0.3–0.5% pullulan to an omelette during preparation prevents water or drip loss when the frozen product is thawed, and it maintains the 'freshly cooked' texture.

Pullulan acts as a humectant and binder and it improves the physical properties of bakery foods by retaining moisture and preventing the retrogradation of starch. The addition of pullulan to the ingredients of sponge cake helps it to retain moisture, shape and appearance.

14.5.3 Binding agent

The strong adhesive properties of pullulan are used to bind food pastes. For example, an aqueous pullulan solution (1.0–5.0%) can be used to strongly fix sesame seeds or a sheet of laver (seaweed) to the surface of a rice cracker. Pullulan is also used to prepare novel snack foods such as sticks or sheets made from fish, beef or pork. Pullulan is quite adhesive and, when sprayed as a solution, acts as an effective food binder, especially for dried foods. A solution of 3–8% gum is easily sprayed because of its relatively low viscosity, resulting in easy granulation for materials that are difficult to granulate. Powders that have been granulated with pullulan are easily dissolved in water because of their high solubility.

14.5.4 Glazing agent

The low viscosity and adhesive nature of pullulan make it a useful agent for forming a satisfactory ice membrane or glaze on frozen foods, such as fish and shellfish. For example, shellfish such as crab, shrimp and cuttlefish may be dipped in an aqueous pullulan solution (0.5–1.0%) and frozen. The glaze prevents discolouration and cracking during freezing and storage and gives valuable improvements in yield.

14.5.5 Coating agent

Pullulan can enhance the strength of adhesion and interaction between the sugar coating and tablet, and reduce surface cracking. It has a low tendency to browning in heated pan coaters.

14.5.6 Film former

Pullulan film is oil resistant, anti-static, transparent, readily soluble in water and of low oxygen permeability. It is, therefore, a novel and versatile packaging material. For example, one serving of coffee, soup, curry, soy sauce powder, freeze-dried vegetables or meat can be conveniently wrapped or packaged in pullulan film to preserve or retain the flavour or appearance for a prolonged period and then cooked without opening the wrapping or package.

An edible coating can be applied by immersing the food in an aqueous pullulan solution, followed by drying. Coating tea bags with pullulan helps prevent oxidation of the contents and retains the flavour. Similarly, peanuts, cashew nuts, jellies, dried fish, fresh vegetables and fruits and eggs can be coated to prevent oxidation and retain their freshness over a long period. A coating of pullulan provides a gloss to certain foods and increases their commercial value.

Recently, the demand for pullulan for oral care film and hard capsules has increased because of its film-forming characteristics, transparency, solubility and non-gummy mouth-feel.

14.5.7 Tablet binder

Pullulan has superior adhesive properties and tablets can be easily prepared from pullulan powder by direct compression. Strongly bound tablets can be prepared by blending a small amount of pullulan and the active ingredient. A pullulan coating gives a smooth and hard surface and tablets retain their integrity under harsh handling conditions.

14.5.8 Dietary fibre

Since pullulan is largely undegraded by digestive enzymes, it may be used as a low-calorie food ingredient. For example, low-calorie artificial rice or noodles can be prepared by replacing a portion of the wheat flour or starch with pullulan. Also, in bakery products, the appropriate replacement of flour with pullulan provides low-calorie doughnuts, cookies and biscuits resembling conventional products. Pullulan functions as a prebiotic by improving the environment condition of the human intestinal tract, for its selective utilisation by bifidobacterium.

14.6 FUTURE DEVELOPMENTS

Pullulan is used for many applications in foods, pharmaceuticals and cosmetics, primarily due to its adhesive properties, film-formation characteristics and edibility. It can be processed into films, sheets and shaped goods and has been called an 'edible plastic'. Hard capsules using pullulan have some advantages, such as being easy to swallow, dissolving rapidly and being impermeable to oxygen. The initial focus of pullulan in dietary supplements and health food applications will be expanded to include pharmaceutical products. Pullulan is also expected to be used for coating, tableting and granulating foods and pharmaceuticals.

Recently, consumers have shown a preference for natural foods and plant-derived materials. Thus, demand for pullulan, a material of plant origin, is expected to accelerate.

References

Bauer, R. (1938) Physiology of *Dematium pullulans* de Bary. *Zentralblatt fur Bakteriologie, Parasitenkunde, Infektionskrankheiten und Hygiene Abt II*, **98**, 133–167.

Bender, H., Lehmann, J. and Wallenfels, K. (1959) Pullulan, an extracellular glucan from *Pullularia pullulans*. *Biochimica et Biophysica Acta*, **36**, 309–316.

Bender, H. and Wallenfels, K. (1961) Investigations on pullulan II. Specific degradation by means of a bacterial enzyme. *Biochemistry Zeitschrift*, **334**, 79–95.

Catley, B.J. and Whelan, W.J. (1971) Observations on the structure of pullulan. *Archives of Biochemistry and Biophysics*, **143**, 138–142.

Kang, J., Sakai, Y., Yoshida, Y. and Yamamoto, Y. (2003) Effects of dietary pectin and pullulan on calcium absorption and bone mineral contents in rats. *Seikatsu Eisei*, **47**, 95–100.

Nakamura, S. (1985) Production and applications of pullulan film. *Abura kagaku*, **34**, 865–871.

Okada, K., Yoneyama, M., Mandai, T., Aga, H., Sakai, S. and Ichikawa, T. (1990) Digestion and fermentation of pullulan. *Nihon Eiyo Syokuryo Gakkaisi*, **43**, 23–29.

Sugimoto, K. (1978) Production and applications of pullulan. *Fermentation and Industry*, **36**, 98–108.

Tsujisaka, Y. and Mitsuhashi, M. (1993) Pullulan. In: R.L. Whistler and J.N. BeMiller (eds), *Industrial Gums. Polysaccharides and Their Derivatives*, 3rd edn. Academic Press, Inc., San Diego, pp. 447–460.

Ueda, S., Fujita, K., Komatsu, K. and Nakashima, Z. (1963) Polysaccharide produced by the genus *Pullularia* I. Production of polysaccharide by growing cells. *Applied Microbiology*, **11**, 211–215.

Wallenfels, K., Bender, H., Keilich, G. and Bechtler, G. (1961) On pullulan, the glucan of the slime coat of *Pullularia* pullulans. *Angewandte Chemie*, **73**, 245–246.

Wallenfels, K., Keilich, G., Bechtler, G. and Freudenberger, D. (1965) Investigations on pullulan. IV. Resolution of structural problems using physical, chemical and enzymatic methods. *Biochemische Zeitschrift*, **341**, 433–450.

Wolf, B.W., Garleb, K.A., Choe, Y.S., Humphrey, P.M. and Maki, K.C. (2003) Pullulan is a slowly digested carbohydrate in humans. *Journal of Nutrition* **133**, 1051–1055.

Yoneyama, M., Okada, K., Mandai, T., Aga, H., Sakai, S. and Ichikawa, T. (1990) Effect of pullulan intake in humans. *Denpun Kagaku*, **37**, 123–127.

Yuen, S. (1974) Pullulan and its applications. *Process Biochemistry*, **9**, 7–9.

15 Seed Gums

Willem Wielinga

ABSTRACT

The origin of seed gums, guar gum, carob or locust bean gum and tara gum, is described. The gums are produced by removing the outer coating of the seed and grinding the endosperm. The composition and structure of the galactomannans are described and linked to their functional properties. All three seed gums are very efficient thickening agents in water. In addition, guar and tara gums interact synergistically with xanthan to increase viscosity. Carob gum forms cohesive, elastic gels with xanthan gum and it increases the strength and elasticity of kappa carrageenan and agar gels. The main food applications are for thickening convenience foods, dairy products, including frozen products such as ice cream, soft drinks and fruit juices, bread and pastry, fruit preserves, baby food, instant products including puddings, flans and pudding powder, and for dietary fibre in baked goods and pet foods.

15.1 INTRODUCTION

Non-ionic galactomannans act as reserve carbohydrates and are found as storage polysaccharides in the cell walls of various seeds. The endosperm of these seeds develops alongside the germ and completely envelopes it. The endosperm itself is protected by a seed coat. The endosperm contains very little cellulose and no lignin. Gums derived from the seeds of the carob tree (*Ceratonia siliqua*), the guar plant (*Cyamopsis tetragonoloba*) and, to a lesser extent, from the tara shrub or tree (*Caesalpinia spinosa*) are widely used in the food industry. The average annual consumption in food and pet food applications is around 9000–10 000 tonnes of carob gum, also known as locust bean gum, about 55 000 tonnes of guar gum and 2000 tonnes of tara gum.

15.2 RAW MATERIALS

15.2.1 Carob gum

Evergreen carob trees grow around the Mediterranean Sea (Wielinga, 1993), although one species is also grown in the USA. There are nine main species. The largest suppliers of pods are located in Spain, Italy, Portugal and Greece, providing 70–75% of the world's total production. Other suppliers of carob seeds are in Morocco, Cyprus, Turkey, Algeria and some other locations.

Table 15.1 Composition of carob powder.

Components	% By weight
Carbohydrates (low MW)	41–52
Tannins	4–13
Polysaccharides (pectins)	7–9
Protein	3–6
Minerals	2.5–3.0
Fat	0.5–0.8

It takes 8–10 years before the trees bear fruit called pods or St. John's bread. The pod length varies from 10 to 20 cm and the carob seeds, or carob kernels, amount to 8–15% by weight of the pod. The yield of pods amounts to about 2700–2800 kg/ha. Under normal meteorological conditions, about 280 000–300 000 tonnes of pods are collected manually each year.

The pods are broken using a kibbling process, during which the seeds are set free. The seeds are recovered from the kibbled pods by a special screening operation.

The kibbled de-seeded pods are known as 'pulp' and finely ground pulp, also called carob powder, has the composition shown in Table 15.1.

The low-molecular-weight carbohydrates in carob powder consist of 65–75% of sucrose, 15–25% of glucose and fructose and 10–20% of other sugars (Baumgartner, 1986). The tannins in the pulp have molecular weights of 5000–32 000 daltons and the fat content of the pods consists of approximately 50% of saturated fatty acids and 50% of unsaturated fatty acids. The typical mineral content of 100 g of pulp contains 100 mg potassium, 307 mg calcium, 42 mg magnesium, 13 mg sodium, 0.23 mg copper, 1.04 mg iron, 0.40 mg manganese and 0.59 mg zinc.

Finely ground pulp, or carob powder, is used as animal feed for all kinds of livestock, although the tannins have a negative influence on the digestion of proteins, thus lowering the feed value of the pulp.

Carob powder is also used for human consumption and has been employed for many years as a remedy against diarrhoea in children. Roasted pulp can be used as a replacement for cocoa and as a flavour in different foodstuffs. The pulp is also used as a raw material for the production of alcohol, yielding 22–24 L of alcohol from 100 kg of pulp. A small quantity of carob syrup is produced by adding water to the pulp and concentrating the extract to give 'carob honey'.

Official programs have been and are under way to emphasise the ecological value of the carob tree for soil and water conservation, and for providing shade for grazing animals, in addition to the nutritional value of its fruits.

15.2.2 Guar gum

The guar plant (*C. tetragonoloba* (L.)), or cluster bean, is a drought-resistant annual legume that thrives in a hot, dry climate and grows in semi-arid regions. For centuries, it has been cultivated in India. Here the farmers supply about 80% of the total amount of guar produced in the world and 70% of this production is grown in Rajasthan.

Guar pods have a length of 3.5–12 cm, containing 5–12 guar seeds, which amount to about 60% of the gross weight of the pod (Justus Leibig University, 2000). In India, guar pods are cooked as a vegetable for human consumption and are used as cattle feed. The guar plant is also cultivated in Pakistan for food and agricultural purposes.

Since 1944, the guar plant has been grown in the USA in west Texas, south-west Oklahoma, Arizona and New Mexico where irrigation has been beneficial for its success. In recent years, cultivation of the guar plant has been tried in the southern hemisphere in Brazil, Argentina and Columbia, Africa and Australia, so that two crops a year could be harvested. It remains to be seen whether these agronomy projects outside of the Indian subcontinent will be successful in the long run.

The annual guar crop fluctuates due to weather conditions but, on average, about 500 000 tonnes of guar seeds are available.

15.2.3 Tara gum

The tara tree or shrub, *C. spinosa* (L.), is native to the Cordillera region of Bolivia, Peru and northern Chile and also grows in Ecuador, Colombia, Venezuela and Cuba. This tree has been introduced into Morocco as well as East Africa.

Tara trees produce flat orange-coloured pods, which are 10 cm long and 2.5 cm wide. These contain 4–7 large round seeds or kernels, amounting to about 28% of the weight of the pod. The seeds are black when mature and similar in appearance to carob seeds.

The main interest in the fruits of the tara tree is the presence of a very useful tannic acid composition in the pod, which can be employed as a natural tanning agent in the leather industry. The tannins can be extracted from the de-seeded pods by hot water. The extract is then concentrated, dried, ground and sold as tara powder, containing 55–62% of tannins (Markman, 2006; Pebani, 2008).

Tara seeds, therefore, are considered a by-product. The ground endosperm of these seeds is the raw material for the production of tara gum, recently approved by the EU as additive E417 for use as a thickener, stabiliser and gelling agent in food (Del Re-Jiménez and Amado, 1989; Polygal, 2008). It is estimated that 1500–2000 tonnes of tara gum are available each year, mainly for use in the food industry.

Carob seeds, or carob kernels, and guar seeds are shown in Fig. 15.1. The tara seed looks similar to the carob kernel.

15.2.4 Composition of galactomannan seeds

The composition of galactomannan seeds is given in Table 15.2.

The composition of tara seed is approximate and, based on the seed weight, the yield of high-grade tara gum is only 21–22%. However, the EU purity criteria for tara gum (E417) specifies a maximum limit for protein content of 3.5% (N% × 5.7). This is significantly lower than the permitted maximum limit of 7% protein for carob gum/locust bean gum and a 10% limit permitted for guar gum (both N% × 6.25). Thus, food-grade tara gum is clearly the purest seed gum. This stringent protein limit for tara gum automatically leads to a lower yield in production so that the actual botanical endosperm content of the tara seed might be as high as about 28% instead of 22%.

Guar seeds have an outer aleurone layer and certain peripheral cells can contain about 25% of protein. Inside this coating is the high-quality galactomannan-rich endosperm. This surrounds two small cotyledons and the embryonic root of the germ in the middle of the seed, within a very thin layer of protein-rich endosperm. This endospermic film, amounting to about 2.8–3.5% of the dry weight of the seed, can develop a viscosity in water of about

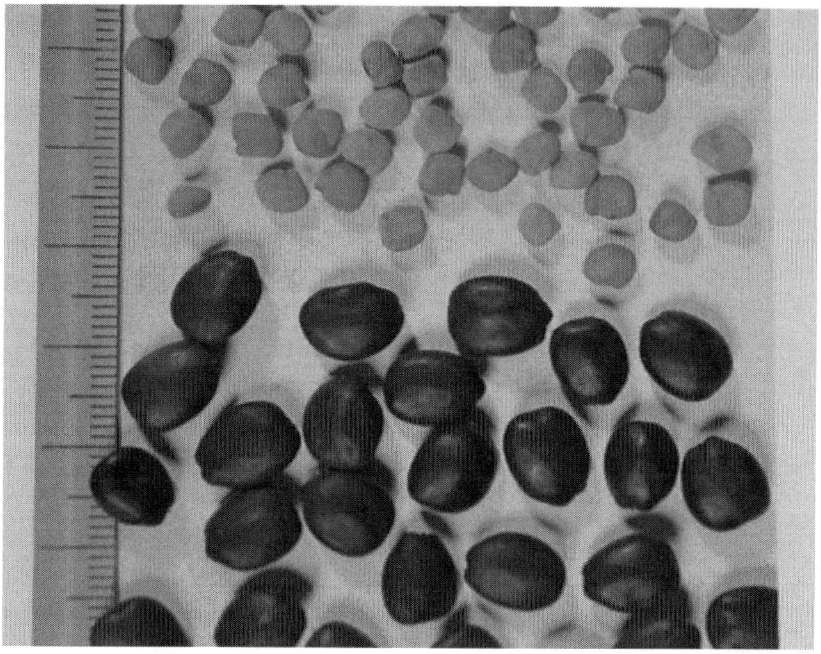

Fig. 15.1 Carob kernels and guar seeds.

1000 mPa s at 1% concentration. Further information on the structure and properties of guar seeds is presented in many reviews (Hui and Neukom, 1964; McCleary *et al.*, 1984; Wielinga, 1984, 1990; Clark *et al.*, 1986).

During germination of the seeds, the endosperm absorbs up to 75% of water, based on its own dry weight. The depletion of galactomannans occurs rapidly during germination of guar seeds and carob kernels, as illustrated in Fig. 15.2. It is clear that to obtain the maximum yield of gum, seeds must be stored to avoid germination (McCleary and Matheson, 1974).

The hard, compact endosperm halves of all three seeds contain more than 88% galactomannan on a dry basis. Endosperm halves are obtained by removing most of the hull mechanically, physically or chemically. The fine, fractured hull fragments, as well as fine endosperm parts and the scattered more-friable germ of the seeds, can be separated by screening. Solutions prepared with pure high-grade guar endosperm can achieve viscosities of 5000–11 000 mPa s at 1% concentration (Brookfield RVT viscometer at 20 rpm and 25°C), after an adjustment for 10% moisture content in the gum.

The histological structure of the guar seed endosperm (Fig. 15.3) shows that the compact aleurone cell layers have a total thickness of about 0.1 mm, showing the highest protein content underneath the hull. The larger inner endosperm cells have significantly less protein of about 1.5–1.6%. The very first cell layer of the aleurone cells consists of cup-like cells in

Table 15.2 The composition of galactomannan seeds by weight.

Component	Carob seed	Guar seed	Tara seed
Hull	30–33%	20–22%	~38%
Endosperm	42–46%	32–36%	~22%
Germ	23–25%	44–46%	~40%

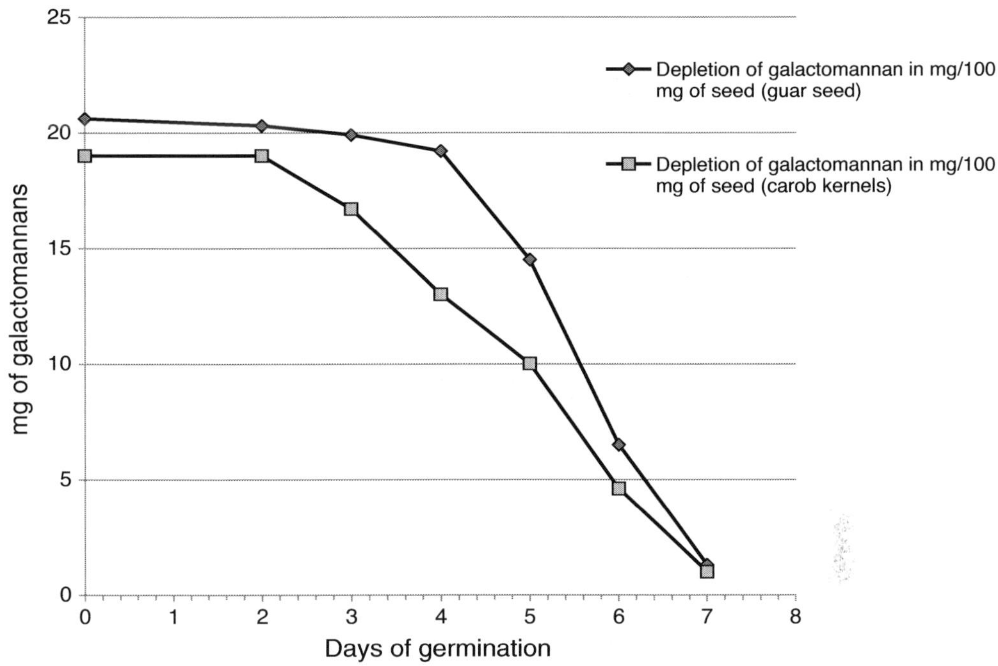

Fig. 15.2 Depletion of galactomannans during germination of guar seed and carob kernels.

which protein bodies are stored. These have a height of about 0.015 mm, with an opening towards the hull. Underneath this first layer, there are two more cell layers with almost the same width as the first one, but longer, each of 0.025–0.030 mm.

The aleurone cells on the convex periphery of the endosperm of *C. tetragonoloba* seeds, called sheath cells in Fig. 15.3, are the only living cell layers which can synthesise enzymes, such as β-mannosidase, EC 3.2.1.25, β-mannanase, EC 3.2.1.78 and α-galactosidase, EC 3.2.1.22. (Mo and Bewley, 2003; Dhugga *et al.*, 2004). The peripheral cells of the aleurone layers are the main cause for the turbidity of aqueous solutions of untreated guar products. These compact cells swell only to a limited extent and are not very susceptible to enzymatic attacks which can take place in the endosperm. These cells either stay as swollen clusters of cells in solution, if sufficient shear and the required temperature have not been applied, or as smaller particles, if sheared. These smaller and larger cell clusters, together with other insoluble matter, can be separated from solution by strong centrifugal forces of 35 000 g or higher.

The histological structures of the endosperms of carob and tara seeds also consist of living cells which can synthesise the enzymes mentioned previously. When examined under the microscope, the convex periphery of endosperm halves of both seeds seems to have a higher concentration of proteins than the rest of the endosperm.

In the future, another seed gum might be available for food applications: cassia gum obtained from the seeds of *Cassia tora/obtusifolia*. The endosperm also contains more than 75% galactomannan with a galactose to mannose (G/M) ratio similar to carob gum, although the fine structure may be different. The purity of the gum, with 5–10% of protein and 7–13% of acid insoluble residue, seems to be lower than that of the other three seed gums. The viscosity of a 5% solution, prepared hot, is 6000–8000 mPa s at 25°C. It can be used in pet food but it has not been approved for use in food in the EU and has no E-number at present.

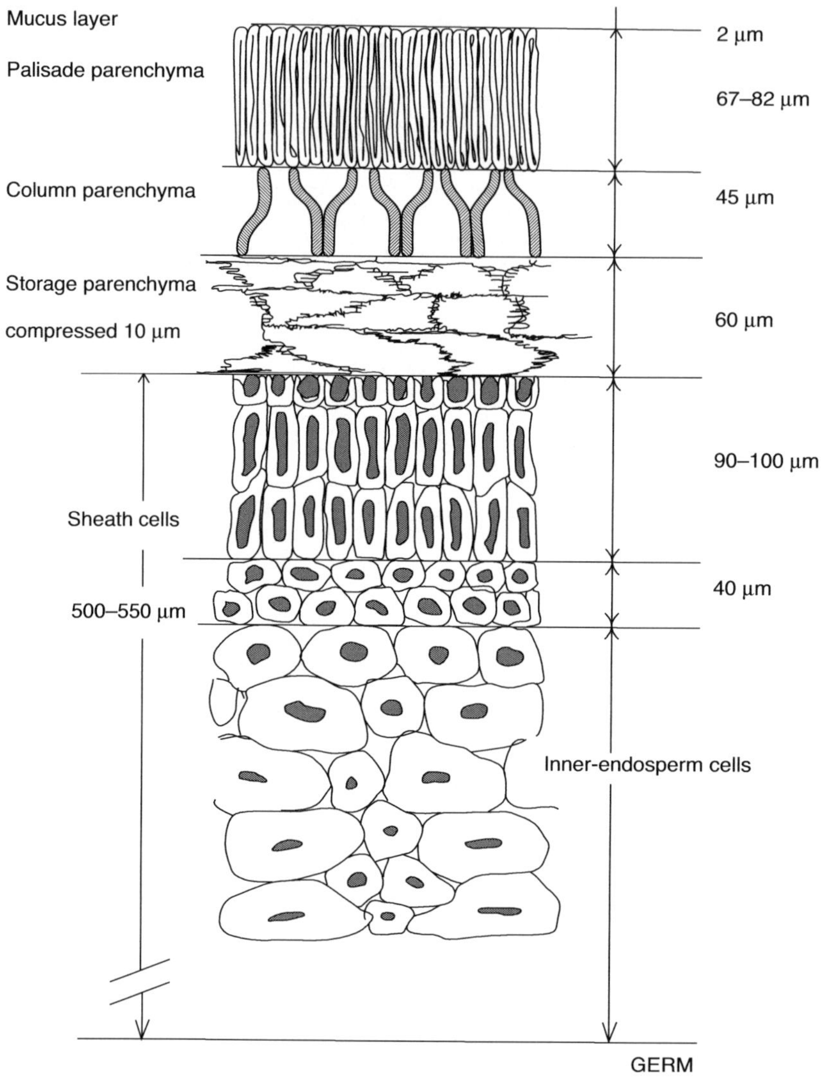

Fig. 15.3 Histological structure of guar endosperm/hull at maximum thickness.

15.3 PRODUCTION

15.3.1 Tara gum and carob/locust bean gum

The hulls of the carob and tara seeds are either carbonised by hot sulphuric acid, followed by intensive and efficient washing with water, or largely removed by roasting in a rotating furnace at temperatures of about 550°C, after which residual hull fragments are rubbed off mechanically. Then the treated remnants of the seeds are split and sifted. The recovered endosperm halves can then be dried to the required degree of hydration and ground into a fine, off-white powder and classified to meet particle size requirements (FAO, 1995; FSANZ, 2006; Unipektin, 2008).

15.3.2 Guar gum

The endosperm halves of guar seeds, also called guar splits, are obtained by feeding whole seeds into an attrition mill, or into any other type of mill having two grinding surfaces travelling at different speeds. The seed is split into endosperm halves covered with hull and finer material, derived mostly from the more fragile germ, which can be sifted off later.

The periphery of the crude crack (endosperm plus hull) must be heated at temperatures in excess of 250°C to dry and preferentially soften the hull. After this, the heated splits are fed into another mill to abrade the partially loosened hull from the endosperm, or into a hammer mill, where the partially loosened hull is shattered and removed. Any remaining germ particles are also pulverised during this step, and after a further sifting, the coarse fraction essentially represents pure endosperm halves. However, a certain number of splits still have residual hull fragments, depending on the number of treatments. Thus, the resulting splits are referred to as single-, double- or triple-purified splits.

On average, a triple purification decreases the largest split size from about 3.4 mm to about 2.7 mm but, of course, the quality of the guar split improves significantly. The gum content gradually increases from single- to triple-purified splits, whereas the protein content and acid insoluble residue (AIR) decrease.

The finer fraction recovered from this cracking/screening process is called guar meal, which contains mainly the pulverised germ, the ground hull and the abraded fine endosperm parts. Guar meal has a minimum protein content of 35% (N% × 6.25) and is marketed as animal feed. If this guar meal is fed to chickens, most of its trypsin inhibitor must be inactivated by heat, which requires a lot of energy, but feeding ruminants with guar meal apparently does not require this costly heat treatment.

To obtain powdered guar gum, the guar splits are subjected to hydration/washing, possibly followed by flaking, and/or drying/milling and screening techniques. For grinding, different types of mills, such as roller mills, hammer mills and pin mills, can be used.

To obtain a fine powder with fast-swelling and high viscosity properties, flash grinding of the hydrated guar splits, for instance in a hammer mill, is recommended. During this process, the swollen splits are dried with a large amount of preheated air, to reduce final moisture content from 45–74% to a final moisture content of 10% (Fig. 15.4) and almost simultaneously reduced to a final particle size by forcing the wet material through a screen with defined small openings (Sattler, 1979; Bepex, 2008; Hosokawa, 2008).

Flaking of wet guar splits has become popular in the last 15–20 years. In this process, the hydrated galactomannan molecules of the flattened splits appear to be compacted, thus facilitating the subsequent grinding operation. The flaked and ground material shows a slightly higher specific weight than a simple flash-ground guar product. The flaked guar gum is, therefore, probably more able to absorb water during solution preparation. An improved rate of hydration is observed in various applications, although the aqueous solutions of these flaked guar gum products are slightly less transparent.

15.4 COMPOSITION

The analysis of these three seeds is rather complicated and the results are not exact. The galactomannans from these three seeds are composed entirely of linear (1→4)-β-D-mannan chains with varying amounts of single terminal D-galactose units linked to the main backbone by (1→6)-α-glycosidic bonds to the 4,6-mannose units. The third residue in the backbone

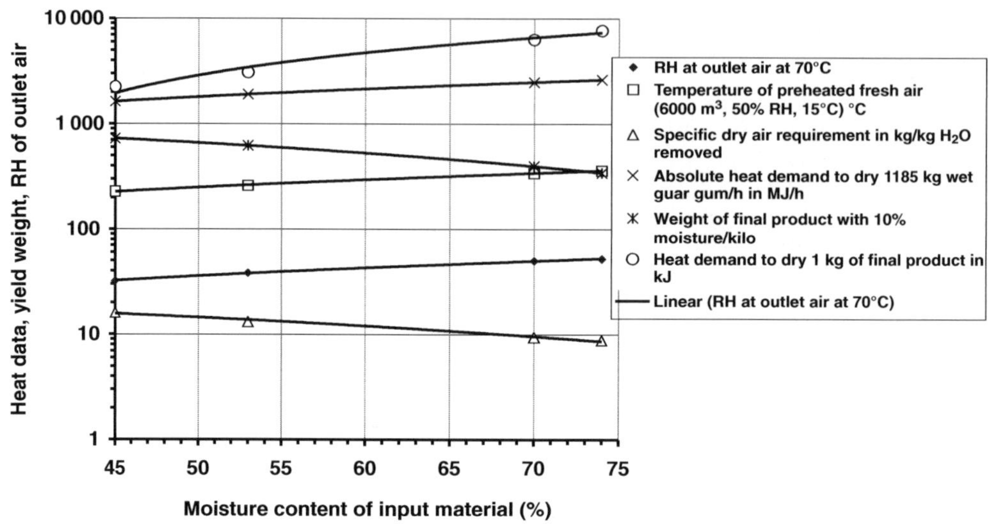

Fig. 15.4 Heat demand per hour to flash grind guar products with different moisture contents to 10% or per kg of final product.

is 4-mannose. These galactomannans can be easily distinguished from each other by their overall mannose–galactose ratios which are between 1.6 to 1 and about 3.5 to 1. The large amount of galactose side stubs, about 20–40% of the galactomannan weight, prevents strong cohesion between the polymer backbones, so that no extensive crystalline regions can be formed. Thus, water at room temperature and above can easily penetrate between the single molecules to hydrate the accessible gum.

The galactose content of guar gum is 33–40%. Therefore, the amount of the 4,6-mannose units is 33–40% as well. The quantity of the 4-mannose units then varies from 20% to 34%. A simplified theoretical building block of a galactomannan of guar gum with 38.4% of D-galactose consisting of 26 hexose units is shown in Table 15.3. The building blocks and main repeating units reflecting the average composition of the galactomannans of tara gum and locust bean gum are shown for comparison.

Two sp^3 hybrid orbitals, each with two paired electrons, are located at all oxygen atoms of the hexose units. The O-atom has a tetrahedral structure. Numerous orbitals will be involved in intra- and inter-hydrogen bonds. Figure 15.5 illustrates the relationship between the viscosity of aqueous solutions of a high-grade carob bean gum and different shear rates at various concentrations. The hydrogen bonds play an important role in the flow behaviour of both carob and guar gum. Figure 15.6 shows the relationship between viscosity at different concentrations and shear rates for very high-viscosity guar gum. The graphs clearly demonstrate that the thickening power of the gum in water depends on the shear rate.

Table 15.3 The average building blocks in galactomannans.

Guar gum:	$(G)_{10}$ (4, 6-M)$_{10}$ (4-M)$_6$	G/M ratio = 1.0:1.6
Tara gum:	$(G)_1$ (4, 6-M)$_1$ (4-M)$_2$	G/M ratio = 1.0:3.0
Carob gum:	$(G)_2$ (4, 6-M)$_2$ (4-M)$_5$	G/M ratio = 1.0:3.5

G, galactose; M, mannose.

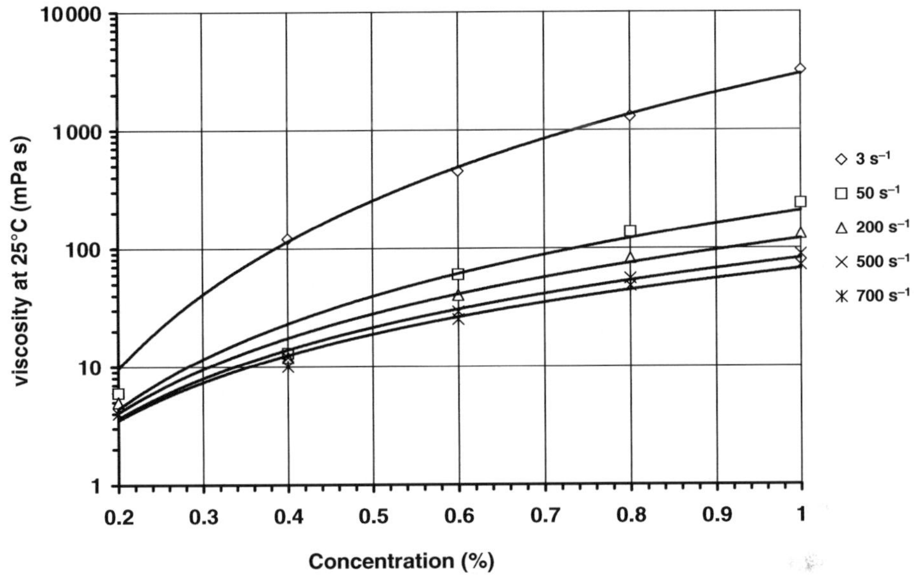

Fig. 15.5 Viscosity of carob bean gum solutions at different concentrations and shear rates.

A twofold increase in the concentration of guar gum in solution gives a viscosity 9.4 times higher at 4 s^{-1} but only 2.2 times higher at 10 000 s^{-1}.

The fine structure of these galactomannans can be quite irregular with respect to the distribution of the galactose units. As many as 5 unsubstituted mannose units in a row may occur in certain galactomannans of guar gum, and as many as 10 unsubstituted mannose units in carob bean gum. Substitution of the mannan chain by more than 12% of galactose makes the galactomannan water soluble.

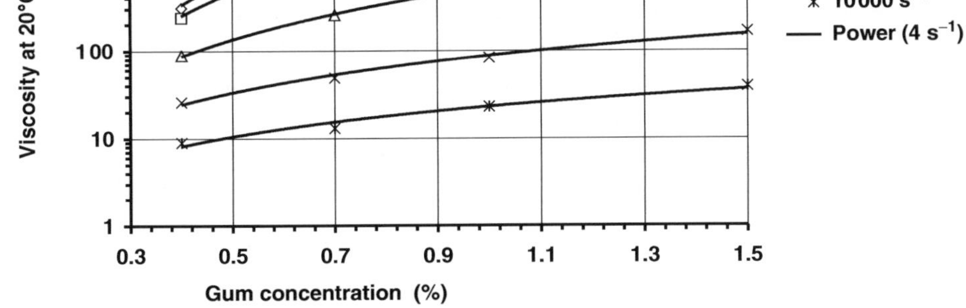

Fig. 15.6 Viscosity of high-viscosity guar gum at different concentrations and shear rates. (Guar gum intrinsic viscosity 14.6 dL/g; solutions hydrated at 86–89°C for 10 min, viscosity measured at 20°C.)

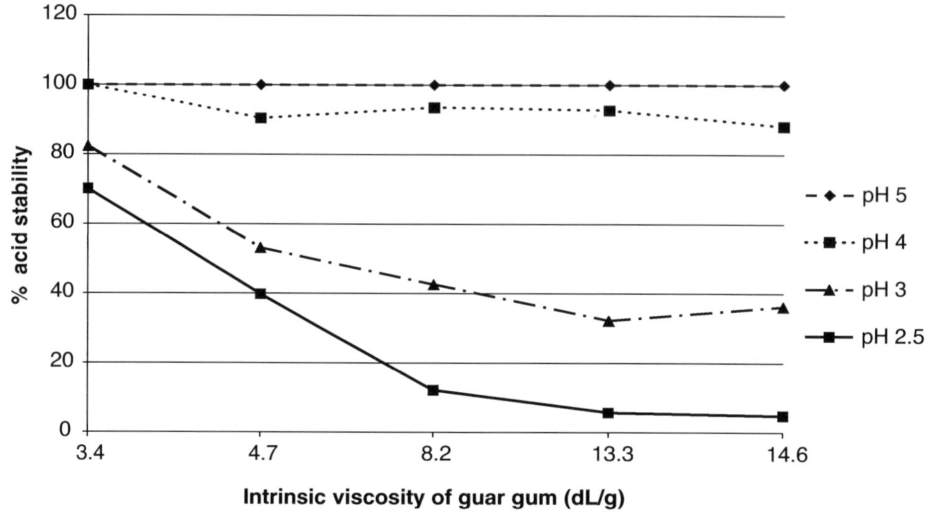

Fig. 15.7 Acid stability of solutions of guar gum of different viscosity expressed as per cent of viscosity at pH 5. (1% solutions stirred whilst hydrating at 86–89°C for 10 min at pH values of 5, 4, 3 and 2.5.)

Certain hot water-soluble fractions of carob gum, with a G/M ratio of 1 to 4.3–4.9, self-associate forming weak three-dimensional gel networks, especially after a freezing process (Lozinsky *et al.*, 2000). It is possible that nano-crystalline regions of 3–5 nm are formed during this process of cryogelation. Self-association does not occur when using guar gum processed under the same conditions and solutions of guar gum are freeze–thaw stable.

Galactomannans are susceptible to degradation by organic acids, such as citric, acetic and ascorbic acid, so that depolymerisation may occur to different extents when applying these polysaccharides, especially in heated acidic food systems with pH values below 3.5, as shown in Fig. 15.7.

The different galactomannans of the three seed gums show a wide distribution of molecular weights. Tara gum has not yet been assigned a specification for its molecular weight range for use in food, but this is expected to be similar to that of carob gum. For the EU and JECFA specifications, the molecular weights of the galactomannans in carob gum or locust bean gum have to be within the range of 50–3000 kDa and in guar gum within the range of 50–8000 kDa. This means that the degree of polymerisation (DP) of the galactomannans has to be in the range of 300–18 500 hexoses for carob bean gum and 300–49 400 hexoses for guar.

The average length of the shortest and longest galactomannan molecules, based on the length of 0.515 nm/mannose unit, varies from 120 to 7410 nm for carob bean gum and from 95 to 15 656 nm for guar gum.

If carob endosperm halves of different provenance, with known acid-insoluble residue, ash and protein content, are dissolved in hot water at 85°C under high shear, the hot water solubility of certain proteins can be determined. The insoluble material in these aqueous solutions can be separated by centrifuging at 35 000 g for 30 min and then dried at 105°C.

Table 15.4 Composition of carob endosperm of different origin.

Carob endosperm type	Intrinsic viscosity, 1% w/v, mPa s	Insoluble matter, by centrifu-gation	Acid insoluble residue	Ash	Protein total	Protein soluble	Protein insoluble	Gum content
		% dry wt						
A	3770	11.8	1.62	0.75	6.21	**1.23**	**4.97**	91.42
B	3520	11.8	1.82	0.77	6.83	**2.29**	**4.54**	90.54
C	3850	8.8	1.46	0.64	5.64	**2.16**	**3.48**	92.26
D	2630	13	1.89	0.75	6.79	**1.67**	**5.12**	90.57

The insoluble protein content indirectly determines the soluble protein and the results are summarised in Table 15.4.

The EU purity criteria for galactomannans published in 1998 (EU, 1998) are shown in Table 15.5.

Galactomannans are susceptible, of course, to strong acids and to alkali in presence of air, to strong oxidising agents and to specific enzymes (Goswami, 1999; Swartz *et al.*, 2000). These properties are utilised to depolymerise seed gums. Very strongly depolymerised guar products, sold as food ingredients under trade names such as Benefiber, Fiberon and Sunfiber, do not meet the specifications for E412 and have no E-number. These products have molecular weights of about 20 kDa.

Table 15.5 EU purity criteria for guar, carob and tara gums with JECFA microbiological limits.

Material	Carob gum	Guar gum	Tara gum
E number/INS number	**E410**	**E412**	**E417**
Loss on drying (105°C, 5 h), % max.	15	15	15
Ash content (800°C, 3–4 h), % max.	1.2	1.5	1.5
Proteins (N% × 6.25), % max.	7	10	3.5 (N% × 5.7)
Acid insoluble residue, % max.	4	7	2
Galactomannan (gum) content, % min.	75	75	
Starch	ND	ND	ND
Arsenic, ppm max.	3	3	3
Lead, ppm max.	5	5	5
Mercury, ppm max.	1	1	1
Cadmium, ppm max.	1	1	1
Heavy metals (expressed as Pb), ppm max.	20	20	20
Ethyl alcohol or iso-propyl alcohol, % max.	1		
Molecular weight, kDa	50–3 000	50–8 000	
Einecs number	232-541-5	232-536-0	254-409-9
	JECFA	JECFA	
Total Plate Count, cfu/g max.	5 000	5 000	
E. coli, cfu/g	Negative by test	Negative by test	
Salmonella, cfu/g	Negative by test	Negative by test	
Yeasts and moulds, cfu/g max.	500	500	

15.5 FUNCTIONAL PROPERTIES

The functional properties of the galactomannans are mainly due to their great ability to change the rheology of aqueous systems. All three seed gums are very efficient thickening agents in water. The gums are able to interact with xanthan gum to form gels, or to fortify three-dimensional stabilising structures. In particular, carob/locust bean and tara seed gums interact with agar–agar, carrageenan and Danish agar, a traditional term for furcellaran and now re-classified as carrageenan (Del Re-Jiménez and Amado, 1989). The thickening power of the galactomannans depends, of course, upon the size and frequency of interactions and of hyper-entanglements between these macromolecules and, therefore, on their apparent molecular weight.

These thickening and gelling agents are widely used in the food industry to make the products more appealing and attractive to the consumer by:

- improving shelf life by binding water,
- controlling the texture,
- influencing crystallisation,
- preventing creaming or settling,
- improving the freeze–thaw behaviour,
- preventing syneresis,
- preventing the retrogradation of starch products,
- maintaining turbidity in soft drinks and juices,
- acting as dietary fibre, and
- stabilising foam.

This means that these food additives find their main applications in convenience food, dairy products, including frozen products such as ice cream, soft drinks and fruit juices, bread and pastry, fruit preserves, baby food, instant products including puddings, flans and pudding powder, as dietary fibre, and in pet foods (Wielinga, 1977). These seed gums also find application in pharmaceutical and cosmetic products.

Current commercial guar products have intrinsic viscosities of 0.7–15.0 dL/g, which correspond to apparent viscosities of about 5–100 000 mPa s for aqueous solutions at 2% concentration measured with a Brookfield RVT viscometer, 20 rpm at 25°C. Carob bean gum shows an intrinsic viscosity of 11.8 dL/g and fine-mesh carob bean gum, with a faster swelling rate, one of 11.4 dL/g.

The three seed gums have no restrictions on acceptable daily intake (ADI) values and may be used *quantum satis*, the level necessary to achieve the required technological effect in food applications (EU, 1998). The rate of addition for all three seed gums in food products varies from below 0.1 to 2.0%, depending on the application. As clouding agents, the gums are used at less than 0.1%, and in moulded confectionery, up to 2.0% is used in gumdrops and jelly candy. The most common rate of addition lies between 0.2% and 0.5%.

All three seed gums thicken aqueous systems very efficiently, thereby controlling the mobilisation of water. This influences the consistency, body and shelf life of aqueous food systems and the stabilisation of oil-in-water and water-in-oil emulsions.

These seed gums, in combination with minor amounts of carrageenan and blended with the appropriate emulsifier, allow the production of ice cream with a shelf life of at least 18 months, under well-controlled storage conditions, without much change in quality. These

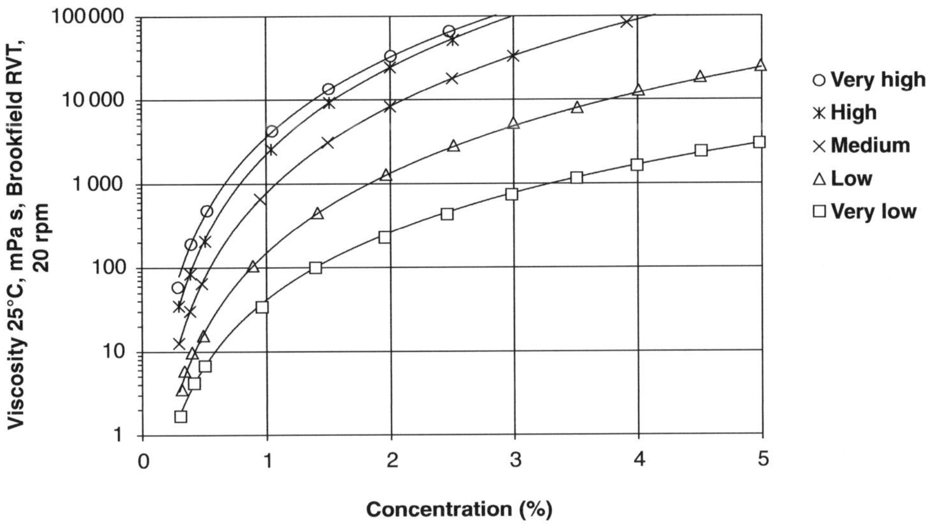

Fig. 15.8 Relationship between viscosity of solutions of guar gum of different degree of polymerisation at different concentrations, prepared at 25°C. (Solutions prepared and measured at 25°C with Brookfield RVT, 20 rpm.)

gums can control the shape retention and meltdown of frozen products and protect them against heat shock.

Hot water-soluble galactomannan fractions of carob gum, with a G/M ratio of 1 to 4.3–4.9, form a network through self-association and by synergistic interaction with carrageenan, especially during the freezing process (Lozinsky *et al.*, 2000). In addition, the three-dimensional molecular structure of carrageenan with kappa-casein is fortified by the network of galactomannans to improve the shelf life of the ice cream significantly.

The cold water-soluble galactomannans in carob bean gum may be separated in a special process so that about 70% of cold-swelling carob bean gum dissolves at 25°C, compared to about 31–33% at the same temperature for standard carob bean gum. This means that the cold-swelling type can stabilise sensitive proteins better at lower temperatures than regular carob bean gum.

Tara gum hydrates more readily with 80% soluble in water at 25°C.

Figure 15.8 demonstrates the Brookfield viscosity of aqueous solutions of guar products of various DP, after hydrating and measuring at 25°C. For comparison, Fig. 15.9 shows the viscosity of aqueous solutions of guar products at concentrations below 0.5% prepared at 86–89°C before cooling and measuring at 25°C. Both preparation methods give identical solution viscosities for the same type and level of guar gum showing that this polymer fully hydrates at room temperature. These graphs demonstrate the almost linear flow behaviour of dilute guar gum solutions (Fig. 15.9). At higher concentrations, chain entanglement results in changes in rheology and thixotropic flow is observed (Fig. 15.8). The intrinsic viscosity of very high-viscosity guar is 14.6 dL/g. High-, medium-, low- and very-low-viscosity guars show limiting viscosities of 13.3, 8.2, 4.7 and 3.4 dL/g, respectively.

Another series of trials compared the behaviour of these gums in water and in pasteurised milk. The gums were blended with 5% of sucrose to facilitate their dispersion into the liquids,

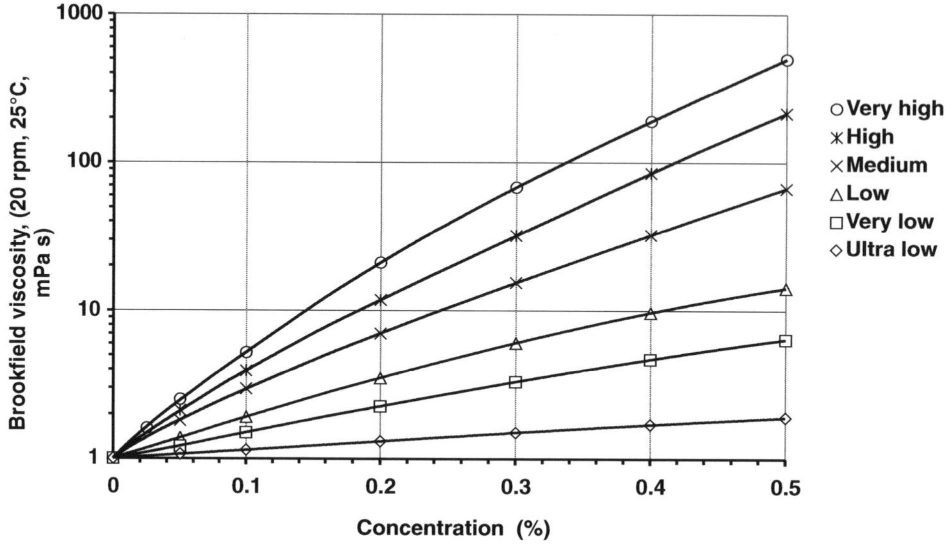

Fig. 15.9 Viscosity of aqueous solutions of guar gum of different degree of polymerisation at concentrations up to 0.5%. (Solutions stirred at 86–89°C for 10 min, cooled and measured at 25°C.)

which were stirred and heated to 86–89°C, held at this temperature for 10 min and then cooled to room temperature. Two concentrations for each product were investigated (Table 15.6). It was found that 0.5% of the two products with the highest viscosities caused wheying-off in milk after 24 h at 25°C. At a rate of 1%, the thickened milk solution showed viscosities between 5800 and 6800 mPa s after 24 h with no visible separation due to the high viscosity.

Seed gums are polydisperse and commercial products contain a range of different molecular weight polymers. Figure 15.10 illustrates the influence of the different molecular weight distributions, such as peak molecular weight (MW), weight average molecular weight (Mw) and number average molecular weight (Mn) of different cold-swelling carob bean gums and the 1% solution viscosity.

The functional properties of these seed gums, blended with carrageenan and/or other biopolymers, are extensively utilised in pet food and dairy products.

Table 15.6 Viscosity of guar gum in water and milk after 1 and 24 h.

Guar gum viscosity grade	Concentration (%)	Viscosity in water		Viscosity in milk		Stability in milk
		1 h	24 h	1 h	24 h	
Very low	1.0	22	22	121	241	nd
	2.0	138	144	835	900	nd
Low	1.0	70	72	417	683	nd
	2.0	1 080	1 200	4 590	4 685	nd
Medium	1.0	731	762	2 340	2 425	nd
	2.0	9 500	10 413	22 525	23 015	nd
High	0.5	305	324	562	nd	Whey separation
	1.0	3 365	3 540	5 065	5 845	Stable
Very high	0.5	464	508	718	nd	Whey separation
	1.0	4755	5015	6790	6 850	Stable

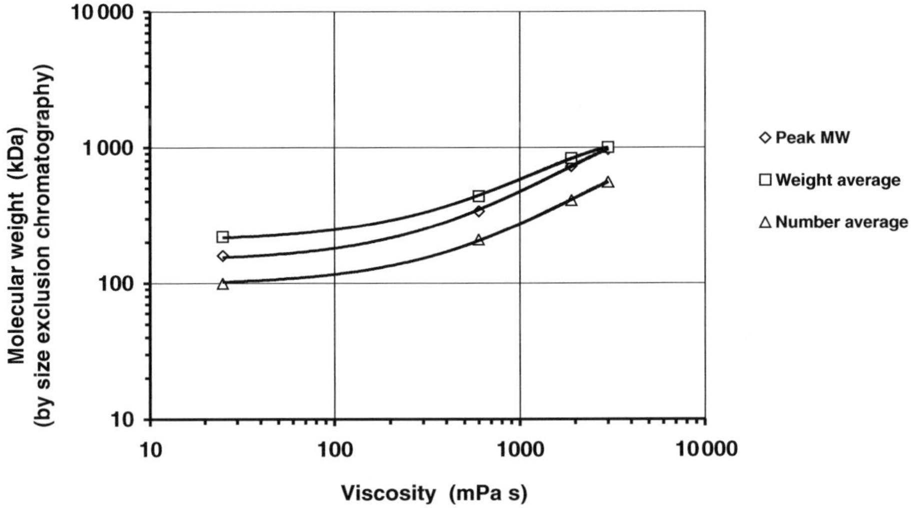

Fig. 15.10 Relationship between viscosity of carob bean gum and different molecular weights distributions. (1% solution prepared at 86–89°C and measured by Brookfield RVT viscosity at 20 rpm, 25°C.)

The retrogradation of starches and modified starches in bakery products can be retarded or even prevented. Efficient stabilisers are obtained by blending these seed gums with unmodified or modified starches in convenience foods, such as salad dressings and prepared meals. Emulsification is enhanced in convenience foods if these gums are blended with native and/or modified starches, as well as with proteins, such as sodium caseinate, or with protein-containing products, for example skim milk powder and whey powder.

Depolymerised guar products are employed, where almost no increase in viscosity is desired, for protecting certain proteins in acidic media, for example in pasteurised cultured milk products such as quark and processed cheeses. Highly depolymerised guar products are applied, for instance, for spray embedding sensitive pharmaceuticals.

15.6 FURTHER DEVELOPMENTS

Recent work was initiated to improve the light transmission of guar solutions. Selected splits were treated under alkaline or acidic conditions and then neutralised. Tap water was used to extract unwanted salts such as sodium sulphate or sodium acetate and, if required, dehydrated with alcohols. Immediately after the treatment, the swollen splits were put into a high-speed mixer to prepare the aqueous solution, which was then analysed for viscosity and light transmission. The patented technology to produce guar products that give clear aqueous solutions comprises a simple alkaline treatment of the selected semi-wet splits, using 8–10% of alkali in presence of a minor amount of hydrogen peroxide, at an elevated temperature in a sigma blender (Yeh, 1996; Wielinga, 2003). The peripheral cell layers are rubbed off and can then be recovered by screening. The purified splits are washed with water, dehydrated with alcohol and ground.

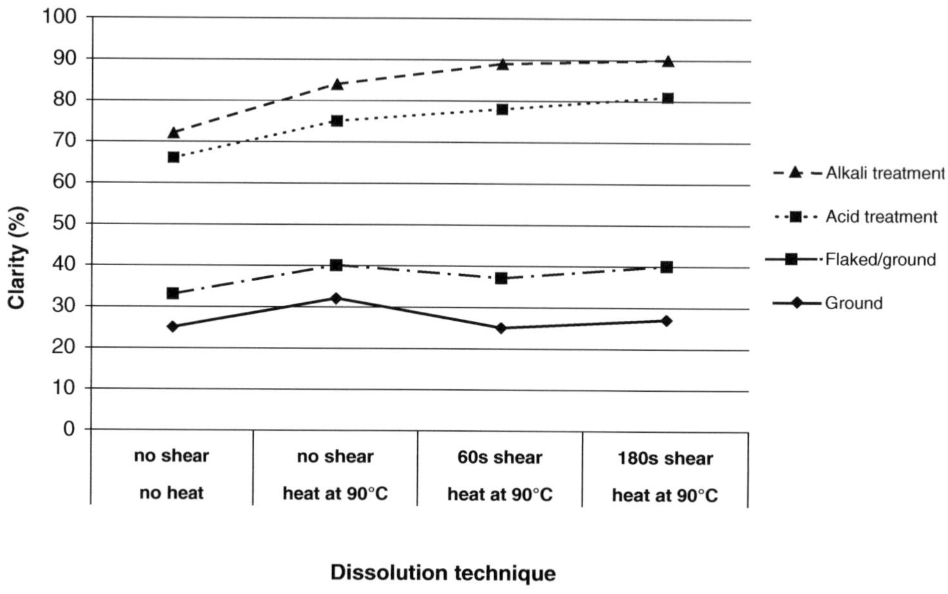

Fig. 15.11 Clarity of different guar solutions. (Concentration 0.5%, 1 cm light path at 500 nm.)

Figure 15.11 compares the light transmission of 0.5% solutions at 500 nm of two standard guar gums, one flash-ground and the other flaked and ground, with two purified guar products manufactured according to these patents. It is evident that these treatments significantly improve solution clarity.

Wet splits of known moisture content were dissolved in water in a high-speed mixer. It was found that the hydration procedure had a pronounced effect upon the final viscosity and the amount of insoluble material. Three hydration techniques were compared using high-shear mixing for different periods, hydrating hot (90°C) or a combination of both. Viscosity readings were made after 24 h, using a Brookfield RVT viscometer at 25°C.

Guar gums with different particle size, a higher viscosity material and a pure guar gum were tested using different hydration procedures. The results are shown in Table 15.7 and clearly show the outstanding properties of pure guar gum with reduced insoluble material and, hence, higher clarity.

Table 15.7 Effect of heat and shear on guar viscosity and insoluble material.

Guar properties			Viscosity			Insoluble material		
Purity	Viscosity	Particle size	No heat No shear	Heat at 90°C No shear	Heat at 90°C 60 s shear	No heat No shear	Heat at 90°C No shear	Heat at 90°C 60 s shear
Standard	Medium	Coarse	4000	5130	5600	21.9	13.5	9.2
Standard	Medium	Fine	3085	3465	4250	25.2	16.3	13.1
Standard	High	Standard	6160	6600	7750	20.4	17.1	9
Pure	Medium	Standard	3810	5200	5300	6.6	1.7	0.8

Such pure guar gums can be produced with viscosities as low as 2 mPa s and as high as 8000–9000 mPa s at 1% concentration. Under the heat and shear conditions mentioned above, the clarity of the aqueous solutions can be as high as 90–96%.

15.7 DERIVATISED SEED GUMS FOR TECHNICAL APPLICATIONS

More than 90% of the available carob bean and tara gums currently used are no longer chemically derivatised. However, the hydroxyl groups of galactomannans can be derivatised and the seed gum manufacturers provide the market with non-ionic, anionic, cationic and amphoteric derivatives.

Usually, it is assumed that the reactivity of the primary and secondary hydroxyl groups has practically the same reactivity which, however, is not the case. Each of the terminal galactose side stubs has only three secondary OH-groups and one primary OH-group, whereas the substituted 4,6-mannose unit has only two secondary OH groups and the unsubstituted 4-mannose possesses two secondary and one primary OH. The maximum average degree of substitution (DS), therefore, is 3. Introducing more substituents to one hydroxyl group leads to a molar substitution.

Both mannose and galactose contain vicinal secondary *cis*-hydroxyl groups, for mannose at C2 and C3 and for galactose at C3 and C4. These vicinal *cis*-hydroxyl groups form cyclic complexes with appropriate reagents, such as borax. Semi-dry processes allow the production of derivatives and modified products, which can be water washed, if required.

The new techniques allow the recovery of the fine fraction consisting of the aleurone cell layers (20–25%), and the coarse fraction of the inner endosperm (75–80%). The latter can then be taken as the starting material for the manufacture of pure modified products. These new technologies enable the production of ionic, non-ionic and amphoteric guar derivatives which hydrate in water to give crystal clear solutions, even at a very low DS of about 0.1 (Chowdhary, 1998). Thus significantly smaller amounts of etherifying reagents are required to produce water-soluble products of high clarity, compared to cellulose derivatives.

Guar is derivatised for technical applications to provide hydroxypropyl and hydroxyethyl guar, carboxymethyl-hydroxypropyl guar, carboxymethyl guar and cationic guar, as well as cationic hydroxypropyl guar and hydrophobic guar, or with guar phosphates. These technical products are used in the cosmetic, textile, paper, explosive, drilling, oilfield and chemical industries.

References

Baumgartner, S. (1986) 'Untersuchungen an der Kohlehydratfraktion von Johannisbrot, *Ceratonia siliqua*' (Investigations of the carbohydrate fraction in carob seed *Ceratonia siliqua*) Carbohydrates in Carob Seed. Dissertation, Eidgenössische Technische Hochschule Zürich (Swiss Federal Institute of Technology Zurich), No. 8103.

Bepex (2008) *Bepex Particle Processing Technology*. Pulvocron dryer, Bepex International, Minneapolis. Available at http://www.bepex.com/pulvocron.htm (accessed 5 August 2009).

Chowdhary, M.S. (1998) Derivatized guar gum composition including nonionic and cationic groups which demonstrate excellent solution clarity properties, US Patent 5,756,720.

Clark, H., Dea, I.C.M. and McCleary, B.V. (1986) The effect of galactomannan fine structure on their interaction properties. In: G.O. Phillips, D.J. Wedlock and P.A. Williams (eds), *Gums and Stabilisers for the Food Industry 3*. Elsevier Applied Science Publishers, London.

Del Re-Jiménez, B.L. and Amado, R. (1989) Comparative study of the chemical composition of germ meals from carob, guar and tara seeds. *Food Hydrocolloids*, **3**(2), 149–156.

Dhugga, K.S., Barreiro, R., Whitten, B., Stecca, K., Hazebroek, J., Randhawa, G.S., Dolan, M., Kinney, A.J., Tomes, D., Nichols, S. and Anderson, P. (2004) Guar seed β-mannan synthase is a member of the cellulose synthase super gene family. *Science*, **303**, 363–366.

EU (1998) Commission Directive 98/86/EC. *Official Journal of the European Communities*, **334**, 11–13, 17.

FAO (1995) Seed gums. In: *Gums, Resins and Latexes of Plant Origin, Non-Wood Forest Products* **6**, pp. 152. Available at http://www.fao.org/docrep/v9236e/V9236e06.htm (accessed 5 August 2009).

FSANZ (2006) Final Assessment Report, Tara gum as a food additive. *Food Standards Australia New Zealand*, Application A546, pp. 1–44.

Goswami, A. (1999) Production of galactomannan products by enzymatic reaction on guar splits. *World Intellectual Property Organisation* WO1999/004027.

Hosokawa (2008) Hosokawa Micron flash dryer, Hosokawa Micron Ltd. Available at http://www.hosokawa.co.uk/flashdryer.php (accessed 5 August 2009).

Hui, P.A. and Neukom, H. (1964) Some properties of galactomannans. *Tappi*, **47**, 39–42.

Justus Leibig University (2000) Guarbohne (*Cyamopsis tetragonolobus*[L.] Taub. [= *C. psoralioides* DC.]) Available at http://bibd.uni-giessen.de/gdoc/2000/uni/p000003/guarbohn.htm (accessed 5 August 2009).

Lozinsky, V.I., Damshkaln, L.G., Brown, R. and Norton, I.T. (2000) Study of cryostructuring of polymer systems. XIX. On the nature of intermolecular links in the cryogels of locust bean gum. *Polymer International*, **49**(11), 1434–1443.

Markman, C.D. (2006) Vegetable tanning extracts/others. Christian D. Markmann GmbH, Hamburg, Germany. Available at http://www.cdm-impex.de/others.html (accessed 5 August 2009).

McCleary, B.V., Dea, I.C.M. and Clark, A.H. (1984) The fine structure of carob and galactomannans. In: G.O. Phillips, D.J. Wedlock, and P.A. Williams (eds), *Gums and Stabilisers for the Food Industry 2*, Pergamon Press Ltd., Oxford, pp. 33–44.

McCleary, B.V. and Matheson, N.K. (1974) α-D-Galactosidase activity and galactomannan and galactosylsucrose oligosaccharide depletion in germinating legume seeds. *Phytochemistry*, **13**, 1747–1757.

Mo, B. and Bewley, J.D. (2003) The relationship between β-mannosidase and endo-β-mannanase activities in tomato seeds during and following germination: a comparison of seed populations and individual seeds. *Journal of Experimental Botany*, **54**, 2503–2510.

Pebani (2008) *Pebani Products, Others*. Pebani, Lima, Peru. Available at http://www.pebani.com.pe/others.html (accessed 5 August 2009).

Polygal (2008) *Raw Materials and Technology*. Polygal A.G., Märstetten, Switzerland. Available at http:// www.polygal.ch/site/content.asp?typ=Raw±materials±%26±Technology&category=01%24Raw±Materials&subcategory=06%24Tara&lang=EN (accessed 5 August 2009).

Sattler, K. (1979) Thermische Trennverfahren, Aufgaben und Lösungen, Auslegungsbeispiele, Vogel Verlag, Würzburg (Thermal separation process, Problems and solutions, design examples, Bird-Verlag, Würzburg).

Swartz, W.E., Hoppe, C.A. and Elfstrum, J.T. (2000) Reduced viscosity, low ash modified guar and process for producing same, US Patent 6,048,563.

Unipektin (2008) Production, locust bean gum, Unipecktin Ingredients AG. Available at http://www. unipektin.ch/pages/template_textbild.asp?pid=26&pmid=144 (accessed 5 August 2009).

Wielinga, W.C. (1977) Verdickungsmittel aus Guar- und Johannisbrotkernen. *ZFL* **28**(1), 16–22.

Wielinga, W.C. (1984) Application of gum-based stabilization systems in ice cream and fruit ices. In: G.O. Phillips, D.J. Wedlock and P.A. Williams (eds), *Gums and Stabilisers for the Food Industry 2*. Pergamon Press Ltd., Oxford, pp. 251–276.

Wielinga, W.C. (1990) Production and applications of seed gums. In: G.O. Phillips, D.J. Wedlock and P.A. Williams (eds), *Gums and Stabilisers for the Food Industry 5*. IRL Press at Oxford University Press, Oxford, pp. 383–403.

Wielinga, W.C. (1993) Evaluation of carob from trees, pods and kernels. Private communication to Institut Europeen des Industries de la Gomme de Caroube (INEC), Bruxelles.

Wielinga, W.C. (2003) Process for producing pure guar seed flour, US Patent 6,664,381.

Yeh, M.H. (1996) Guar gum composition and process for making it, US Patent 5,489,674.

16 Starch

Paul Sheldrake

ABSTRACT

Starch is one of the most widely used, functional and flexible food stabilisers for both thickening and gelling. This chapter introduces the food scientist to the range of different sources for these many and varied raw materials and the potential they have to offer in texturising products. After a short introduction, the chapter is broadly split into three main segments: the examination of the starches from the molecular through to the macroscopic level, how starches can be controlled to create a whole range of different thickening or gelling systems and, finally, how these different textures can be utilised to their best effect in different product applications. A series of tables and figures is used to exemplify the different aspects and clearly show how to achieve certain textures and effects. The tables and figures can be used to compare and contrast the different properties of starches and to establish how to create the most innovative solutions for the consumers. The aim is to show how to create a 'twist in the tail' by moving away from the traditional thickening and gelling properties through the use of technologies to develop new innovative starch-based products.

16.1 INTRODUCTION

Chefs or food scientists often make a classic choice when they are developing their latest innovative creation: they know exactly how they are going to delight the consumer and where to source all the ingredients in order to do this. Alternatively, they go to their store shelves and see what they have in stock and work out how to create their latest innovation from the materials at hand. The former is exciting and potentially unlimited where their creativity can produce totally new solutions by combining the latest and best materials. The latter has the same potential but it is more about being innovative and creative with existing ingredients and obtaining value by pushing the limits of the functionality of these already-available raw materials. Both of these two routes are the purpose and expected delivery of this chapter on starch, a truly flexible and functional thickening and gelling agent.

For the food scientists that are new to the industry, this guide will explain the role of starches in many food systems whilst being an opportunity to further develop the understanding of those scientists who already have a general grounding in the technology. The chapter brings this all together as a solid resource that explains the latest innovations, developments and trends and gives an insight into how to achieve the textures that food scientists need when developing new products for the food industry. In all cases, the format will allow the food scientist to dip in and out and check things through the use of helpful tables, diagrams and figures.

The twist is in the tail! Through a deeper understanding of the different functional properties of the core raw materials, food scientists will gain a solid grounding in the basic properties, know the questions to ask in order to maximise the benefit of the starches in use, be stimulated to want to know the more in-depth parts that can really enhance their work and, finally, create that unique dish. When the consumer is delighted in the end-creation and they think to themselves, 'Who would have ever thought of that combination?' you know that the twist has done the trick.

On a practical note, the scope of this chapter is specifically around the gelling and thickening properties of starch products, which is only part of the range of the totally flexible and diverse functionalities that can be gained from the application of starches.

16.2 RAW MATERIALS

Hydrated carbon – *carbohydrate* – is the energy block for life and this makes it one of the world's largest agricultural crops. The diversity of carbohydrate content, in terms of both availability and functionality, has allowed a specific industry to develop whose primary focus is the extraction of the starch component for more than its simple calorific value. The diversity and flexibility of the texturising properties has resulted in starch, a water-soluble, digestible polymer, becoming one of the largest and most widely used functional ingredients across the spectrum of many global industries: food, industrial, medical and cosmetics. There are many complex issues that have to be taken into account when sourcing a particular product to ensure that it meets the specific needs of the end-application. In simple terms, these issues can be categorised into three inter-related core aspects that need to be evaluated together in order to develop the right texturising solution: functionality, availability and cost.

16.2.1 Sources

Industrially, the largest crops that are cultivated for starch extraction are cassava (tapioca), corn (maize), potato and wheat, with each starch generally grown in regions that are particularly suited to that type plant from an environmental point of view. The next group of commercially available starches, rice, pea, sweet potato and mungbean, are extracted on a more limited industrial scale. Figure 16.1 and Table 16.1 show the production statistics of the largest commercially grown crops by region.

Once extracted, starches can be distributed globally to where they are needed for their specific texturising or other functional properties. It is important to note that there is a strong agronomy and bio-engineering knowledge base and expertise that has been built up over many years associated with all these crops. This is of paramount importance to ensure that new breeds and varieties are constantly being developed to improve the resistance to disease or climate change or to improve other functional benefits of the starches.

16.2.2 Market changes

Once the raw starch has been extracted, the diversity and flexibility of the material mean that it is used in many different industries and, in the ever-changing market place, it is important to monitor the market trends when looking for new opportunities. There is a constant change

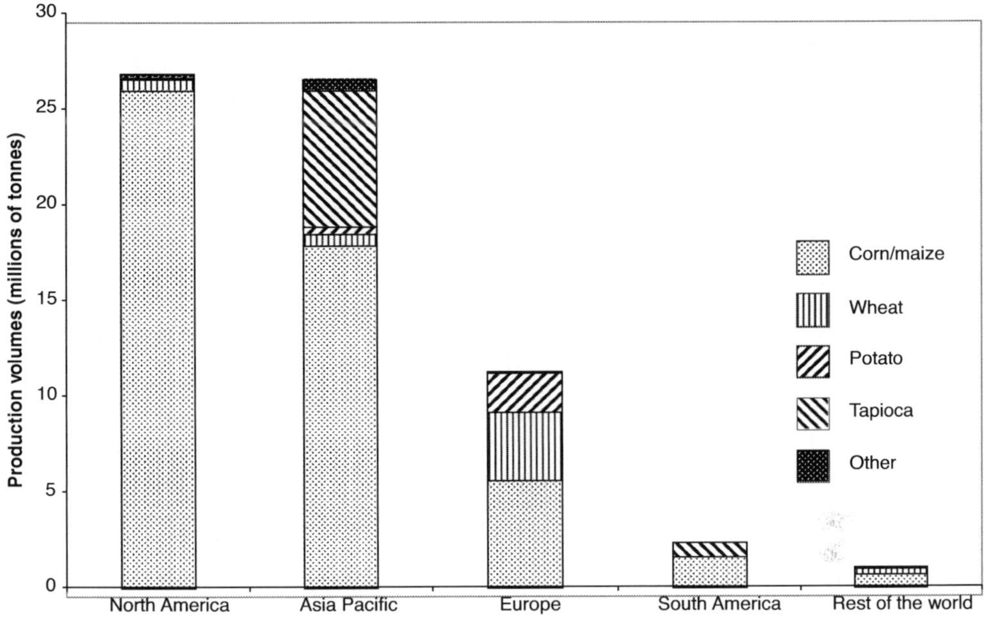

Fig. 16.1 Global starch production volumes by region.

in the supply and demand for each starch material in different market segments due to factors such as the overall demand versus the pricing (profitability) versus the need for specific application-related functional properties. New market situations also have to be taken into account, such as the recent trend to produce bio-ethanol from starch for fuel, which, due to the different and potentially higher demand, affects the overall availability, and thus pricing, for other market segments.

By monitoring specific countries or regions, the flow of the different raw materials can be followed around the world through trade and other market statistics.

16.2.3 Environmental changes

The issues of global warming and crop yields through improved biotechnology continue to be major issues in the marketplace. Warmer weather with less rain has an effect on the growing

Table 16.1 Global starch production volumes 2006.[a]

	Corn/maize	Wheat	Potato	Tapioca	Other starches
Rest of the world	0.64	0.27	–	0.03	0.02
South America	1.49	0.03	–	0.76	–
Europe	5.55	3.57	1.99	–	0.10
Asia Pacific	17.83	0.54	0.44	7.12	0.54
North America	25.92	0.56	0.12	0.01	0.20
Total	54.43	4.96	2.56	7.92	0.86

Source: GIRACT (2006).
[a]Values are in million tonnes.

conditions of many starch products and with the consequential reduction in availability there is often an effect on pricing.

The development of GMO strains of crops is an emotive issue, particularly in Europe, and contrasts with a greater acceptance in other regions. However, GMO crops have been developed to be the most cost-effective and to produce higher yields under current climate conditions. The availability and traceability of the GMO products are issues that have potentially significant cost implications.

16.2.4 Safety

All starches have good safety and toxicological records with the industry having the capability to manage the growing and extraction of the raw materials to very high manufacturing standards and in conditions consistent with good manufacturing practice. The two main regulatory bodies that most of the global community follow for these matters are the FDA (USA) and the European Community who both assess the suitability and safety of these food ingredients. As in all cases, the legislation is under constant review and, thus, it is important to keep up to date with any significant changes that may occur.

16.3 PROCESSING

In general, there are some basic processing steps to extract the native starch from the starchy raw materials: harvesting, extracting, drying, bagging and storage. Water plays a major role in all starch extraction processes as the primary extraction fluid but, more importantly, due to the actual water content of the starchy feed materials, it explains the main differences between the various harvesting methods, production periods and locations. Table 16.2 lists the main crops and shows the differences in starch, moisture and other physical differences.

For cereal-based starches, the harvested crops, such as wheat and corn or maize, have a relatively low-moisture content and thus they can be stored in the unrefined form until the point of extraction. This contrasts with tubers or roots, such as potato and cassava (tapioca), which have a much higher moisture content and, consequently, the starch needs to be extracted as soon as practicable to prevent spoilage. As a result, these high-moisture

Table 16.2 Differences between the starchy raw material supplies (collated from Atwell and Thomas, 1997; Imeson, 1999; Phillips and Williams, 2000).

	Source	Moisture in source (% by weight)	Starch in source (% by weight)	Starch on dry substance (% by weight)
Corn/maize	Cereal	16	60	71
Waxy corn/maize	Cereal	20	57	71
High amylose corn/maize	Cereal	20	57	71
Potato	Tuber	78	17	77
Waxy potato	Tuber	78	17	77
Rice	Cereal	12	65	89
Tapioca	Root	66	26	77
Wheat	Cereal	14	64	74

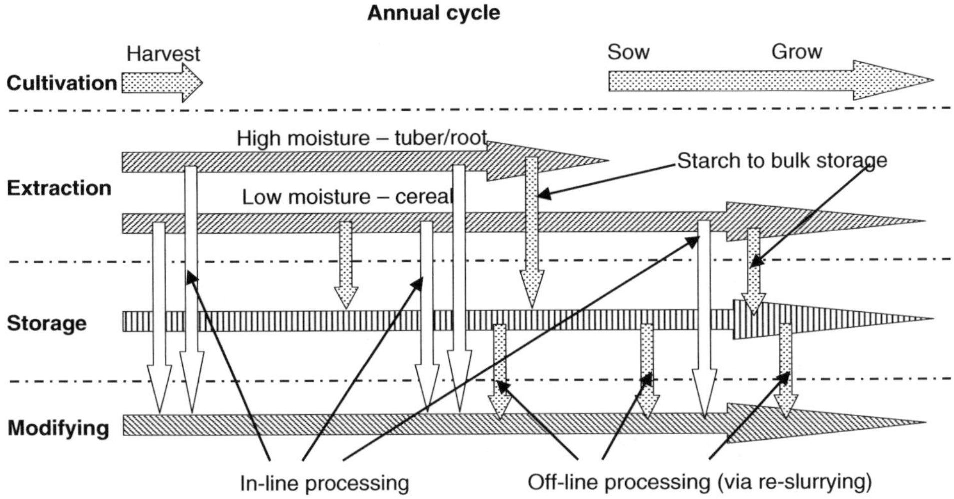

Fig. 16.2 Starch extraction and annual processing cycle.

starches are generally extracted in the vicinity of their growing locations whilst the cereal starches, subject to transportation costs, may be extracted further afield and at later times. The specific details of the different processes are well documented but the overall operational differences are not often compared.

Figure 16.2 shows a schematic diagram of the different operations within the starch-processing industry. The first part is the cultivation process which, for most manufacturers, begins with the harvest of the new crop to ensure the continued supply of the feed material. During the rest of the cycle there are the typical sowing and growing phases.

The next step is extraction and, as pointed out earlier, there is a difference between high- and low-moisture content starchy foods. The high-moisture products need to be extracted as soon as practicable and, thus, they run for a limited time in the year, commonly known as a campaign. The use of technology has allowed this campaign to be extended for several months but at some point there is no more feedstock and the extraction process stops. For the low-moisture content products extraction can go on throughout the year as there are generally no problems in storing the raw materials.

There is then a choice of routes for the extracted starch. One is to storage, for example in bulk silos, whilst the other is to take some of the freshly extracted starch directly to further in-line modification processes that need to be carried out. Great energy efficiency is gained by not first drying the starch to put it into storage and later re-slurrying it for further modification in an off-line or de-coupled process.

In the shorter campaigns for the higher moisture raw materials, once there is no fresh in-line feed available, the native starch from the silos has to be re-slurried to continue the production of modified starch for the period outside the extraction campaign.

There is one further place where starch is collected, although this is actually seen as a by-product of another industry. This is known as side streaming. For example, in the further processing of potatoes in the snacks industry potatoes are often chopped or peeled, which releases some of the entrapped potato starch. Through filtering methods, this starch can often be recovered and used as a raw material.

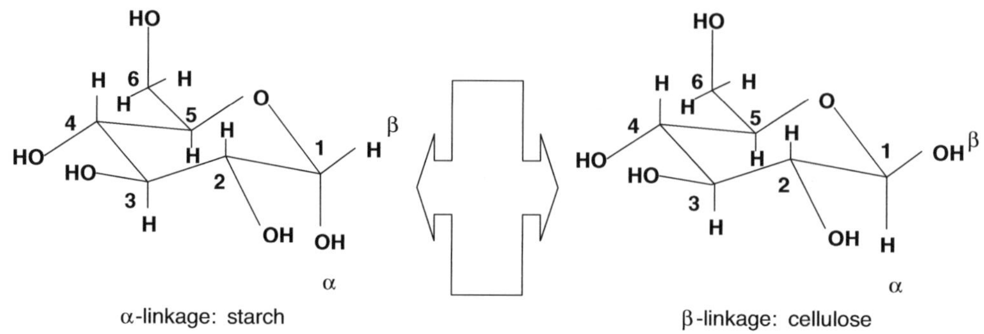

Fig. 16.3 α-Linked versus β-linked glucopyranose units.

16.4 COMPOSITION AND STRUCTURE

There are three levels of composition that need to be considered within the starch world: molecular, microscopic and macroscopic. Each builds on the previous level to develop the overall functional and texturising properties that are utilised in food products.

16.4.1 Molecular composition

In simple terms, starches are long chain polymers of D-glucopyranose units which are differentiated from each other through the growing processes in the different botanical crops. This affects the number and ways these units link together and gives rise to the variety and functional differences between the starches. In fact, cellulose is made up of exactly the same monomer units except for one major difference; starch has α-linked units whilst cellulose has β-linked units (Fig. 16.3). Why such a major difference? α-Linked units are digestible by animals whilst β-linked units are not and, because of this, there are structural differences that give different, useful functional properties (Galliard, 1987).

For polymerisation reactions there are three reactive sites: C1, C4 and C6. In starches, the most common link is the α-1,4 linkage, which results in the elongation of the polymer chains to create the starch backbone (Fig. 16.4). The three-dimensional backbone that is created is

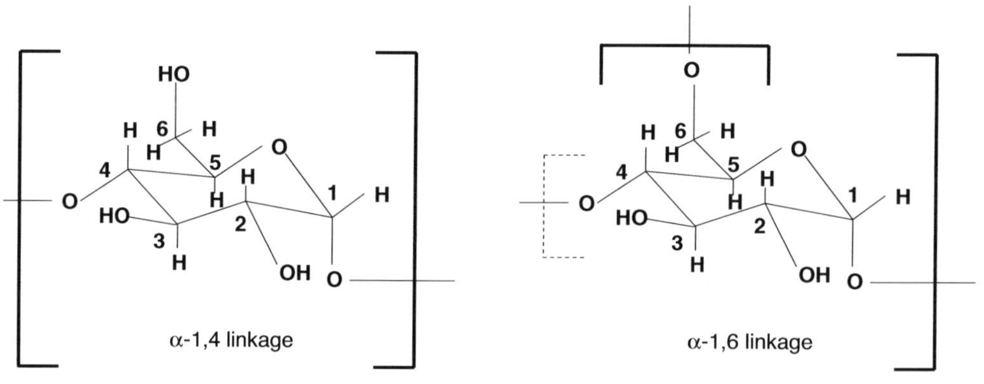

Fig. 16.4 α-1,4 and 1,6 linking units.

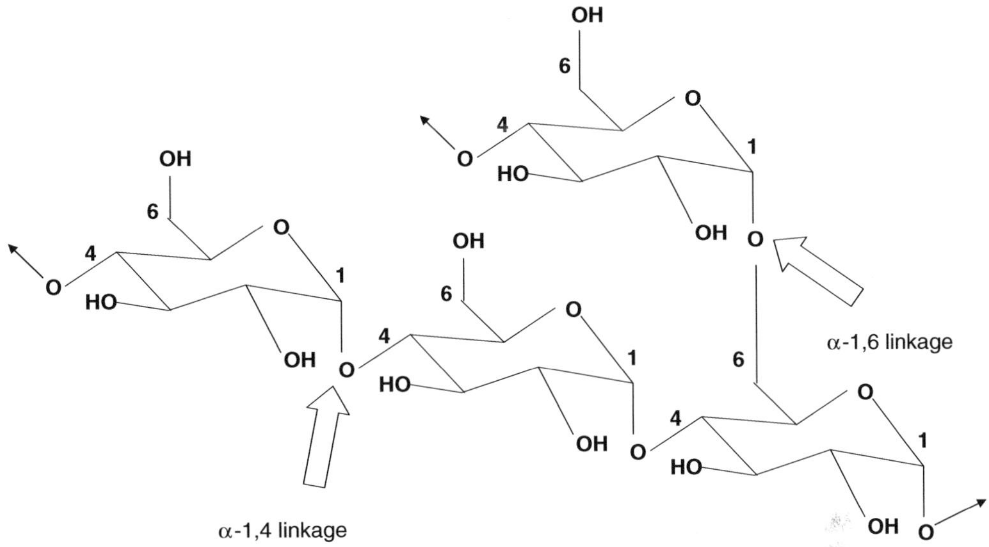

Fig. 16.5 α-1,4 linear and α-1,6 branching points for the polymers.

of a helical nature and contrasts with the linear and sheeting formation that is created from the β-1,4 linkages of cellulose.

The second most common link is the α-1,6 linkage and this has the effect of creating a branching point where a new linear chain can be formed (Figs 16.4 and 16.5).

There are many 1,4 linkages which create the backbone but the real functional differences between the starches can be seen in the way the 1,6 linkages are created. Although it is, in theory, possible to create many different combinations of 1,4 and 1,6 starches, which in turn would give an infinite number of different products, in general only two main types are created in bio-synthesis, amylose and amylopectin. The former is the more linear product with few side chains whilst the latter is a much more highly branched structure (Fig. 16.6).

Table 16.3 shows the different ratios of the two components in a range of commercially available starches. For each of the components, the number of D-glucopyranose units that are in each chain is commonly known as the degree of polymerisation (DP). The differences in the DP and the ratio of amylose and amylopectin both contribute to the different functional properties between the starch raw materials. As starch is a natural polymer, it is important to note that there is always a range of chain lengths and some natural variation in the actual composition of each chain and, thus, the make up of each starch granule.

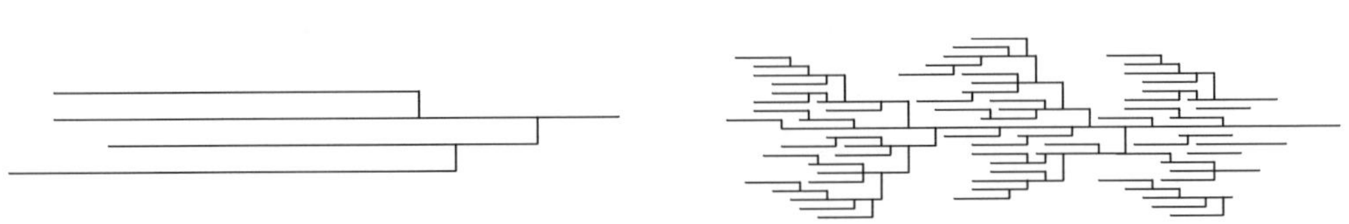

Fig. 16.6 Long chain amylose and highly branched amylopectin.

Table 16.3 Differences between the amylose and amylopectin contents of different starch raw materials.

	Amylose (%)	Amylopectin (%)	DP amylose	DP amylopectin
Corn/maize	25–28	75–72	2 000	2 000 000
Waxy corn/maize	<1 (5)	>99 (95)	–	2 500 000
High amylose corn/maize	55–70 or greater	45–30 or less	1 300	2 300 000
Potato	19–21	81–79	10 000	3 000 000
Waxy potato	<1	>99	–	3 000 000
Rice	17–19	83–81	1 000	2 000 000
Waxy rice	<1	>99	–	2 000 000
Tapioca	17	83	5 000	3 000 000
Wheat	25	75	8 000	2 500 000
Pea (smooth)	35	65	1 300	2 600 000
Pea (wrinkled)	63–75	37–25	1 100	2 700 000

16.4.2 Microscopic structure

From the molecular level, the polymer chains group together to form the microscopic structures (Fig. 16.7) that can be examined and show some of the first measurable physical differences between the starches (Whistler *et al.*, 1984). The ratios of amylose and amylopectin and the different bio-synthetic pathways in the different species give rise to the different types of starch granules. Again there is a range of starch granule sizes and structures that give a further effect in the end-application. Table 16.4 highlights some of the key differences in the granular compositions that are created. Within these granules are other residual components that are not generally removed, including proteins, fats, salts and moisture.

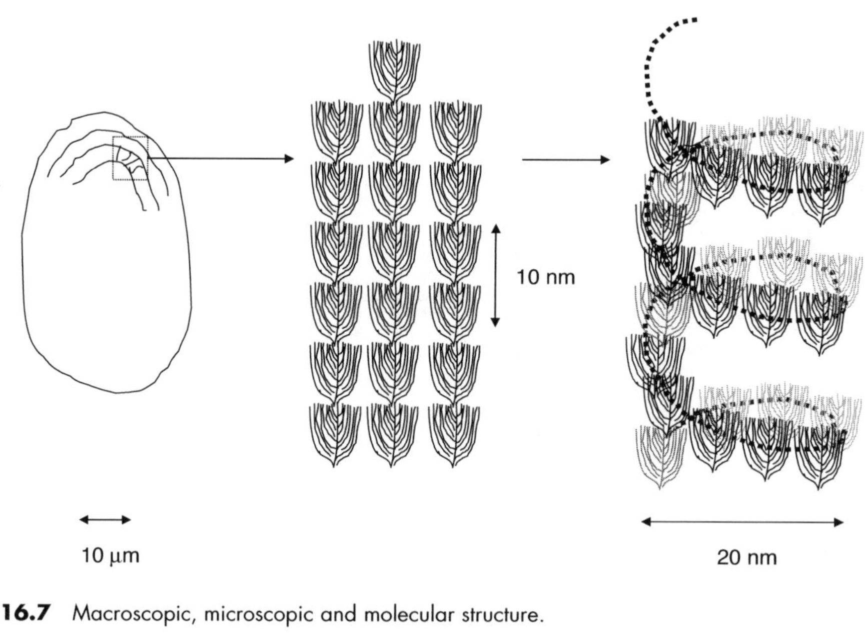

Fig. 16.7 Macroscopic, microscopic and molecular structure.

Table 16.4 Differences between starch raw materials.

	Moisture at 65% RH 20°C (% by weight)	Protein (% by weight on dry substance)	Fat (% by weight on dry substance)	Granular size distribution (μm)	Gelatinisation temperature	Shape
Corn/maize	13	0.8	0.35	3–26	75–80	Polygonal round
Waxy corn/maize	13	0.2	0.25	3–26	63–72	Polygonal round
High amylose corn/maize	13			3–36	63–92	Polygonal round, irregular, elongated
Potato	19	0.1	0.1	5–100	56–69	Oval, spherical
Waxy potato	19	0.1	0.1	5–100	56–69	Oval, spherical
Rice	12			3–8	68–78	Polygonal, spherical, compound granules
Tapioca	13	0.1	0.1	4–35	62–73	Oval truncated, 'kettle drum'
Wheat	13	0.9	0.4	1–40	80–85	Round lenticular
Pea	8	1.0	0.1	30–40	65–70	Reniform

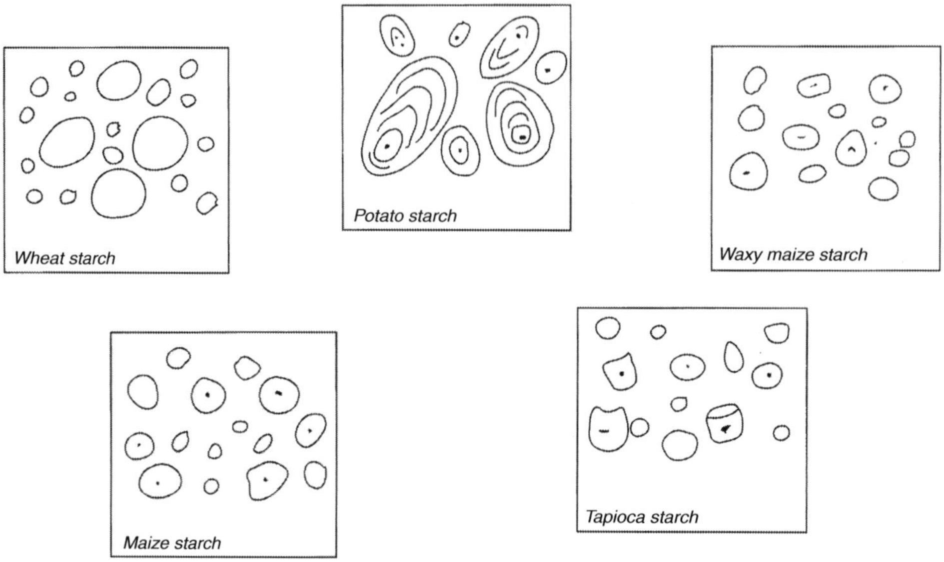

Fig. 16.8 Granular shapes of different starch raw materials.

16.4.3 Macroscopic organisation

When the range of starch granules is looked at in bulk (Fig. 16.8), the overall or average effects of that particular material can be seen and it is the effects at this macroscopic level that the food scientist generally uses in order to create the desired effects in developing the food product. This is where the attention starts to focus on the application of starches as thickening and gelling agents.

16.5 THICKENING AND GELLING PROPERTIES

The main focus and purpose of this chapter is to understand the thickening and gelling characteristics of starches so that the food scientist can understand how to maximise the functional properties in their specific application area. This is the focus of the final part of this chapter but there are two stages that need to be understood first so that these applications can be put into context and the benefits maximised.

Initially, it is important to understand the flexible and great range of textures that can be achieved with starches that can act as both thickening and gelling agents under different circumstances. Then we see how these properties can be controlled, changed, adjusted and improved to meet the requirements for particular applications before looking at the specific thickening and gelling characteristics in the varied applications in which starches are used.

16.5.1 Changes during processing

In order to understand these thickening and gelling properties, the first step is to examine what happens to a native starch when it is processed.

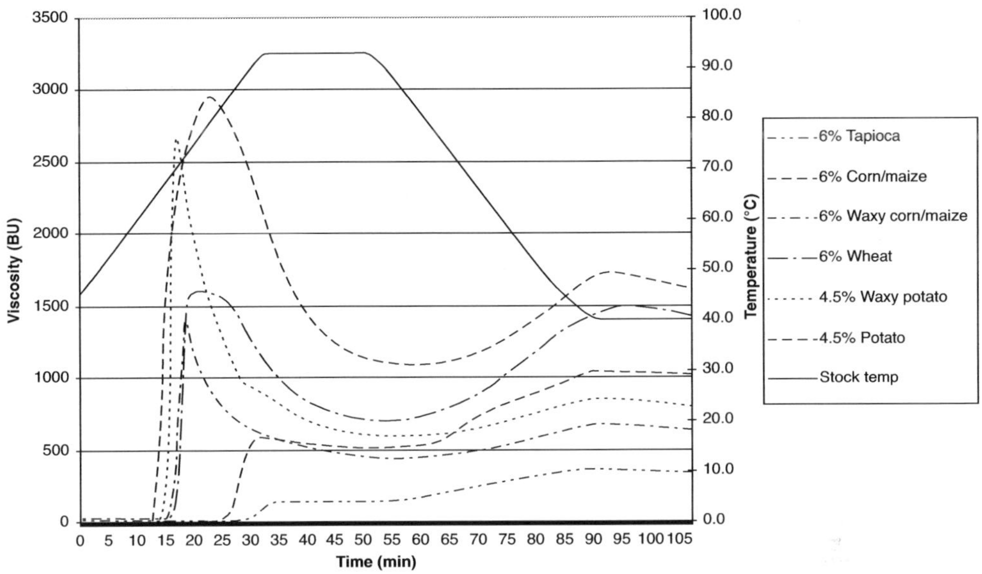

Fig. 16.9 Brabender Viscograph profiles of a range of commercially available starches.

Typically, the starch granules are distributed and suspended in an aqueous solution before heat is applied. As this suspension is heated, it passes through the gelling point of the starch, which is the temperature at which the starch starts to swell (T_g). As the starch granules swell they absorb water until they reach their maximum size. At this point the maximum thickening viscosity is achieved. As heating continues, the starch granules cannot swell any further and so they start to break down. If this process is continued long enough all the granules break down and the amylose and amylopectin polymers are released into the solution. When this solution cools, it tends to set to give a gel-like structure.

During this process, in theory the food technologist has the ability to 'stop' the starch and capture the texture associated with that point along the cycle. In fact, this is the key target that scientists are trying to achieve in order to obtain the functional properties they need for their specific application.

The whole process can be measured and monitored in a number of ways, with the use of either a Brabender Viscograph or a Rapid Visco Analyser being the most common in industry. By plotting the process time, temperature and shear against the viscosity, profiles for a range of starches can be obtained as shown in Fig. 16.9. The state of the starch granules can also be monitored by examining them under a microscope and an experienced starch scientist is able to make interpretations of the state and nature of the granules and product structure.

At this point, it is worth looking at some of the key differences between these different starches as it has an effect on the choice for the end-application. The first point to look at is the gelatinisation temperature, which is the point when the starch viscosity starts to increase rapidly. Those with the lowest values are typically the potato starches and the highest are the wheat starches. The second point to look at is the peak viscosity – the highest viscosity developed during heating. Potato starches, both regular and waxy potato, have the highest weight-for-weight viscosity – note the lower concentration in these graphs compared to the others. The lowest peak viscosity is for wheat starch whilst the others are intermediate. The third point is known as the breakdown viscosity where the native starch breaks down. The potato starches show the greatest percentage breakdown whilst there appears to be no reduction in viscosity for the maize starch. The final stage is during the cooling phase where there is gelation and 'setback'. Here, the key difference is between starches containing amylose versus the high amylopectin (waxy) starches, such as waxy maize or waxy potato. Starches with high amylose content tend to set back and gel more than those with lower amylose content. This is an important factor in designing products for liquid systems and a significant reason why waxy starches, such as waxy maize, which dominates this sector, or the newly introduced waxy potato or waxy wheat starches, have such a high share of these market areas. The key point is to choose and use the right starch for the application.

As we move towards the applications, it is important to look at the microscopic level of the starches as this is where much of the texture is influenced. Initially, we will look at the two 'extremes' and by understanding these we will be able to cover the range of textures.

16.5.2 Thickening starches

The greatest thickening characteristic of starch occurs when the starch granules attain their maximum size and retain this throughout the life of the product. In simple terms, the swollen granules act like 'balloons' which fill a space or volume and it is the inability to move around easily that gives viscosity (Fig. 16.10).

The flexible nature of the balloons means that other ingredients can be located in the gaps either in the continuous phase, such as dissolved sugars, salts or hydrocolloids, or as discrete 'particles', such as emulsified oil droplets or food particles such as seeds or herbs.

Of course, there are often more specific interactions within the food system that also have to be taken into account. For example, steric, ionic, covalent or electrostatic effects may cause the ingredients within the mixture to interact with each other and give different effects.

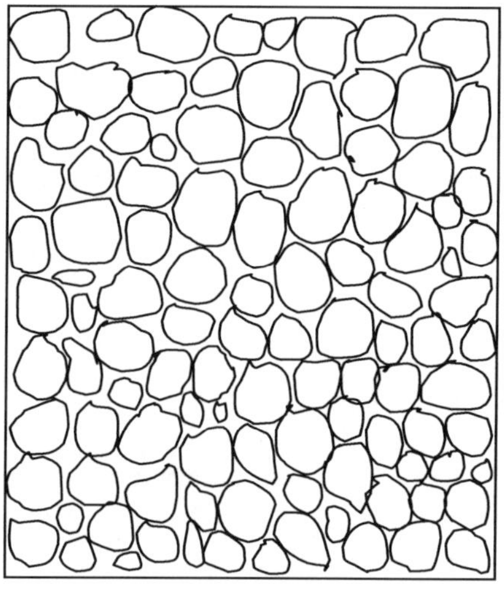

Fig. 16.10 The space-filling nature of swollen starch granules.

16.5.3 Gelling starches

At the other extreme, starches can act as gelling agents. Here, the starch granules have broken down and the amylose and amylopectin chains are free in solution and can associate to form a gel. The type of interactions, chain length and interaction with other ingredients has an effect on the gel properties. There are no discrete starch granules and the free starch polymers are dissolved within the continuous phase.

In hot solutions these polymers have a certain degree of freedom to move around and that allows them to start to interact with each other and to create associative networks. It is, in fact, during the cooling phase that these interactions increase and a stronger gelling network is formed. Care should be taken when these are created. Generally with fairly rapid cooling a gelled network is formed but if the cooling is done more slowly, or if there is a higher concentration of starch or low levels of other ingredients that can interact, a process called retrogradation can occur where the starch polymers align to give small, discrete, insoluble particles. This precipitate is generally undesirable for gelling and thickening but important for other food properties such as the creation of insoluble fibrous starch material (Fig. 16.11). Starches with a higher amylose content are more likely to gel or retrograde but these processes can be avoided by using the 100% amylopectin starches.

This shows the two extremes that can be achieved and takes us to the next important step: how to achieve the required thickening or gelling characteristics.

16.5.4 Controlling thickening and gelling properties

The control of thickening and gelling characteristics is a major target for all starch manufacturing companies. It is the predictable nature of achieving consistent results that is most important to their customers and consumers and this is the driver for 'modifying starches'.

It would be possible to review in full the technical details of every type of starch with every type of modification but in a text of this size it would not be possible to do this comprehensively, especially when this is covered in other publications.

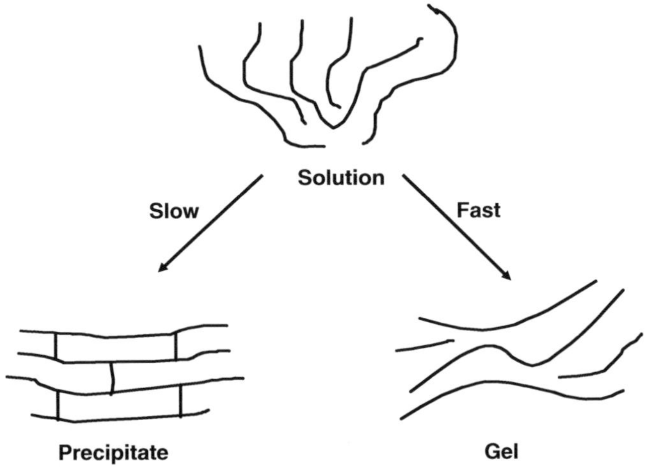

Fig. 16.11 Fast setting to give a gel compared to slow gelling to give a precipitate.

It is more important to focus on creativity and innovation and how, by controlling starch behaviour, food scientists are able to maximise the functionality of this raw material. As a starting point, it is important to examine some of the traditional methods used to modify the properties of starch. The current modification routes can be categorised into three main groups: granular strengthening, weakening and chain length reduction. The future brings chain length extension!

16.6 STARCH MODIFICATION

There are many different routes to modifying starches which are well documented. This overview will try to explain the purpose of these modifications.

16.6.1 Granule strengthening by cross-linking and substitution

Cross-linking and substitution of starches forms a very large part of the modified starch market. There are three sites on each glucopyranose unit where a reaction can take place: C2, C3 and C6 (Fig. 16.12) and thus, in theory, every single unit could be reacted three times. In practical and legal terms it is significantly lower.

Cross-linking with agents such as adipic anhydride, to form starch adipates, or phosphorous oxychloride, to form starch phosphates, strengthens the starch granules by chemically bonding different parts of the chains together in order to keep the granules intact for as long as possible throughout processing and to maintain the thickening characteristics.

Substituting starch has the primary purpose of 'holding the gains' and preventing the swollen starch from collapsing or retrograding and losing the thickening characteristics. The different types of substitution, esterification in acetylated starches or etherification in hydroxypropylated starches, have an effect on the stability, clarity and mouthfeel characteristics. In fact there is a natural substitution as the amylopectin chains also act partially as substituting agents. This natural enhancement is the reason why waxy starches are often preferred to starches containing amylose in liquid systems.

With cross-linking and substitution, the different levels and ratio have the effect of increasing stability to shear, pH and heat through cross-linking and increasing freeze–thaw

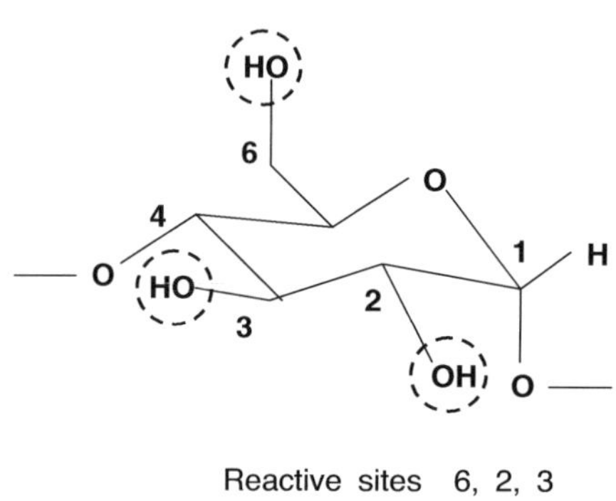

Fig. 16.12 Reactive sites on the glucopyranose unit.

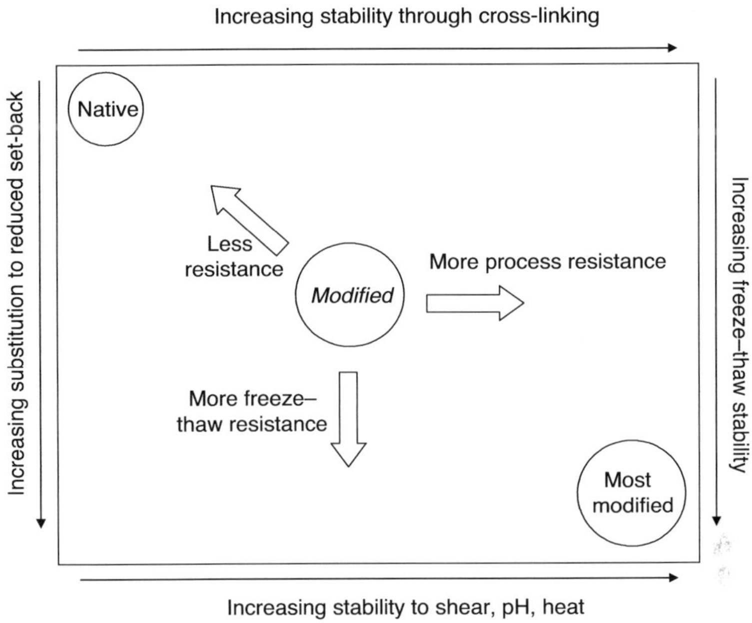

Fig. 16.13 Map showing the effects of modifying starches on stability attributes.

stability and reducing retrogradation through substitution. These two factors can often be plotted on a graph, as in Fig. 16.13, to help the application scientist determine the correct starch choices for different processes.

16.6.2 Granule weakening – oxidation and hydrolysis

At the other extreme, if the scientist is trying to increase the rate of gel formation they can achieve this by weakening the granules through oxidation or hydrolysis reactions. The reactive site is the link between the glucopyranose units (Fig. 16.14). By breaking these bonds, shorter or more disrupted chains can be made that are able to gel more easily.

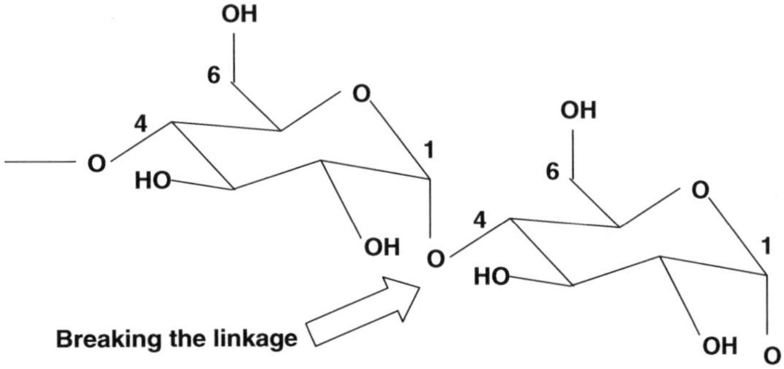

Fig. 16.14 Oxidation/hydrolysis sites on the glucopyranose unit.

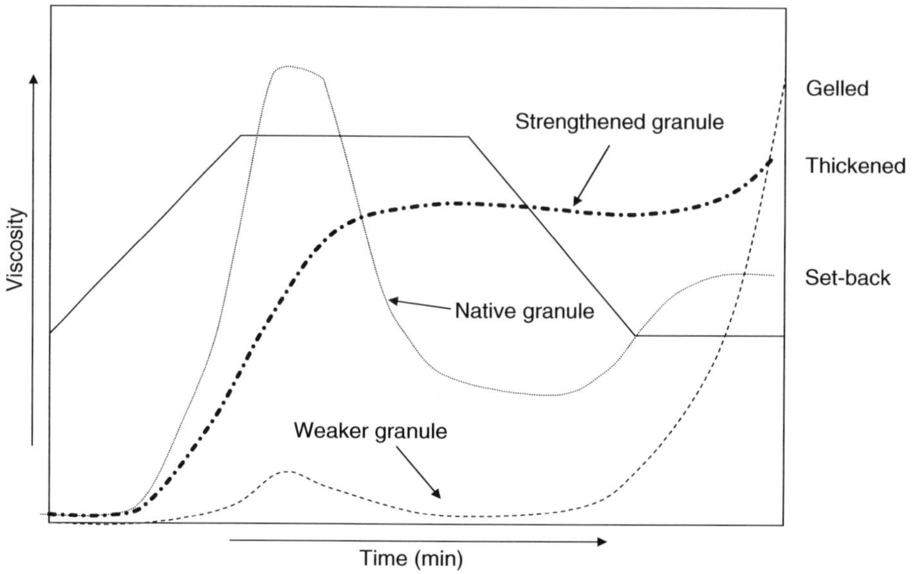

Fig. 16.15 Brabender Viscographs of regular, strengthened and weakened granules.

The contrast in starch properties is shown in the Brabender Viscographs in Fig. 16.15. The native granule follows the expected route whilst the strengthened granule remains intact and maintains its viscosity. The weakened granule rapidly loses viscosity and only reaches the gelled viscosity once the product is cooled.

16.6.3 Chain length reduction

The main products with reduced chain length are maltodextrins, spray-dried glucoses and dextrose. A maltodextrin is formed when the chain is cut by an enzymatic action and this increases the DE or dextrose equivalence – a measure of the reducing power of the starch system. Starch has a DE of one whilst dextrose or glucose has a DE of 42. From a texturising point of view, maltodextrins, especially the low DE maltodextrins, have some specific characteristics where they can form weak gel structures that have some very interesting and useful properties. Different properties are a function of the DE as exemplified in Fig. 16.16.

16.6.4 Creating a 'twist'

The 'twist' in starch developments is in the truly exciting innovations that the starch manufacturers have developed in recent years. Starch scientists have been able to challenge and change some of the properties of the basic raw materials and 'divert from the expected' to give specific and innovative functional properties. In many of these cases, these unique differences have been recognised through intellectual property rights being granted. Figure 16.17 illustrates this point. A traditional starch would be expected to work in one way but, by use of new methods, different and interesting properties can be gained that add to the overall flexibility and versatility of the starch conferring new gelling and thickening characteristics.

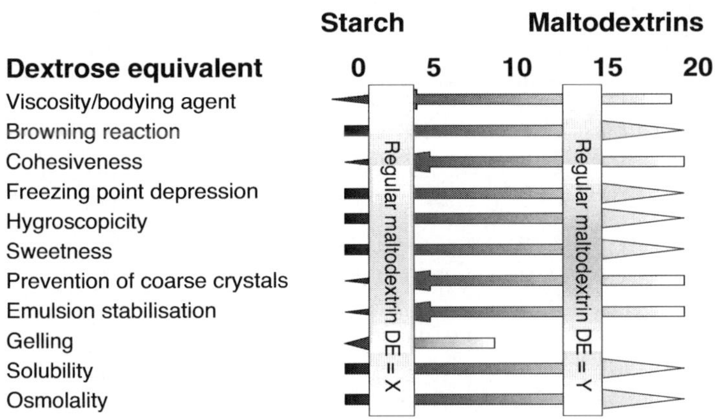

Fig. 16.16 Schematic effects of dextrose equivalent (DE) on textural properties.

Hence, it is always exciting to be able to review all the latest innovations in starch texturising, thickening and gelling properties and to demonstrate how these have been achieved. It will also become clear from the review how these innovations are now leading to some of the most successful applications in the marketplace.

16.6.5 Modification routes to new products

Starting with the premise that there are four basic starch raw materials that are grown on a very large commercial scale, corn or maize, potato, tapioca and wheat, there is the potential to be able to create four thickened and four gelled textures which can be extended further with a range of textures in between. In fact there are many different textures and structures that have been achieved by modifying the raw materials.

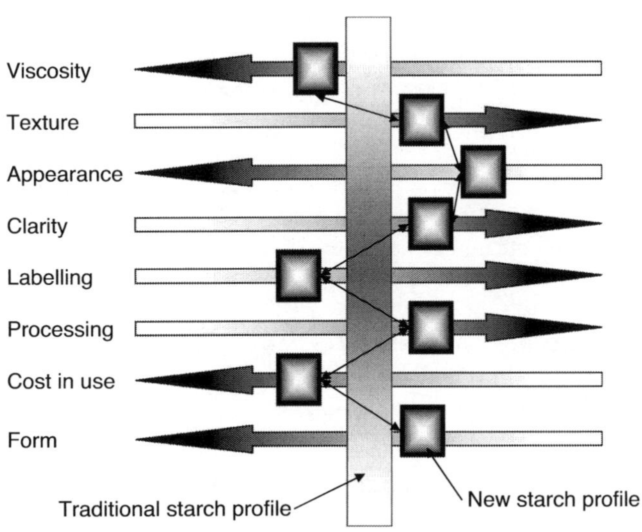

Fig. 16.17 Schematic effects of a 'twist in behaviour' to give a new product.

In simple terms, there are four principal routes that have been taken: modifying the raw material and using different chemical, physical and enzymatic modifications. By considering each of these areas in detail, the benefits that have been brought to the market in recent years can be seen.

16.6.5.1 *Raw material modification*

Starches are grown by plants, and advances in agronomy mean that new starches have become available from new sources. In terms of new crops, for example, there is an increasing demand for pea proteins and, as a by-product of this process, the natural starches in peas are now being separated and made available to the food market. Here is an example of a new source that is available because it is a useful by-product from another industry and pea starch offers a new source of raw material with different functional characteristics.

Natural breeding or modern biotechnology can preferentially select or adjust the amounts of amylose or amylopectin that are generated within the starch granule as it grows resulting in starches having different properties. For example, in potato starches natural breeding techniques have produced a potato with only the amylopectin component developed to create a new waxy potato starch. The same is the case for wheat starches where there is now a waxy wheat starch. These examples compare to the corn and (amylopectin only) waxy corn starches that have been available for many decades. Increasing the amylose content has been achieved in products, such as corn starches. In all cases, by changing the ratio and content of amylose and amylopectin, new starches that offer new functional benefits have been prepared from traditional raw materials.

16.6.5.2 *Chemically modified starches*

The word 'modified' on many starch labels has become a major issue within certain food segments, customers and regions and thus some clarity in this area is always needed.

In general, when the term 'modified starch' is seen on an ingredient declaration it is referring to a chemically modified starch, such as those described for strengthening or weakening the granule. These offer many technical and functional benefits that cannot always be achieved in other ways or as efficiently as through chemical modification. The number of chemicals that can be used in preparing these products is tightly controlled and regulated to ensure that the safety of the end-product is not in doubt. There are, in principle, two major routes to develop new products.

Firstly, using the approved chemicals in a new way or varying the processing routes or methods can lead to changes that can allow products to be developed with new functional behaviours. For example, changing the order or levels of the modification ingredients can affect the speed at which the starch granules may swell under certain conditions to give a new product.

The second is to introduce new chemicals to modify the starches. This is a much longer term prospect as the full safety and toxological effects would have to be proven before approval could be gained and new benefits offered.

16.6.5.3 *Physically modified starches*

There are a number of ways in which starches can be physically modified in order to adjust their functionality. These can be subdivided into different categories:

- Pregelatinisation so that the starch is functional in cold-water systems without the end-user needing to apply heat.
- Changing the form, for example by agglomeration or making finer or coarser grades.
- Particle or grain selection to adjust the particle distribution.
- Moisture reduction, particularly to extend shelf life in dried products.
- Other heat, moisture and pH treatments to change the swelling behaviour of the starch.

It is in this last area that there has been greatest interest with the market trend to products that have 'clean label declarations'. A broad range of products has been developed by applying these technologies to a wide range of raw material bases.

16.6.5.4 *Enzymatic treatments*

Historically, this category focused on chain length reduction but, in the latest developments, chain length is maintained or extended to change the functional properties of the end-product. To make starches with shortened chains there are a number of enzymes that can be used to cleave the starch backbone in different ways to give different functional properties. For chain lengthening, enzymes selectively attach chains to specific sites on the existing amylopectin fraction to create products with new functional properties.

Figure 16.18 shows the range of different processes and how starches with new functions can be developed through the various routes. What must also be remembered is that these different modification routes are often used together or in different combinations to develop an even wider range of starches.

Although not a totally comprehensive list, Table 16.5 shows some of the more recent examples of innovations from the starch manufacturers. Those with brand names tend to be associated with products that have been protected with intellectual property and thus are proprietary to one starch manufacturer.

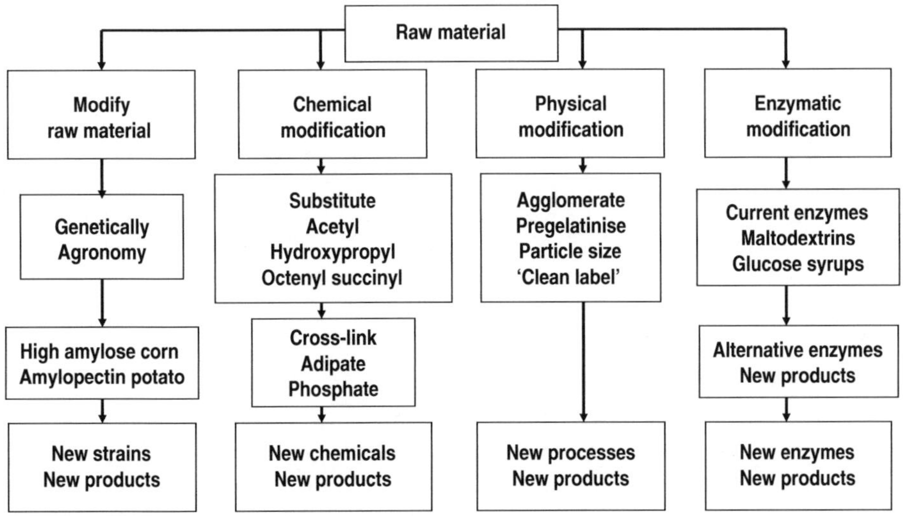

Fig. 16.18 Schematic diagram of the four main routes in starch modification.

Table 16.5 Examples of recent starch innovations.

Modification area	Starch functionality	Specific/unique brand
New/more widely available starch raw materials	Pea starch	Various
	Amylopectin potato starch	ELIANE™
	Amylopectin wheat starch	HOME CRAFT™
Chemical modifications	Adjusting processability/swelling/gelling	Various
Physical modifications	Agglomerated	Various
	Clean label	NOVATION®
	Spray cooked	Various
Enzyme modifications	Low DE maltodextrins	Various
	Chain lengthened	ETENIA™

16.6.5.5 *Physical interactions of starches*

In closing this section, it should be noted that starches generally are not used in isolation and hence it is important to know the effects of other ingredients. There are two general effects that must be considered: physical interactions, such as bulking or blocking, and electrostatic interactions.

The first is seen where there is another ingredient in the system that has the effect of reducing the space for the starches to interact. This bulking effect can affect the overall viscosity and texture.

The second effect is much more important and can be split into a number of areas. Introducing other water-soluble solids such as sugars has the effect of reducing the water in the system which is available for the starch to hydrate and develop its functional properties. The addition of more water, or changing the order of ingredient addition to allow the starch to hydrate first, can have different effects. The extra solids also tend to increase the gelling temperature. When the water-soluble solid is charged, for example salts, there are also interactions between the starch and these salts through ionic interactions. This generally has the effect of reducing the swelling power of the starch.

There are strong interactions between starches and other hydrocolloids, usually through hydrogen bonding. One of the most common is that between starch and xanthan gum. These interact synergistically so that the viscosity of the combination is greater than the sum of the components. In other systems, such as with gelatin, at certain concentrations a de-mixing effect can occur where one part of the mixture creates discrete particles within the continuous phase of the other. Each interaction occurs under certain conditions and must be understood in the specific application in order to maximise the benefit from it.

16.7 FOOD APPLICATIONS

This next section focuses on applications and will take a journey from the thickening to the gelling characteristics of starches (Fig. 16.19)! In all of the applications it is the effects of the starch interactions, particle size and shape that are being considered in order to get to the right end-product.

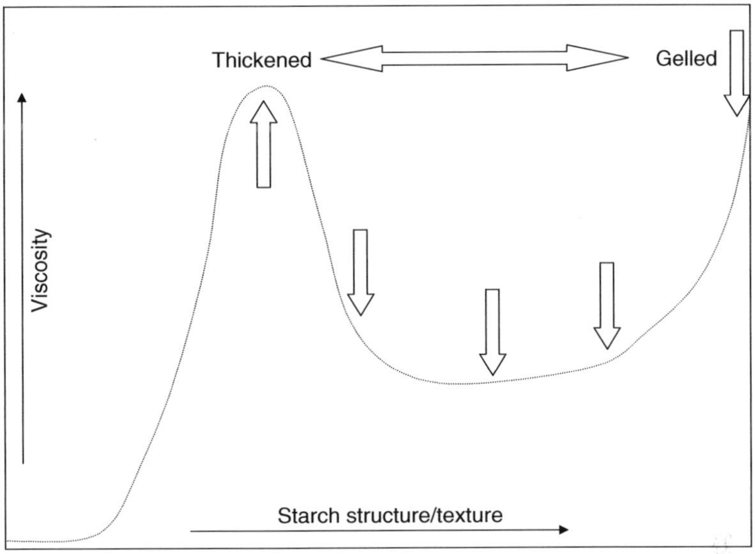

Fig. 16.19 Schematic diagram of the range from thickened to gelled characteristics.

16.7.1 Soups and sauces

The first aim is to maintain the granular structure of the starch to obtain the maximum thickening power. The soups and sauces market is the first application as thickening is the main reason for using starches in these products.

There are many different subcategories within the total soups and sauces market which relate to the end-use, applications and textures that are required. For example, there are both emulsified (cream-type) and regular soups and sauces and sometimes the product has been developed with the versatility of being a soup or sauce if required. This involves a more detailed look into the function of all the materials within the product rather than the specific starch properties. Ultimately, when consumed, the product is liquid so the aim is to get the granules to the right swollen state or form at this point (McKenna, 2003).

Figure 16.20 shows the way that starch granules change their structure during the preparation of the soup or sauce. This market is complicated by the many and varied routes that can be used to achieve the end-result. The different starting points are to do with the type of soup or sauce, for example instant, simmer, pasteurised, UHT, retorted or sterilised, and each will be looked at separately. In all cases, the target is to deliver a soup or sauce where the granules of starch are intact at the point of consumption. The differences are generally related to the way the starch has been *modified* and the *form* that is used.

In powdered soup or sauce applications, there is only a minimum amount of processing required to attain the maximum viscosity and thus the starches need limited or no modification. Here the more important factors are the gel or pasting temperature, ease of dispersion, flavour and viscosifying power. Ideally a lower pasting temperature is preferred as less heat is required to thicken the starch. Ease of dispersion is also important as the starch should disperse without lumping in order for it to swell optimally. This is often the reason why agglomerated starches are chosen as they disperse more easily and thicken faster. Maximising the swelling power of the starch is also invaluable because less starch is required to

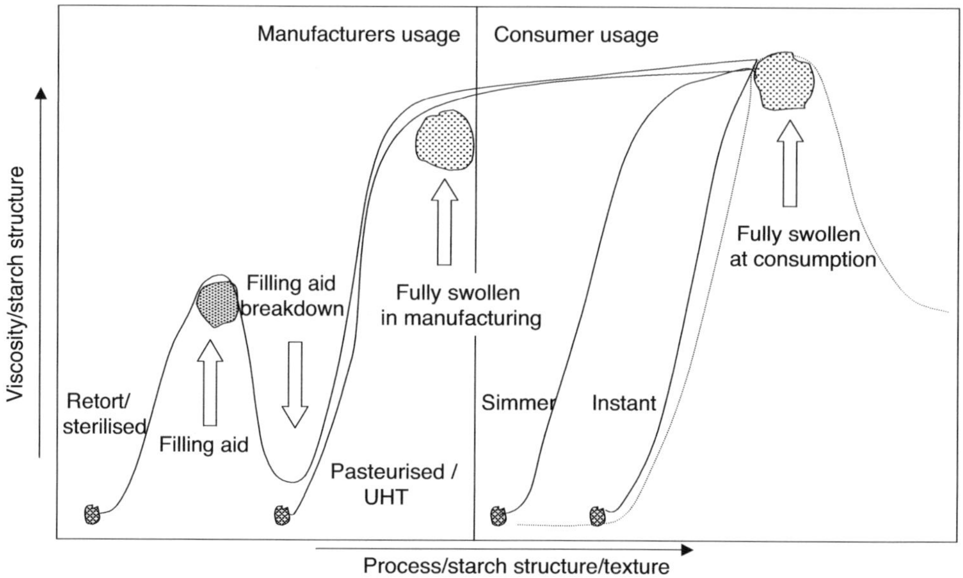

Fig. 16.20 Changes in the starch granule during the preparation of soups and sauces.

achieve the maximum viscosity. In many cases, the starches are also dried to reduce moisture migration between the starch and other ingredients in the mix.

As we progress from the basic instant soup to a product that needs to be heated and then held for a slightly longer time, such as a simmer soup, some level of modification is often carried out. This is for two reasons: firstly to ensure that the granules remain intact during the more-extended holding period and, secondly, to prevent some retrogradation or setback that often leads to skin forming on top of the product.

The next areas to cover are generally in the manufacturing environment where a range of different processes can be chosen to fully swell the granules before products are despatched to the consumers. This can be divided into two key areas for the texturising systems, thin–thick and thin–thick–thin–thick systems.

Pasteurisation or UHT processes are examples of thin–thick systems. The ingredients are put into the pre-blending vessel as a thin mixture which is then heated to swell and thicken the starches before it is then transferred into the storage containers. Once cooled, the product is in the required consumer-ready format and the starch granules are at their most swollen. Due to the extreme processing conditions, there is need for starches that are more resistant to pH, shear and heating and generally starches that have been more extensively modified are used. The target is to hold the starch granules for as long as possible in the maximum swollen state to give a thickened texturised product.

The second major class is the thin–thick–thin–thick systems. Starting with a thin mixture, all the ingredients are premixed and then thickened to distribute all the particles throughout the batch to enable the product to be filled into the final packaging. These packs need to be sterilised, and in order to do this, it is desirable to have the maximum thermal conductivity provided by a thin liquid. This contrasts with the need to have the starch thickened to suspend and distribute the ingredients in the pre-mixing stage. Hence, the design of more specialised and sophisticated systems where two starches are used. The first starch thickens long enough

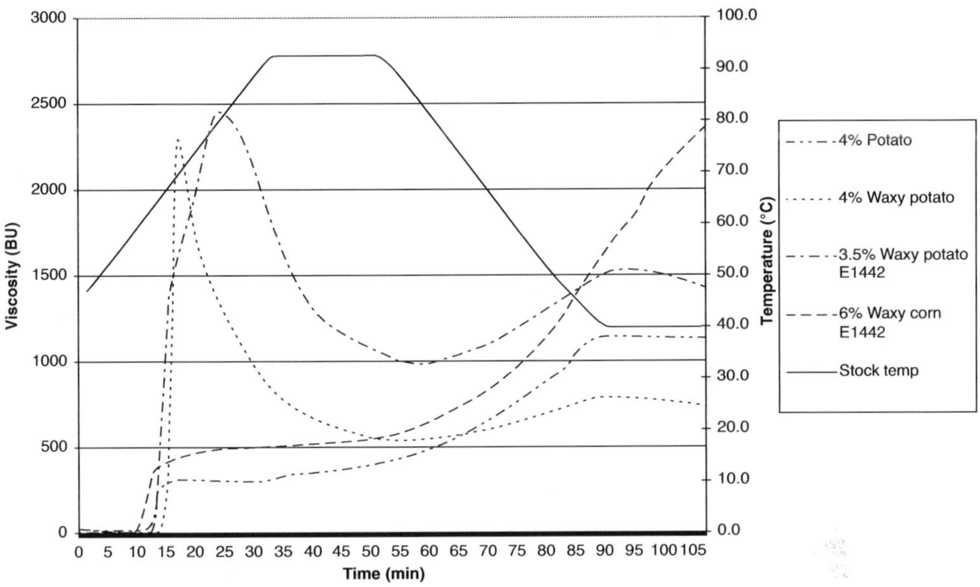

Fig. 16.21 Brabender curves of different starches.

to evenly distribute the mixture in the first part of the process before breaking down to allow good thermal conductivity. A second starch thickens later in order to give the right consistency for the end-product.

Figure 16.21 shows the profile of two starches used in this type of system. The first starch thickens rapidly to give a high peak viscosity to distribute the premix. Once this has been transferred into the separate containers, it can break down to give minimal residual viscosity. Then the second starch takes over to give the end-viscosity.

The choice of the first starch is based on the following:

- Low dosage (and cost) to maximise the filling aid properties.
- Low gelatinisation temperature (T_g) so as to thicken before the second starch.
- Rapid breakdown during sterilisation to allow good thermal conductivity.
- Limited or no setback at the end.

The second starch should develop viscosity later or have latent viscosity to allow maximum heat penetration.

Figure 16.21 shows examples of potato and waxy potato starch used as filling aids as these tend to have the ideal parameters for this application. The comparison shows how the waxy potato starch does not set back and thicken later to the same level as regular potato starch and it breaks down more rapidly and to a greater extent to increase the thermal conductivity.

The second starch tends to be a waxy starch, such as waxy maize or waxy potato, as these have the good thickening characteristics that are required for the end-application.

One new trend seen in the market is the use of liquid concentrates that will thicken further when used by the consumer. These systems need a starch that will bind the product together as a liquid but have enough latent thickening power so that, when the consumer dilutes some into a sauce, it thickens further during cooking to give the final viscosity and texture.

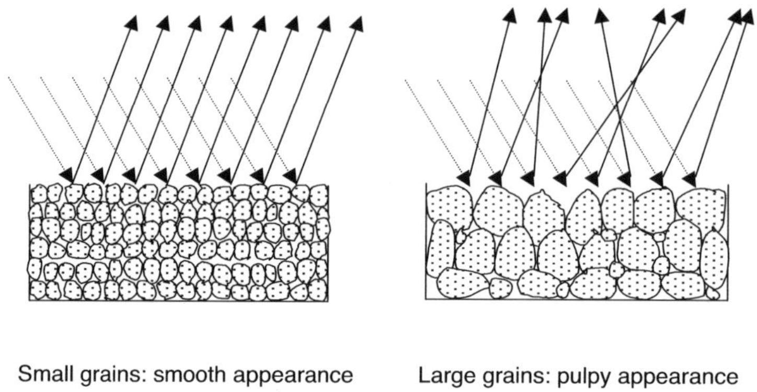

Fig. 16.22 Different surface effects from different starches.

16.7.2 Dressings and condiments

The dressings and condiments category has a number of similar attributes to that of soups and sauces. In some ways the simplest way to differentiate them is to define this section as sauces that are ready to be consumed with no further preparation, for example table sauces such as tomato ketchup or mayonnaise.

For hot prepared sauces, such as ketchup or a dipping sauce, they generally have the same demands as for pasteurised soups and sauces but, as they tend to be low pH products, the starch granules need to be strengthened to tolerate the low pH and heat of the process. Thus some modification is required in order to withstand the processing conditions.

The choice of starch is related to the same factors as any product where the maximum granular integrity is needed. Each different starch has different factors that must be considered in order to achieve different effects and end-results.

Using different starches in these products, one of the most visible differences can be seen on the surface. A change in the appearance from pulpy to smooth is due to the difference in starch granule size (Fig. 16.22). This is particularly evident when making tomato-based sauces where large grains from potato starches give a pulpy appearance whilst smaller grains from maize starches give a smoother appearance.

For dressings, it is critical to fully understand the technology of the interactions between the starch granules so that the route ahead can be seen. Many of these products are emulsified oil-in-water sauces and the first product we review is the basic emulsified sauce, mayonnaise. In mayonnaise, oil droplets are packed together and the surface interactions between the droplets hold the system together. These oil droplets act like discrete particles – very much like intact starch granules. Thus with the trends to healthy products, and also to reduce costs, starches were able to 'replace' the oil droplets to give a structure that represented the mayonnaise in texture and appearance (Fig. 16.23).

Mayonnaise manufacture involves the use of high shear and starches must remain intact through this process in order to deliver the end-structure. Mayonnaise is also a partly gelled product and it was found that some breakdown of the granular structure is needed to give the right product consistency, although if the product is over-sheared the granules break down too much and the product thins. Figure 16.23 shows the granules of starch that represent a typical mayonnaise. In fact this can be made with either a cook-up or a pre-gelatinised starch, depending on the processing route, but the amount of process tolerance does vary.

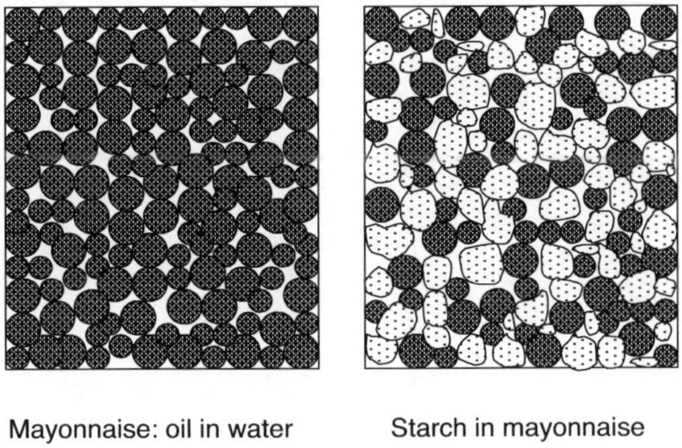

Fig. 16.23 Diagram of mayonnaise with and without starch.

Cook-up starches need to be pre-heated with the water in order to thicken. If they swell too quickly they break down, but if they do not swell enough then they will not attain the target viscosity.

For pre-gelatinised starches, a whole range of different technologies exists to get granules into different states before shearing to create the emulsion. Spray-cooked starches have the highest level of process stability for a given level of cross-linking and tend to give the best results. Figure 16.24 shows the starting point/swelling point for different starches in order to reach the end-goal. The further along the process curve they are the less tolerance they have to shear during manufacture.

Here, smooth glossy texture and appearance giving a rich creamy mouthfeel are generally the key characteristics that are required. Waxy starches dominate as they have the stability and range to be able to create different products. Each starch will impact the product with different flavours and colours and, again, this affects the choice.

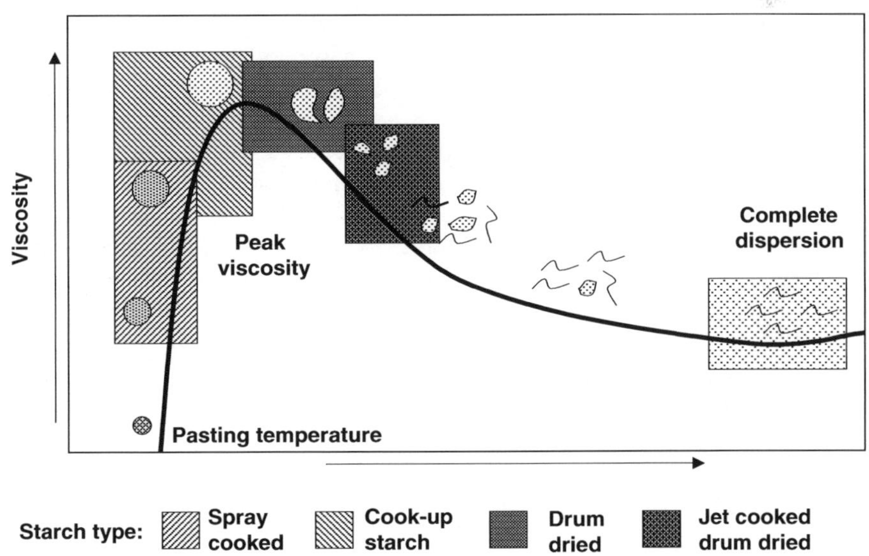

Fig. 16.24 Diagram of gelling network and Brabender profile.

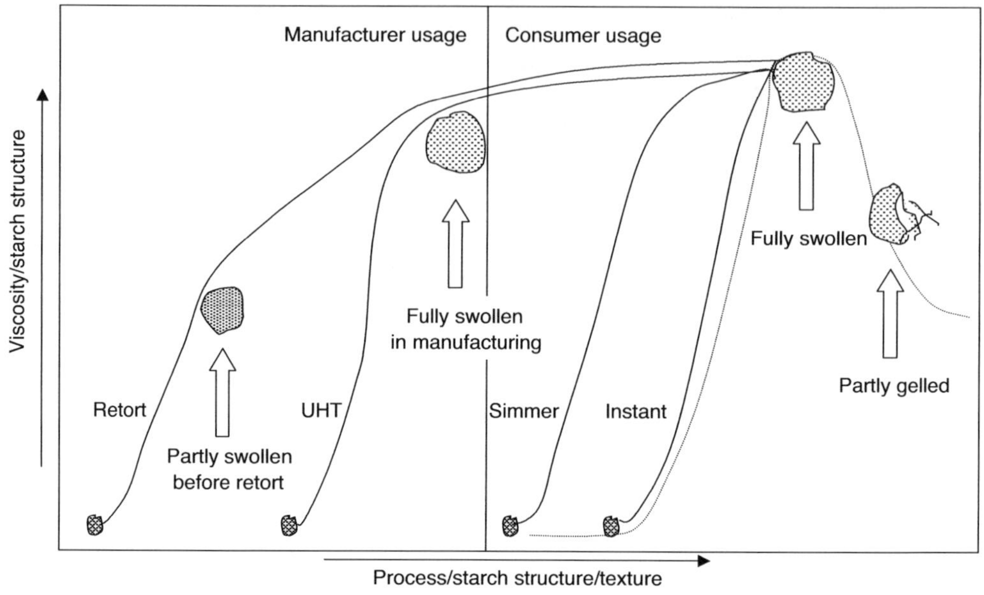

Fig. 16.25 Diagram of gelling network and Brabender profile.

16.7.3 Dairy

Dairy products are the next group of products along the profile from thickening to gelling: a wide range of different textures are observed depending on the area of application. In principle the first division that can be made is between non-fermented and fermented products, for example a ready-to-serve dessert versus a ready-to-eat yoghurt. In the former, all the key gelling and thickening characteristics come from the texturising agent whilst in the latter the culture and how it reacts with the proteins also play an important role. There is a wide range of textures from just-thickened to thickened, soft gelling to more firmly gelled to formed and moulded desserts.

Most ready-to-serve or ready-to-eat desserts are generally smooth, glossy, rich and creamy through the contributions from the dairy or dairy-alternative components and the thickening or gelling system. A range of starches is needed to cope with the range of conditions from instant products through to more highly modified starches that tend to be used for high-level processing, such as UHT, that is needed to give safe, neutral pH end-products if they have a long shelf life. Thus the same profiles exist for the desserts as the savoury soups and sauces.

Figure 16.25 shows this range and highlights one of the key differences: although many desserts are thickened, more have a gelled character. The move towards a gelled structure is achieved by releasing more of the amylose and amylopectin. In order to achieve this, the products rely on a degree of breakdown of the starch granule to release the gelling components. As the dessert moves more towards a gelled texture, there is also the potential to change from the waxy starches to those with more amylose, such as potato or corn starches, to give firmer gels. Hence, lightly cross-linked systems are used for those desserts that are prepared at home compared to the most highly linked for a UHT factory. Again in retorted products there may be a need to use two starch systems. For instant desserts there is a range of starches that can go from thickened through to gelled in nature if required.

Cultured products can be thickened, as in a stirred yoghurt, through to gelled, with a more set-type yoghurt, by using the same principles for the desserts.

16.7.4 Bakery

Bakery products often have to go through one more stage during processing that adds to the stability requirements and that is, of course, a baking step. The requirements for thickening and texturising characteristics in this market are generally for a filling or cream-type product where there is a strong role for the starch. Again the range is from a thickened through to a gelled system. For the former there is a need to hold the granules in their integral form to create a cream or perhaps for use in an acidic fruit filling. In some products there is also the need to have a degree of gelling in order to create structure for the cream filling.

The same is also the case for product made up cold, such as bakery creams and, again, the breakdown of the granular structure, to free the amylose to help in gelling, is an important application for these types of products.

16.7.5 Spreads

The properties of the starch products are rapidly moving towards those with more gelling characteristics and the intact granular structure has now all but gone. This is the realm of weakened granules that have stronger gelling properties. A spread is a gel with a high level of complexity and its production is a specialised business for the major skilled manufacturers.

In these systems another important property needs to be managed – the inversion of the emulsion from an oil-in-water to a water-in-oil emulsion. This has the effect of forming starch gels into more discrete particles rather than the continuous phase. In fact this is why the gelling properties are important as these particles add to the overall structure of the product. This product has moved from an oil-in-water emulsion to a water-in-oil emulsion where the starches play a key role in creating gelled particles that mimic the properties of full fat systems.

Figure 16.26 shows some of the key processes. Firstly, the water phase forms discrete droplets that are thickened to give the right consistency to the end-product. They need to be thick and gelled enough to ensure that some structure remains and that the water within this part of the material cannot escape but soft enough so that on spreading the gel particles spreads smoothly rather than staying as solid particles within the mix. Weakened starches which have a higher amylose content, such as potato or tapioca starches, have found much use in this application because they can form these spreadable gels. In fact low DE maltodextrins, at high concentration, have similar properties and can be used in the same way.

By this stage, the granular structure has gone and the starch properties that are more important are the gel strength, thermoreversibility of gelling, set temperature and ease of processing. This is also the one area, as discussed earlier, where the interactions between different hydrocolloids have been used to make lower fat systems and where de-mixing of hydrocolloids has been used to achieve new textures.

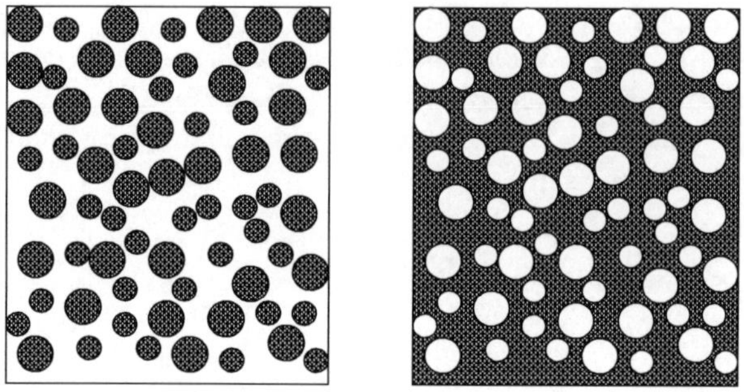

Fig. 16.26 Oil in water (mayonnaise) versus water in oil (spread) emulsions.

16.7.6 Confectionery

The confectionery market is at the furthest end of the spectrum in the journey from thickening through to gelling and it is in the area of sugar confectionery that there are the most applications for the gelling properties of starches. Within this category there are a number of subdivisions that need to be considered.

16.7.6.1 *Moulded confections*

In moulded confections, the end-product tends to be a relatively clear, highly elastic confection that has been formed by pouring a molten solution into a mould which is then cooled and often dried to reduce the moisture content to the desired level. The gelling properties for this product can be achieved by the use of starches. Success has been achieved with a whole range of starches but it is those with a higher amylose content that have proven most effective. In many cases, the goal has been to replace other hydrocolloids, such as gelatin, to allow vegetarian products to be produced in a cost-effective way.

The starches have to tolerate the various processes shown in Fig. 16.27. Initially all the ingredients are blended, heated and hydrated. Here, it is important that the starch dissolves and mixes easily and retains fluidity at the high temperatures used in moulding. This fluid mix is poured into moulds and then it must gel rapidly and give an elastic texture. There is always a balance between fluidity at high temperatures and forming good gels at cooler temperatures. In many cases, a further drying step is required to achieve the moisture content before the product is ready to pack. One of the major issues for these products is to retain stable characteristics throughout the shelf life. By careful selection of starches these gelled products have been developed.

16.7.6.2 *Chewy confections*

The same approach is taken for chewable products that need to be gelled and also stretch. Again the higher amylose starches are suitable. After moulding the product is stretched in preparation for the final forming and packing so the product must remain elastic for part of

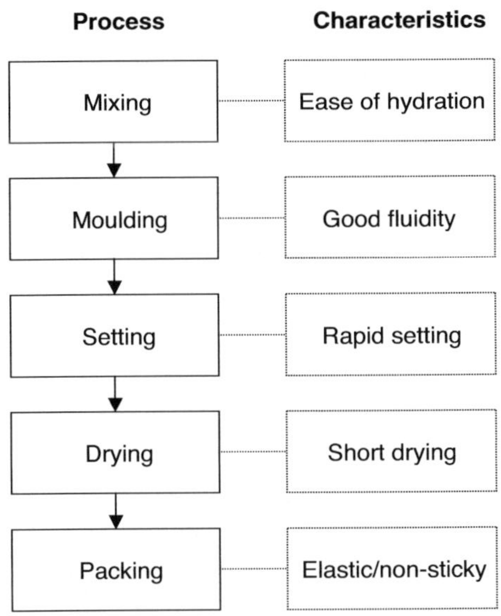

Fig. 16.27 Steps and viscosity needs for starches in a moulded confectionery product.

this period. In the final form they should not be too sticky to touch but should easily soften when chewed in the mouth.

16.7.6.3 *Formed confections*

In a slightly different area of confectionery, but also where the starch does not have integral granules, the starches can also act as good binders in products where a thickened paste is

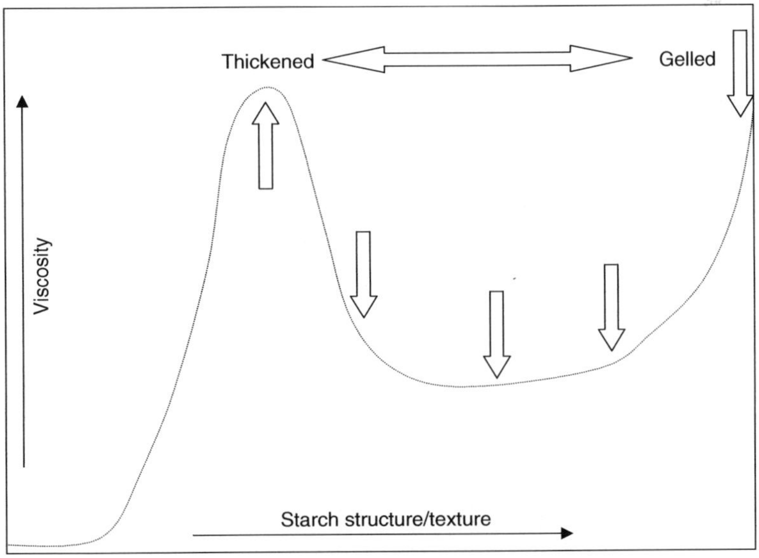

Fig. 16.28 Thickening to gelling in starches – a full spectrum of textures.

Table 16.6 A comparison of the physical, rheological and process stability of different starches.

	Viscosity w/w	Visual appearance	Native texture	Modified texture	Clarity	Flavour	Processing	Thickening	Gelling
Maize	Medium	Smooth	Short	Short	Opaque	Cereal	OK	**	****
Waxy maize	Medium	Smooth	Medium	Short	Cloudy	Some cereal	Very good	*****	**
Potato	High	Pulpy	Long	Short	Clear	Clean	Good	***	*****
Waxy potato	High	Pulpy or smooth	Medium	Short	Very clear	Very clean	Very good	*****	**
Waxy rice	Medium–low	Smooth	Medium	Short	Cloudy	Rice/cereal	Good	*****	**
Tapioca	Medium	Smooth	Long	Medium	Very clear	Very clean	Good	***	*****
Wheat	Medium	Smooth	Short	Short	Opaque	Strong cereal	OK	***	*****
Waxy wheat	Medium	Smooth	Short	Short	Opaque	Slight cereal	Good	****	**

Relative scale of thickening or gelling where ***** = greater extent than **** = greater than *** and so on.

formed and then sometimes further dried. The ability to bind in the cold and absorb enough moisture to hold the system together before the final drying step is important.

16.8 CONCLUSIONS

So, through this brief introduction to these main applications it can be seen that the thickening and gelling characteristics of starches can be applied across the whole spectrum of products from those that are fully thickened through to those that are fully gelled. Hence, starch is one of the most versatile texturising products as it can cover such a widespread range of applications and processes.

Figure 16.28 shows again where many of the products fit on the curve but it is only a guide as many products can be used in different applications and it is impossible to cover every single combination.

Starches exhibit many different properties as shown in Table 16.6. By looking at the key characteristics for each product, the food scientist will be able to choose the appropriate starch for their application. The table is not meant to be totally comprehensive but a guide or starting point to act as a springboard for new ideas.

The final twist is in the tail. These are the 'standard' properties that the starches exhibit but, by discussing the needs for a particular project, starch manufacturers will be able to offer those latest innovations that offer the latest and best twists.

This presents the scientist with an overview on how the benefits can be twisted to create new solutions. It is most important to remember that this is done to create products that delight the consumer and this should always be an important question in the food scientist's mind: what does the consumer need?

Acknowledgements

I would first of all like to thank Alan Imeson for asking me to contribute to such an excellent and broadly based volume that is an essential handbook for all good food scientists in need of gelling, thickening and texturising agents. I must thank all my colleagues at AVEBE FOOD but with special mention to the members the Food Innovation Centre, excellently guided by Dr Piet Buwalda, and for particular inspiration from Dr Jack Bergsma and Dr Cindy Semeijn who were able to explain many of the in-depth starch processes to me. In the preparation of the text I am indebted to Ronald Apeldoorn, who provided the general starch market data, and to my good colleagues Tommy Anzelius and Lars Svensson, who both bring the subject of starch and its application alive. I cannot close without giving my final thanks to Dr Rhian Davies, who has had the patience to support me during the stimulating and yet demanding processing of preparing the text.

Dedication

I would like to dedicate this chapter to my two sons, Rhodri and Aled.

References

Atwell, W.A. and Thomas, D.J. (1997) *Starches*. American Association of Cereal Chemists, St. Paul, MN.

Galliard, T. (ed.) (1987) *Starch; Properties and Potential*. Society of Chemical Industry, John Wiley and Sons, Chichester, UK.

GIRACT database (2006) www.giract.com

Imeson, A. (ed.) (1999) *Thickening and Gelling Agents for Food*, 2nd edn. Blackie Academic and Professional, London.

McKenna, B.M. (ed.) (2003) *Texture in Food Volume 1: Semi-Solid Foods*. Woodhead Publishing, Cambridge.

Phillips, G.O. and Williams, P.A. (eds) (2000) *Handbook of Hydrocolloids*. Woodhead Publishing, Cambridge.

Whistler, R.L., BeMiller, J.N. and Paschall, E.F. (eds) (1984) *Starch Chemistry and Technology*, 2nd edn. Academic Press, Orlando, FL.

17 Xanthan Gum

Graham Sworn

ABSTRACT

Xanthan gum is a high-molecular-weight polysaccharide secreted by the microorganism *Xanthomonas campestris* and produced commercially in a batch fermentation process. It hydrates in cold water to give a viscous solution with pseudoplastic flow behaviour. This gives excellent suspension and cling at low shear and excellent mouthfeel and pouring properties at high shear. The xanthan gum molecule has a cellulosic backbone with side chains that wrap around the backbone protecting it and conferring excellent stability across a wide pH range and tolerance of high salt concentrations and ingredients such as glycerol and alcohol. The rigid backbone helps to maintain viscosity during heating. Xanthan gum shows synergistic thickening with guar gum and forms very elastic cohesive gels with locust bean gum and konjac mannan. Non-food uses include oil field, personal care, pharmaceutical and home care products. Typical food applications include sauces and dressings, baked goods, beverages, desserts and ice creams.

17.1 INTRODUCTION

Xanthan gum is a high-molecular-weight extracellular polysaccharide secreted by the microorganism *Xanthomonas campestris* and is produced commercially in a fermentation process. It is soluble in cold water and has a very wide range of applications. It was first discovered in the 1960s and commercialised in the 1970s. With an annual sales volume of approximately 46 000 metric tonnes, xanthan gum applications are split approximately 50/50 between food and non-food. Non-food includes oil field, personal care, pharmaceutical and home care. Typical food applications include sauces and dressings, baked goods, beverages, desserts and ice creams. The total textural ingredients market for food is currently estimated to be worth 2.8 billion US dollars and xanthan gum represents approximately 11% of this.

17.2 PRODUCTION

Xanthan gum is produced at the cell wall surface by the bacterium *X. campestris* during its normal life cycle (Harding *et al.*, 1995). Commercial production is carried out batchwise by submerged fermentation with strong agitation. The sterile medium contains carbohydrates, a nitrogen source, magnesium sulphate and other trace minerals. After initial inoculation with the selected strain, fermentation is continued for approximately 3 days at 30°C. At the end of the fermentation, the broth undergoes a sterilisation treatment to eliminate any viable microorganisms. The xanthan gum is then recovered by precipitation with alcohol. After separation of the fibres by centrifugation, they are dried and milled before packaging.

17.3 CHEMISTRY

The primary structure of xanthan gum is shown in Fig. 17.1 and consists of a cellulosic backbone of β-(1,4) linked D-glucose units substituted on alternate glucose residues with a trisaccharide side chain. The trisaccharide side chain is composed of two mannose units separated by a glucuronic acid (Jansson *et al.*, 1975; Melton *et al.*, 1976). Approximately half the terminal mannose units are linked to a pyruvate group and the non-terminal residue usually carries an acetyl group. The carboxyl groups on the side chains render the gum molecules anionic. Xanthan gum has a molecular weight of about 2×10^6 daltons with a narrow-molecular-weight distribution compared to most polysaccharides.

X-ray diffraction studies on xanthan gum fibres have identified a right-handed, fivefold helix conformation (Moorhouse *et al.*, 1977). In this conformation, the side chains are aligned with the backbone and stabilise the overall conformation. In solution, the side chains wrap

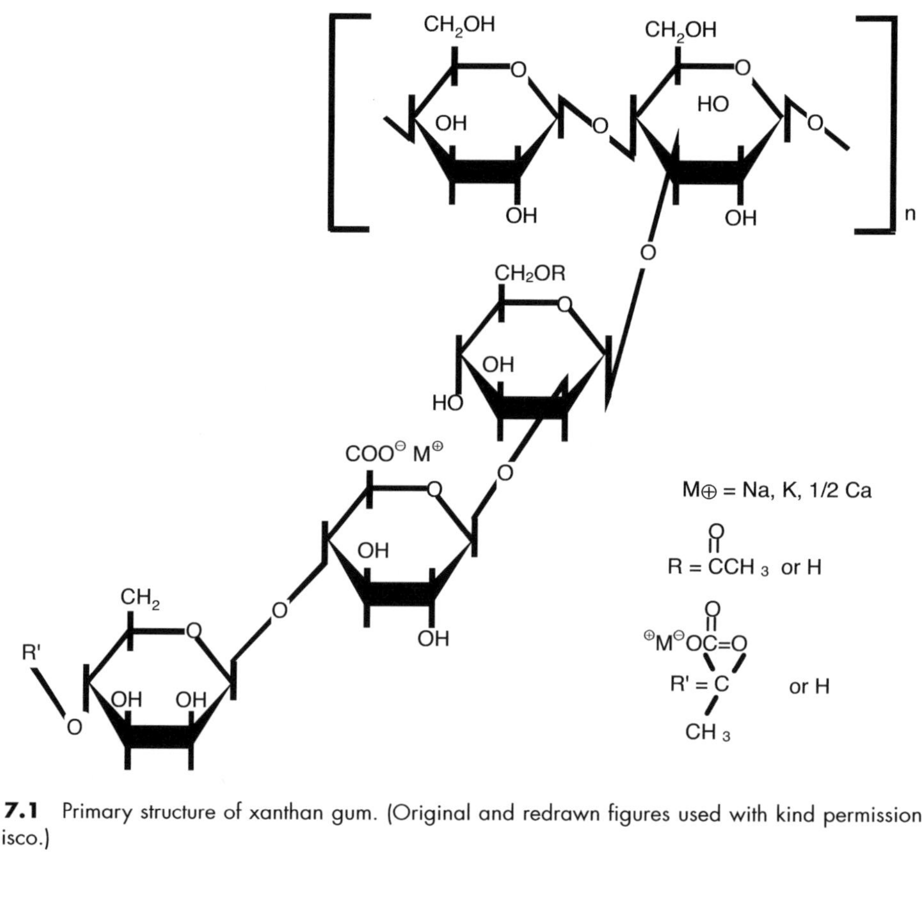

Fig. 17.1 Primary structure of xanthan gum. (Original and redrawn figures used with kind permission of Danisco.)

around the cellulose-like backbone thereby protecting it. It is thought that this is responsible for the excellent stability of xanthan gum under adverse conditions.

Solutions of xanthan gum undergo a conformational transition during heating which is believed to be associated with the change from a rigid ordered state at low temperature to a more flexible, disordered state at high temperatures. This conformational change was first observed as a sigmoidal change in viscosity (Jeanes *et al.*, 1961). Measurements by optical rotation, calorimetry and circular dichroism have shown that this conformational change coincides with the viscosity change (Morris *et al.*, 1977; Kawakami *et al.*, 1991). The temperature at which the conformational transition occurs is primarily dependent on the ionic strength and on structural features such as the pyruvic acid and acetic acid contents of the xanthan molecule (Morris, 1977; Morrison *et al.*, 2004). At xanthan concentrations up to approximately 0.3% in deionised water, the thermal transition occurs at around 40°C, however, in the presence of low levels of salt, typical of those used in food products, the thermal transition occurs at temperatures above 90°C. The presence of low levels of salts helps to maintain the rigid ordered conformation of xanthan gum and the relative insensitivity of the viscosity to additional salt and elevated temperatures is a result of the molecular rigidity (Morris, 1977).

17.4 SOLUTION PREPARATION

Dispersion and hydration are the first steps in all applications of hydrocolloid thickeners. To achieve the optimum functionality of any hydrocolloid, it is important to ensure that the product is properly hydrated before use. The main factors that affect the hydration of xanthan gum are dispersion, mixing speed, particle size and the composition of the solvent.

Xanthan gum is a fast-hydrating water-soluble hydrocolloid which can be dissolved at room temperature. Achieving full viscosity development requires:

- a minimum quantity of water and
- a uniform dispersion of the gum in the water.

Before discussing the preparation of xanthan gum solutions, a distinction needs to be made between solubility and dispersibility:

- Dispersibility is the ease of separation of the individual gum particles as the xanthan gum is put into the liquid.
- Solubility is the ease with which the individual particles hydrate and dissolve.

A very fine powder is difficult to disperse, but, once dispersed, it is quick to hydrate, whereas a course mesh powder is easily dispersed, but hydrates more slowly.

Hydration time depends on:

- the effectiveness of the dispersion,
- the mesh size of the xanthan particles,
- the type of solvent and
- other ingredients in the recipe.

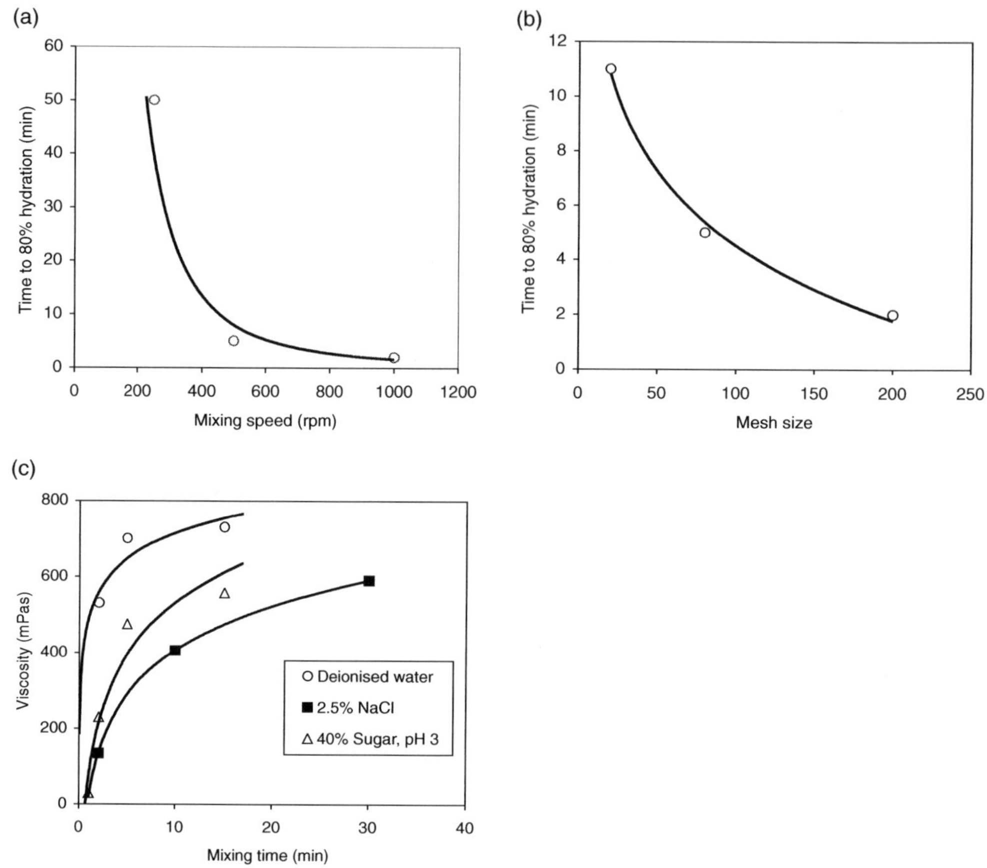

Fig. 17.2 Effect of (a) mixing speed, (b) mesh size and (c) solvent on the hydration of 0.3% xanthan gum. (Original and redrawn figures used with kind permission of Danisco.)

For xanthan gum to hydrate efficiently, the individual gum particles must be well dispersed in the solvent. Poor dispersion leads to lumping of particles during mixing, which results in the formation of swollen lumps (sometimes referred to as 'fish eyes'). Severe lumping prevents complete hydration which reduces functionality. Generally, the larger the particle size the better the dispersibility of the xanthan. Dispersion can also be improved by pre-blending with other ingredients such as sugar, starch or oils. For example, pre-blending 1 part xanthan with 5 parts sugar will assist effective dispersion of the gum.

Figure 17.2a shows the effect of mixing speed on the rate of hydration of 0.3% xanthan gum in 1% sodium chloride solution. As mixing speed is increased, the hydration time is reduced. These tests were carried out on well-dispersed samples. Particle size will have an influence on both dispersion and hydration. As the particle size increases, the xanthan becomes easier to disperse but slower to hydrate. Figure 17.2b illustrates the effect of particle size on the hydration rate of 0.3% xanthan gum in 1% sodium chloride solution. Figure 17.2c shows the effect on hydration of some typical solvents for food products such as salt and sugar. Generally, high ionic strength or high-solids slow-down hydration. In poor solvents, the gum will be easier to disperse but slower to hydrate. As a general guide, hydration takes between 15 and 30 min with a good high-speed mixer and up to 1 h with a slow-speed mixer. It is recommended, where possible, to hydrate the xanthan gum in water before the addition of other ingredients such as salt and acid.

Some suppliers offer a range of xanthan gum products designed for specific dispersion and hydration needs. The type of properties required will depend on the process and application. For example, in poor mixing conditions, products with larger particle size are preferred for better dispersibility. In dry mix applications, such as baking and beverages, where rapid hydration is required, products with a small particle size are preferred. In these applications, dispersion is not normally a problem due to the presence of other ingredients such as sugar, salt or proteins, which act as dispersing agents.

In some applications, the amount of water available for hydration of the ingredients is limited. In these cases, xanthan gum can be prepared as a concentrated solution (2–3%) which can then be added in the correct proportion to the final recipe.

17.5 RHEOLOGY

The flow behaviour of xanthan gum solutions can be modelled using the Cross equation and this is illustrated schematically in Fig. 17.3.

The solutions exhibit Newtonian viscosity (η_0) at very low shear rates, followed by a pseudoplastic region as shear rate increases and, finally, an upper Newtonian viscosity (η_∞) at very high shear rates. This flow behaviour is a result of intermolecular association among xanthan polymer chains which results in the formation of a complex network of entangled rod-like molecules. At very low shear rates, the disentanglement due to shear is slower than the formation of new entanglements and so viscosity is constant. At higher shear rates, these weakly bound aggregates are progressively disrupted under the influence of applied shear resulting in the pronounced pseudoplastic flow of xanthan gum solutions. At very high shear rates, no further disruption of the entanglements can take place and the viscosity is again constant. In addition, this highly ordered network of entangled stiff molecules results in xanthan solutions having the viscoelastic characteristics of a weak gel. This is illustrated in Fig. 17.4, which shows that the xanthan solution has a dominant elastic response to frequency, whereas guar, for instance, which does not form an ordered network in solution, has a dominant viscous response.

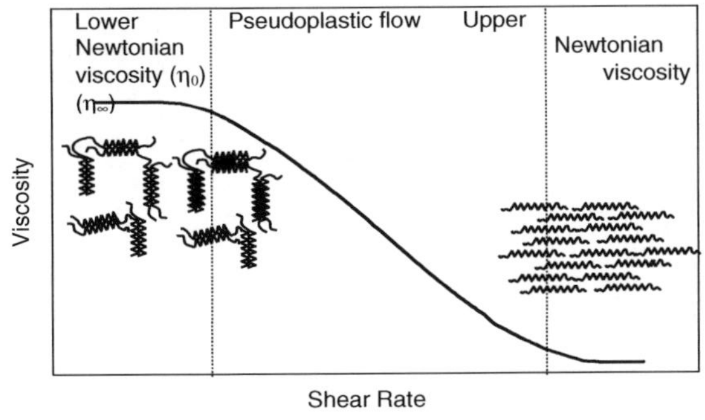

Fig. 17.3 Schematic representation of the molecular origins of the flow behaviour of xanthan gum solutions. (Original and redrawn figures used with kind permission of Danisco.)

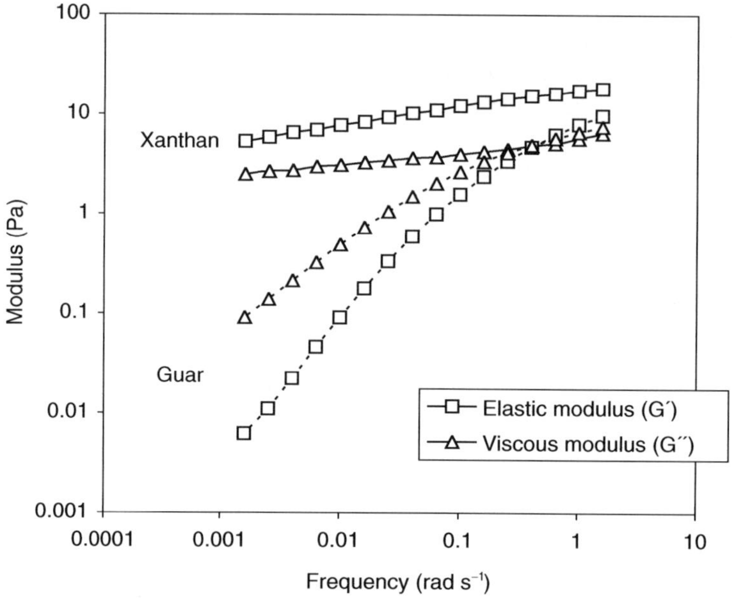

Fig. 17.4 Viscoelastic properties of xanthan gum compared to guar gum (0.6% gum in tap water). (Original and redrawn figures used with kind permission of Danisco.)

In practical terms, these rheological properties translate into the very effective suspending and flow properties exhibited by xanthan gum solutions. High viscosity at low shear rates provides excellent stabilising properties, whilst the highly pseudoplastic flow provides good mouthfeel and pouring qualities. Xanthan gum has the ability to develop an extremely high viscosity even at low concentration. The viscosity/concentration relationship is shown in Fig. 17.5.

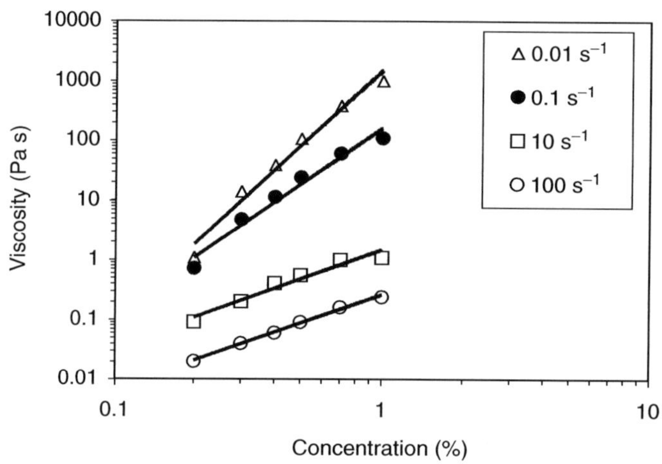

Fig. 17.5 Effect of xanthan gum concentration on viscosity in 2.5% sodium chloride solution, pH 3.5 measured at various shear rates. (Original and redrawn figures used with kind permission of Danisco.)

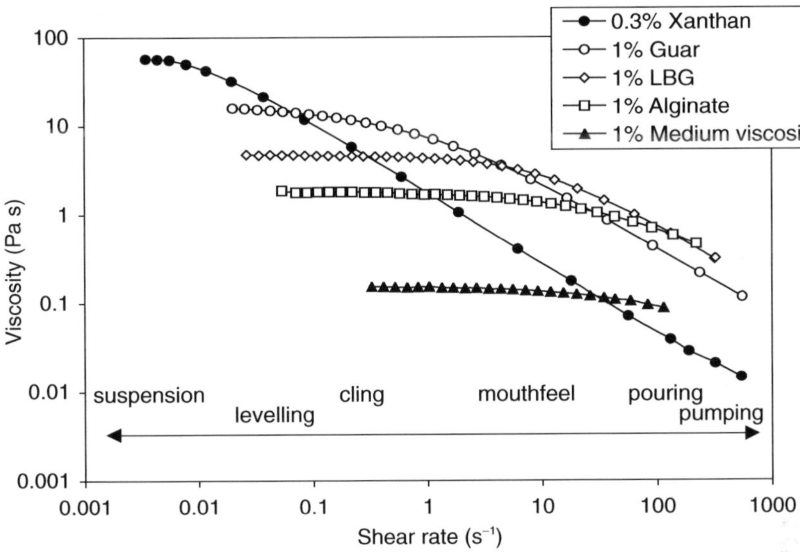

Fig. 17.6 Comparison of the flow behaviour of xanthan gum compared to other thickeners at shear rates typical of some food processes. (Original and redrawn figures used with kind permission of Danisco.)

Xanthan is a very effective thickener and stabiliser compared to other hydrocolloids. It has a higher low-shear viscosity at lower concentrations and is more pseudoplastic at the shear rates typical of food processes compared to other hydrocolloid thickeners (see Fig. 17.6).

17.6 STABILITY AND COMPATIBILITY

17.6.1 Temperature

The temperature stability of xanthan gum compares favourably with other thickeners. Xanthan solutions exhibit exceptional stability during heating, even in the presence of salts and/or acids. As temperature increases, the viscosity of the xanthan solution decreases, but this viscosity is recovered upon cooling. Figure 17.7 illustrates the effect of a temperature cycle between 20°C and 70°C on the viscosity of xanthan gum. It shows that although viscosity is lower at 70°C compared to 20°C, the majority of the viscosity is recovered when the solution is cooled again. When severe heat treatments are applied, such as pasteurisation or sterilisation up to 130°C (266°F) for a few minutes, the solution viscosity remains practically unchanged after cooling. Many other commonly used thickeners lose their viscosity at high temperature and do not recover this on cooling (Urlacher and Dalbe, 1997).

17.6.2 pH

The viscosity of xanthan gum solutions is independent of pH over a wide range.

Figure 17.8 shows that the flow curves of 0.3% xanthan in 1% sodium chloride solution are super-imposable between pH 10 and 4. Only extreme pH conditions (pH 11–12, pH 1–2)

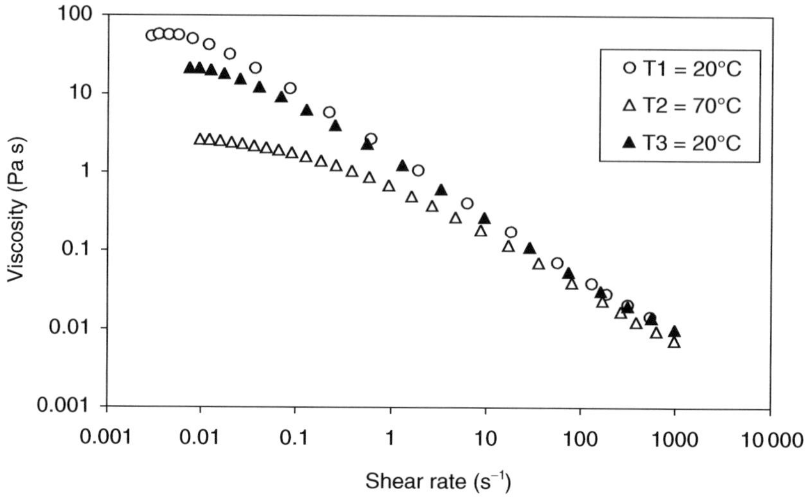

Fig. 17.7 Effect of a temperature cycle between 20°C and 70°C on the viscosity of 0.3% xanthan in 1% sodium chloride solution. (Original and redrawn figures used with kind permission of Danisco.)

affect viscosity. For example, the viscosity of xanthan solutions decreases below approximately pH 3.0, but this viscosity is recoverable when the solution is neutralised. Xanthan gum is compatible with most organic acids including acetic, citric, lactic, tartaric and phosphoric acids, and it is more stable than other commonly-used thickeners. The compatibility with various acids and bases is summarised in Table 17.1. Xanthan gum can be hydrated directly in an acidic solution; however, it is recommended to prepare the gum solution first and then add the acid to achieve the best results.

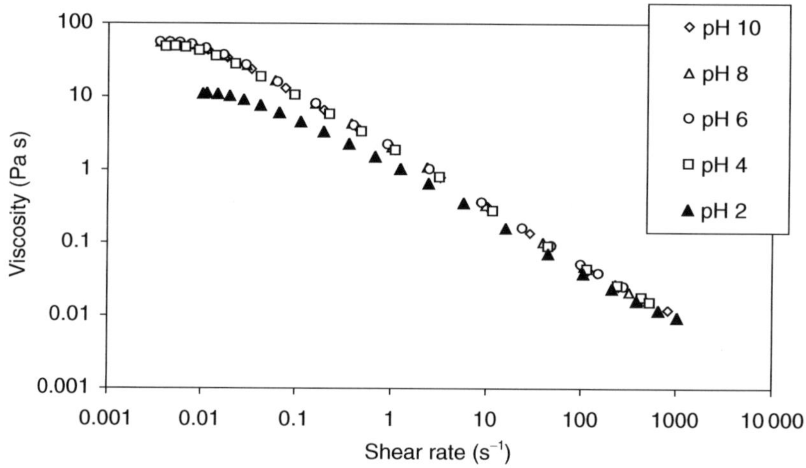

Fig. 17.8 Effect of pH on the flow curves of 0.3% xanthan in 1% sodium chloride solution. (Original and redrawn figures used with kind permission of Danisco.)

Table 17.1 Compatibility of xanthan with acids, bases and chlorinated solutions.

Acid/base	Maximum concentration (%)
Citric acid	20.0
Acetic acid	20.0
Hydrochloric acid	5.0
Sulphuric acid	10.0
Phosphoric acid	30.0
Caustic soda	5.0
Sodium carbonate	20.0
Trisodium phosphate	15.0
Sodium hypochlorite	0.5
Sodium dichloroisocyanurate	0.5

Source: Reproduced with kind permission of Danisco.

17.6.3 Salts

As already discussed in Section 17.4, the hydration rate of xanthan is decreased in the presence of salts particularly above 1–2%. It is recommended to hydrate xanthan in the water before the addition of salt. Once hydrated, xanthan has very good salt tolerance and up to 20–30% salt can be added without adversely affecting the viscosity. Table 17.2 shows the stability of xanthan gum solution viscosity in various salt solutions.

The effect of salts on the viscosity is dependent on the concentration of the xanthan gum. At low concentrations of xanthan, below approximately 0.3%, addition of salts results in a slight decrease in the viscosity. At concentrations above 0.3% addition of salts results in an increase in viscosity. This is illustrated in Fig. 17.9, which shows the viscosity/concentration profile of xanthan gum in deionised water compared to that in 1% potassium chloride solution. The viscosity of the xanthan gum solution is equivalent in the two solvents at 0.3% gum. Below this concentration, xanthan has the higher viscosity in deionised water. Above this concentration, the xanthan has a higher viscosity in the salt solution. The effect of salt on the viscosity occurs at very low salt levels. Typically, this effect will be seen with the addition of as little as 0.1% sodium chloride. Further addition of salt will have little or no further effect on the viscosity (see Fig. 17.10) (Pettitt, 1983).

Table 17.2 1% Xanthan solution viscosity over time in different salts.

	Viscosity over time (mPa s), Brookfield LVT 60 rpm			
Salts (%)	**4 h**	**7 days**	**30 days**	**90 days**
Water	1500	1500	1500	1500
NaCl				
5	2300	2050	2000	1850
30	1700	2300	2800	2800
KCl				
5	2100	2000	2000	1900
30	2900	3000	3000	3000
$CaCl_2$				
5	2650	2350	2300	1900
15	3400	3250	3150	2750
30	Gel	Gel	Gel	Gel

Source: Reproduced with kind permission of Danisco.

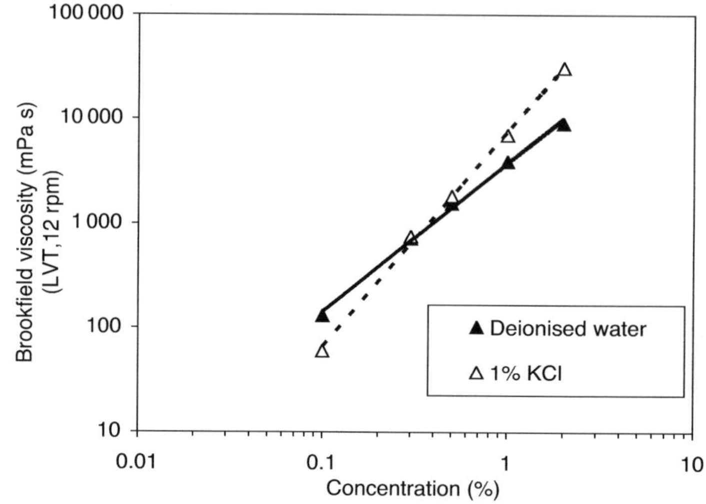

Fig. 17.9 Effect of salt on the viscosity – concentration profile of xanthan gum. (Original and redrawn figures used with kind permission of Danisco.)

17.6.4 Other ingredients

Generally, xanthan gum is not soluble in organic solvents, although it will hydrate directly in glycerol at 65°C (150°F). Nevertheless, after hydrating xanthan gum in water, up to 50% of ethanol or isopropanol can be added without precipitating the gum.

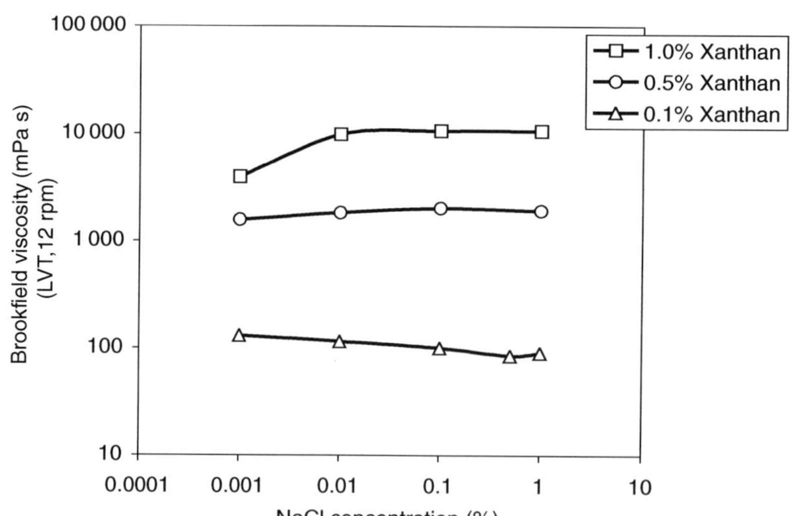

Fig. 17.10 Effect of salt concentration on the viscosity of xanthan gum solutions at various xanthan gum concentrations. (Original and redrawn figures used with kind permission of Danisco.)

Xanthan gum is compatible with all the commonly used thickeners such as starch, carrageenan, pectin, gelatin, agar, alginate and cellulose derivatives. Xanthan does not contain any cellulose-degrading enzymes.

Xanthan gum is very resistant to enzyme degradation. It can be used in the presence of many common enzymes whether they originate from other raw materials or they are added as commercially available products for processing, such as amylases, proteases, pectinases and cellulases.

17.7 INTERACTIONS

17.7.1 Galactomannans and glucomannans

Xanthan gum enjoys strong synergy with galactomannans such as locust bean gum (LBG), guar gum, tara gum and cassia gum as well as with glucomannans such as konjac mannan. Guar gum and LBG are neutral polysaccharides consisting of a linear mannan backbone, the mannose unit being linked together by β-(1,4) glycosidic bonds. Single galactose units are attached as side chains to the mannan backbone by α-(1,6) glycosidic linkages. Galactomannans can be characterised by their galactose content. LBG contains between 17% and 26% galactose, whereas guar typically contains between 33% and 40% galactose. Both galactose-rich ('hairy') and galactose-free ('smooth') regions can be found in guar and LBG molecules, and it is the distribution and number of galactose groups that influence their interaction with xanthan gum.

In solution, the galactose-free (smooth) regions of the mannose backbone form associations with the ordered xanthan helices (Dea and Morris, 1977). These associations result in a synergistic increase in viscosity in the case of guar or gelation in the case of LBG. Generally, the greater the number of galactose-free regions, the greater the synergy with xanthan gum.

Konjac glucomannan forms strong elastic gels after heating and cooling it in mixtures with xanthan gum. Cassia gum exhibits similar behaviour to LBG and tara gum has synergy intermediate to that of guar and LBG.

17.7.1.1 *Guar gum*

Guar gum is the ground endosperm from the seeds of the guar plant *Cyamopsis tetragonoloba*. Synergy between xanthan and guar gum can been characterised by the gain in viscosity and elastic modulus (G') in the mixed system compared to the simple addition of the contributions of the individual components (see Fig. 17.11). The maximum synergy occurs at a ratio of 80:20 guar gum:xanthan gum for both viscosity and elastic modulus (see Fig. 17.12). The maximum synergy should not be confused with the maximum viscosity. Elastic modulus and viscosity of these xanthan/guar blends are always lower than with a 100% xanthan solution (see Fig. 17.11).

The synergy is affected by the quality of the guar gum. For example, a direct correlation can be made between molecular weight of the guar gum and the synergy, in terms of viscosity and elastic modulus (see Fig. 17.13). The structure of the xanthan gum can also influence the synergy with guar gum and LBG (Shatwell *et al.*, 1991a,b; Morrison *et al.*, 2004). Generally, as the number of acetate groups on the non-terminal mannose decreases, the synergy with guar gum and LBG increases (see Fig. 17.14).

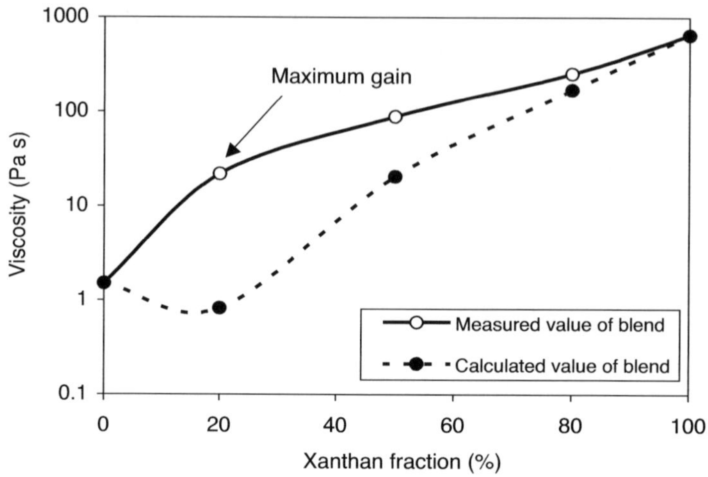

Fig. 17.11 Comparison between calculated values of viscosity from the individual components and the measured values of a mixture of 0.6% xanthan and guar gum in deionised water. (Original and redrawn figures used with kind permission of Danisco.)

17.7.1.2 *Locust bean gum*

LBG is the ground endosperm from the seeds of the carob tree *Ceratonia siliqua* and is also known as carob gum. LBG has a lower degree of galactose substitution than guar gum and, as a result, has a greater synergistic interaction with xanthan gum. LBG is insoluble at

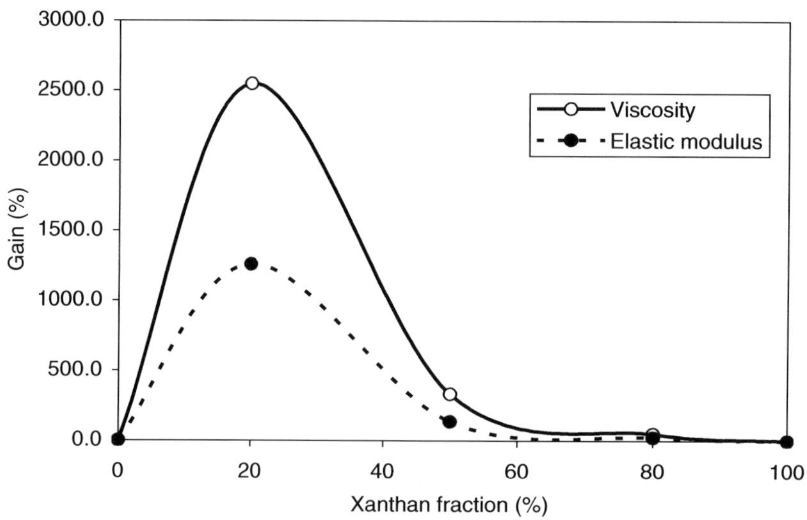

Fig. 17.12 Synergistic gain obtained in terms of viscosity and viscoelasticity as a function of xanthan–guar ratio measured from the comparison of the calculated values for the individual components and the measured values of the mixture (0.6% total gum in deionised water). (Original and redrawn figures used with kind permission of Danisco.)

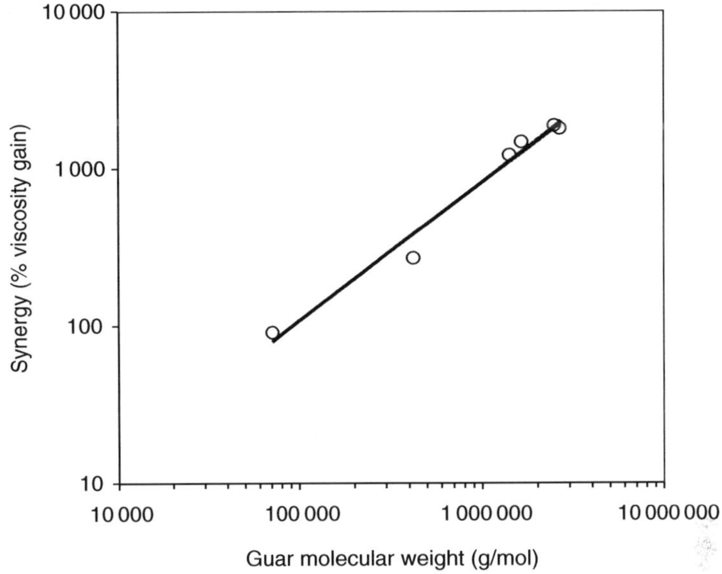

Fig. 17.13 Correlation between guar gum molecular weight and the degree of synergy with xanthan measured from the comparison of the calculated values for the individual components and the measured values of the mixture (0.6% total gum, 80:20 guar:xanthan in deionised water). (Original and redrawn figures used with kind permission of Danisco.)

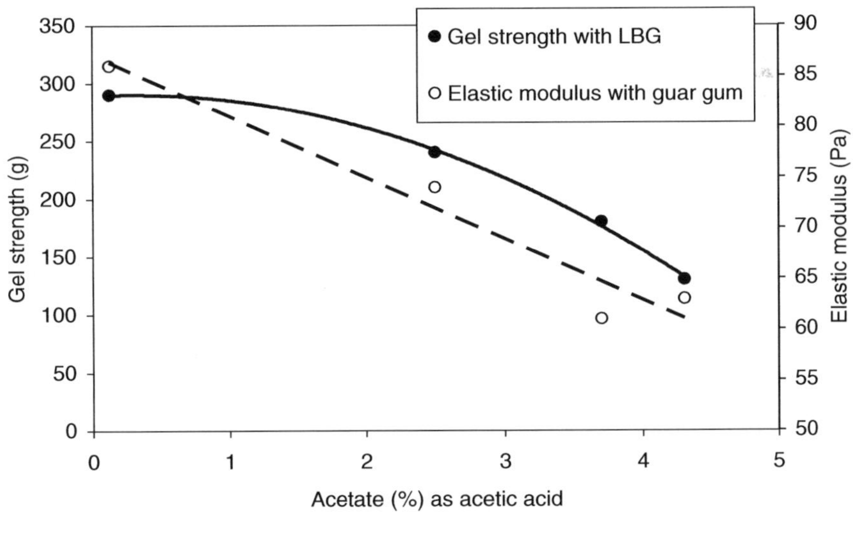

Fig. 17.14 Effect of acetate groups on xanthan synergy with guar gum (1:1 xanthan:guar, 0.5% gum in 5% sucrose) and LBG (1:1 xanthan:LBG, 1% gum in 1% potassium chloride solution). (Original and redrawn figures used with kind permission of Danisco.)

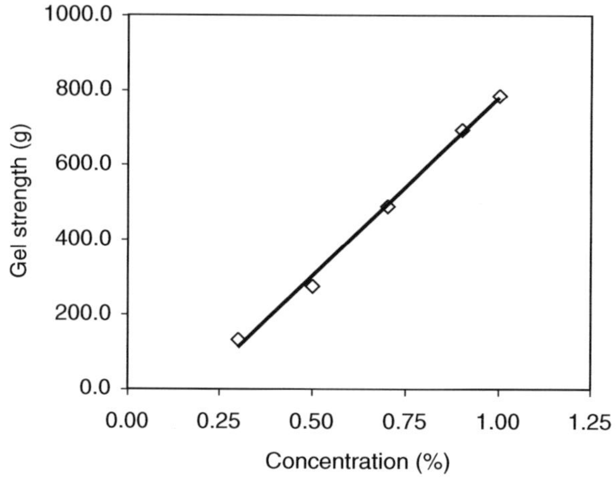

Fig. 17.15 Effect of concentration on the gel strength of xanthan–LBG gels (1 : 1 in 1% potassium chloride solution). (Original and redrawn figures used with kind permission of Danisco.)

room temperature and must be heated to approximately 90°C to fully hydrate. Xanthan is able to form strong, elastic, demouldable gels with LBG at concentrations above approximately 0.3% (see Fig. 17.15) when heated and cooled in mixtures. The optimum gel strength is found at a ratio of approximately 60:40 xanthan:LBG (see Fig. 17.16). At low gum concentration, this synergy can be used to increase the thickening impact of xanthan or LBG alone. The degree of synergy is also influenced by pH and salt content. High salt concentrations and low pH values tend to reduce the interaction between the xanthan and the LBG.

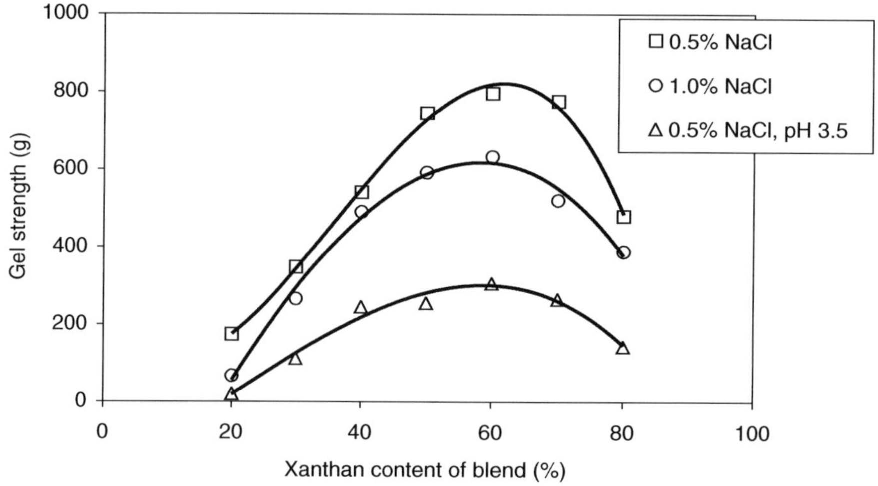

Fig. 17.16 Effect of gum ratio and solvent on the gel strength of 1% xanthan–LBG gels. (Original and redrawn figures used with kind permission of Danisco.)

17.8 APPLICATIONS

Table 17.3 provides a summary of xanthan gum applications, functionalities and the typical use levels required. It also illustrates the diversity of applications for xanthan gum in the food industry. Below is a brief description of some of the main application areas.

17.8.1 Dressings and sauces

This is the largest application area for xanthan in the food industry. Xanthan stability and rheological properties are ideally suited to the demands of the manufacturer and the consumer. Xanthan provides stability to emulsions and can also suspend particles and spices. The pseudoplastic flow properties provided by the xanthan gum make dressings and sauces easy to pour but still with good coating and cling to meats or salads. The pseudoplasticity also enables easy pumping and reduced splashing during filling. Cold solubility enables use in dry mixes and in cold processed dressings. The stability of xanthan to salt, temperature and pH also help to maintain product viscosity and texture during processes such as pasteurisation, ultra-high temperature (UHT), microwave cooking and freeze/thaw cycles, as well as during the shelf life of the product. Texture can also be varied through use of the synergistic interactions with galactomannans such as guar and LBG.

A typical procedure for the preparation of a dressing would be as follows:

- Disperse the xanthan in two to five times its weight of oil and add the slurry to the water with vigorous stirring for approximately 15 min.
- Add the remaining dry ingredients and, if included, any paste-like ingredients such as mustard, eggs or tomato.
- Add the remaining oil and vinegar and homogenise in a colloid mill.

Xanthan should be used at 0.1–0.3% for high oil content (50–60%) and at 0.3–0.5% for low oil (10–20%) dressings.

In sauces and gravies, xanthan gum gives a high viscosity at acidic and neutral pH and in high salt concentrations. The temperature stability is also important to maintain the viscosity when eaten hot. However, due to the thermoreversible transition of xanthan, the viscosity is greatly reduced at sterilisation temperatures ensuring good thermal penetration in UHT and retorted foods.

17.8.2 Bakery products

In cakes, the high viscosity of xanthan at low shear rates helps to improve the even distribution of solid particles, such as fruit and chocolate pieces, by suspending them in the batter before baking. Xanthan can also increase the shelf life and maintain a soft texture in baked products through improved water retention. It also helps to control recrystallisation of the amylose thereby retarding staling. Cakes made with xanthan also show a higher volume and reduced crumbliness. Typically, a 200 mesh, (75-μm) xanthan product will be used since this can be pre-blended with the flour to provide good dispersion and very rapid hydration.

Table 17.3 Summary of xanthan gum applications, functionalities and the typical use levels.

Application / Function		Salad dressings	Gravies, relishes, canned soups and sauces	Instant mix sauces, soups, drinks	Pickles, aspics	Processed meat	Frozen foods	Dairy products, flavoured milk	Whipped cream, mousses	Instant desserts	Ice creams, sherbets, sorbets	Bakery products	Fruit drinks, flavouring concen-trates	Jams and jellies, fruit prepara-tions, toppings
Recommended use level (%)		**0.1–0.5**	**0.1–0.4**	**0.1–0.2·**	**0.1- 0.5**	**0.05–0.2**	**0.2–0.5**	**0.05–0.3**	**0.1–0.3**	**0.1- 0.3·**	**0.1–0.4**	**0.05–0.2**	**0.05–0.2**	**0.2–0.5**
Rheology	Suspension power	◆	◆		◆	◆	◆	◆	◆		◆		◆	◆
	Viscosity	◆	◆	◆			◆		◆	◆				
	Pseudoplasticity	◆		◆				◆	◆			◆	◆	◆
Stability	Salts	◆	◆	◆	◆	◆	◆						◆	
	Acids	◆	◆		◆		◆	◆					◆	◆
	Temperature	◆	◆	◆	◆	◆		◆	◆		◆	◆	◆	◆
Miscellaneous	Synergy	◆				◆		◆		◆	◆	◆	◆	◆
	Cold solubility	◆		◆		◆				◆		◆	◆	
	Syneresis control		◆	◆	◆	◆	◆	◆	◆		◆			◆
	Texture modifier	◆		◆				◆	◆	◆	◆			◆
	Cling	◆	◆	◆	◆		◆							◆
	Crystallisation control		◆				◆				◆			
	Whipping control							◆	◆	◆	◆	◆		

Source: Reproduced with kind permission of Danisco.
Note: ◆ indicates property conferred by xanthan gum in this application.

17.8.3 Dairy products

In dairy products such as desserts, xanthan is most often used in combination with other hydrocolloids such as starch and carrageenan. Low levels of xanthan gum are able to improve the textural properties of the product by improving smoothness and reducing syneresis. In aerated desserts, it helps with air retention, and in layered desserts, it can reduce the migration of colours between layers.

17.8.4 Instant mixes

The very rapid hydration of xanthan gum makes it ideal for instant dry mix products such as beverages, soups and desserts. Good dispersion is achieved through blending with other ingredients such as sugar and proteins. Some special grades of xanthan are available that combine good dispersion with rapid hydration. These can be used in low-calorie instant beverages in which there is often a low level of other ingredients available for the dispersion of the xanthan gum. The xanthan will thicken the product, give body and help with suspension of particles. Typical use levels are between 0.05% and 0.2% in the final product.

References

Dea, I.C.M. and Morris, E.R. (1977) Synergistic xanthan gels. In: P.A. Sandford and A. Laskin (eds), *Extracellular Microbial Polysaccharides*. ACS symposium series no. 45, American Chemical Society, Washington, DC, pp. 174–182.

Harding, N.E., Ielpi, L. and Cleary, J.M. (1995) Genetics and biochemistry of xanthan gum production by *Xanthomonas campestris*. In: Y.H. Hui and G.C. Khachatourians (eds), *Food Biotechnology Microorganisms*. VCH Publishers, New York, pp. 495–514.

Jansson, P.E., Kenne, L. and Lindberg, B. (1975) Structure of the exocellular polysaccharide from *Xanthomonas campestris*. *Carbohydrate Research*, **45**, 275–282.

Jeanes, A., Pittsley, J.E. and Senti, F.R. (1961) Polysaccharide B-1459: a new hydrocolloid polyelectrolyte produced from glucose from bacterial fermentation. *Journal of Applied Polymer Science*, **5**, 519–526.

Kawakami, K., Okabe, Y. and Norisoye, T. (1991) Dissociation of dimerized xanthan in aqueous solution. *Carbohydrate Polymer*, **14**, 189–203.

Melton, L.D., Mindt, L., Rees, D.A. and Sanderson G.R. (1976) Covalent structure of the polysaccharide from *Xanthomonas campestris*: evidence from partial hydrolysis studies. *Carbohydrate Research*, **46**, 245–257.

Moorhouse, R., Walkinshaw, M.D. and Arnott, S. (1977) Xanthan gum – molecular conformation and interactions. In: P.A. Sandford and A. Laskin (eds), *Extracellular Microbial Polysaccharides*. ACS symposium series no. 45, American Chemical Society, Washington, DC, pp. 90–102.

Morris, E.R. (1977) Molecular origin of xanthan solution properties. In: P.A. Sandford and A. Laskin (eds), *Extracellular Microbial Polysaccharides*, ACS symposium series no. 45, American Chemical Society, Washington, DC, pp. 81–89.

Morris, E.R., Rees, D.A., Young, G., Walkinshaw, M.D. and Darke, A. (1977) Order–disorder transition for a bacterial polysaccharide in solution: a role for polysaccharide conformation in recognition between *Xanthomonas* pathogen and its plant host. *Journal of Molecular Biology*, **110**, 1–16.

Morrison, N.A., Clark, R.C., Talashek, T. and Yuan R.C. (2004) New forms of xanthan gum with enhanced properties. In: G.O. Phillips and P.A. Williams (eds), *Gums and Stabilisers for the Food Industry 12*. RSC, Cambridge, pp. 124–130.

Pettitt, D.J. (1983) Xanthan gum. In: M. Glicksman (ed.), *Food Hydrocolloids, Vol I*. CRC Press Inc., Florida, pp. 127–149.

Shatwell, K.P., Sutherland, I.W., Ross-Murphy, S.B. and Dea, I.C.M. (1991a) Influence of the acetyl substituent on the interaction of xanthan with plant polysaccharides – I. Xanthan – locust bean gum systems. *Carbohydrate Polymers*, **14**, 29–51.

Shatwell, K.P., Sutherland, I.W., Ross-Murphy, S.B. and Dea, I.C.M. (1991b) Influence of the acetyl substituent on the interaction of xanthan with plant polysaccharides – II. Xanthan – guar gum systems. *Carbohydrate Polymers*, **14**, 115–130.

Urlacher, B. and Dalbe, B. (1997) Xanthan gum. In: A. Imeson (ed.), *Thickening and Gelling Agents for Food*, 2nd edn. Blackie Academic, Glasgow, pp. 287–289.

Index